Models for Intensive Longitudinal Data

Models for Intensive Longitudinal Data

Edited by
Theodore A. Walls
and
Joseph L. Schafer

2006

OXFORD
UNIVERSITY PRESS

Oxford University Press, Inc., publishes works that
further Oxford University's objective of excellence
in research, scholarship, and education.

Oxford New York
Auckland Cape Town Dar es Salaam Hong Kong Karachi
Kuala Lumpur Madrid Melbourne Mexico City Nairobi
New Delhi Shanghai Taipei Toronto

With offices in
Argentina Austria Brazil Chile Czech Republic France Greece
Guatemala Hungary Italy Japan Poland Portugal Singapore
South Korea Switzerland Thailand Turkey Ukraine Vietnam

Published by Oxford University Press, Inc.
198 Madison Avenue, New York, New York 10016
www.oup.com

Library of Congress Cataloging-in-Publication Data

Models for intensive longitudinal data / edited by Theodore A. Walls, Joseph L. Schafer.
p. cm.
Includes bibliographical references and index.
ISBN-13: 9780195173444
ISBN 0-19-517344-9
1. Social sciences—Research—Statistical methods. 2. Social sciences—Longitudinal studies. 3. Longitudinal methods. I. Walls, Theodore A. II. Schafer, J. L. (Joseph L.)
HA29.M673 2006
300′.72′7—dc22 2005047269

9 8 7 6 5 4 3 2 1
Printed in the United States of America
on acid-free paper

Contents

Contributors

STEVEN M. BOKER
Department of Psychology
University of Notre Dame
Notre Dame, Indiana
USA

CARLOTTA CHING TING FOK
Department of Psychology
McGill University
Montreal, Quebec
Canada

BRIAN R. FLAY
Department of Public Health
Oregon State University
Corvallis, Oregon
USA

CHAD J. GWALTNEY
Department of Community Health
Brown University Medical School
Providence, Rhode Island
USA

DONALD HEDEKER
Division of Epidemiology and Biostatistics
School of Public Health
University of Illinois at Chicago
Chicago, Illinois
USA

MOON-HO RINGO HO
Department of Psychology
McGill University
Montreal, Quebec
Canada

STEPHEN S. INTILLE
Professor
Massachusetts Institute
of Technology
Cambridge, Massachusetts
USA

HYEKYUNG JUNG
Department of Statistics
Pennsylvania State University
University Park, Pennsylvania
USA

JEAN-PHILIPPE LAURENCEAU
Department of Psychology
University of Delaware
Dover, Delaware
USA

RUNZE LI
Department of Statistics
Pennsylvania State University
University Park, Pennsylvania
USA

RANJAN MAITRA
Department of Statistics
Iowa State University
Ames, Iowa
USA

ROBIN J. MERMELSTEIN
Psychology Department
University of Illinois at Chicago
Chicago, Illinois
USA

SARAH M. NUSSER
Department of Statistics
Iowa State University
Ames, Iowa
USA

HERNANDO OMBAO
Department of Statistics
University of Illinois at Urbana
Champaign, Illinois
USA

JAMES O. RAMSAY
Department of Psychology
McGill University
Montreal, Quebec
Canada

STEPHEN L. RATHBUN
Department of Statistics
Pennsylvania State University
University Park, Pennsylvania
USA

TAMMY L. ROOT
The Methodology Center
University Park, Pennsylvania
USA

MICHAEL J. ROVINE
Human Development and Family Studies
The Pennsylvania State University
University Park, Pennsylvania
USA

JOSEPH L. SCHAFER
The Methodology Center and
Department of Statistics
Pennsylvania State University
University Park, Pennsylvania
USA

JOSEPH E. SCHWARTZ
Department of Psychiatry and
Behavioral Science
Stony Brook University
Stony Brook, New York

SAUL SHIFFMAN
Department of Psychology
University of Pittsburgh
Pittsburgh, Pennsylvania
USA

ROBERT SHUMWAY
Department of Statistics
The University of California, Davis
Davis, California
USA

THEODORE A. WALLS
Department of Psychology
University of Rhode Island
Kingston, Rhode Island
USA

Introduction: Intensive Longitudinal Data

Why This Book Is Needed

This volume is devoted to the explanation and demonstration of new statistical approaches for the analysis of *intensive longitudinal data* (ILD). Many excellent references on longitudinal analysis are already available. With few exceptions, however, they tend to focus on examples with no more than about ten occasions. Traditional sources of longitudinal data–panel surveys, clinical trials, studies of human growth or development–were usually able to achieve their goals with only a few waves of measurement. As the capacity of modern computers to store and process information continues to increase, and as new technologies expand the possibilities for data capture in diverse areas of science, longitudinal databases with a much higher intensity and volume of data are becoming commonplace. The ability of researchers to gather ILD has outstripped the capabilities of many data analysts to model them.

Our interest in models for ILD was originally sparked by studies in behavioral science that rely on handheld computers, beepers, web interfaces, and other new tools for data collection. As ideas for this volume developed, we found that data with similar features arise in many disciplines. Moreover, we learned that the kinds of scientific questions that give rise to ILD share much in common. These questions pertain to processes that lead to the expression of time-varying characteristics, to understanding how these processes evolve, and to learning what influences determine the course of these processes. We found that the mainstays of longitudinal analysis–multilevel models with random effects for individual subjects–could address some of these questions if they were creatively applied and extended, but they were clearly inadequate for handling other issues. Other methods, such as time series analysis or spatial analysis, on the other hand, are designed for large numbers of occasions, but typically involve only a single series rather than series from multiple subjects.

In September 2003, a national conference on the science and technology of real-time data capture was held in Charleston, South Carolina. These meetings, which were sponsored by the National Cancer Institute, have resulted in a new volume on the collection of ILD edited by Stone et al. (forthcoming). Numerous grant applications involving intensive longitudinal data collection have been submitted to funding agencies in recent years. We found that applied statisticians, biostatisticians, and methodologists in many fields are already analyzing ILD with varying degrees of success, but they had no common language with respect to common features of these data by which to communicate their successes or learn from one another. We hope that this volume will help to define the modeling of ILD as a distinct emerging field and serve as the first step in filling the void.

For Whom We Are Writing

This volume is addressed to statisticians, data analysts, and methodologically oriented researchers in the social sciences, health sciences, engineering, and any other field where longitudinal data are collected and analyzed. The chapters assume an intermediate knowledge of probability and statistics, including the properties of random variables and vectors, basic matrix algebra, regression, and occasional simple calculus. The material should be accessible to anyone with graduate training in statistics or biostatistics and to those who, through involvement and experience with quantitative research, have gained a working knowledge of the existing tools of longitudinal modeling. We instructed the authors not to shy away from technical material when essential, but to avoid unnecessary details and jargon and to provide enough discussion so that readers may grasp the major points without needing to understand every formula. We have also tried to give each chapter an applied flavor, by motivating and illustrating the methods with interesting and real data examples. Some authors have even supplied programs and source code examples; these are available at the website accompanying this book:

http://www.oup.com/MILD

We believe that this volume will serve as a desk reference for investigators who collect and analyze ILD. Many of the chapters include succinct primers on popular methods for longitudinal analysis, making this book useful to a wide audience of statisticians, modelers, professional researchers, and advanced students. In the classroom, this volume may serve as a primary text for a survey course on the modeling of ILD, or as a supplementary text for courses related to multilevel modeling, time series, functional data analysis, or techniques of dynamic modeling. As a research tool, it will help to inform scientists who lead, design, and carry out ILD-producing studies and help them to develop appropriate analytic strategies. Finally, we hope that this volume will inspire more statisticians and methodologists to become involved in this exciting field, to develop new methods and expand the possibilities for modeling ILD in the years ahead.

The Nature of Intensive Longitudinal Data

Why ILD Are Different

In our usage, intensive longitudinal data or ILD arise in any situation where quantitative or qualitative characteristics of multiple individuals or study units are recorded at more than a handful of time points. The number of occasions may be in the tens, hundreds, or thousands. The frequency or spacing of measurements in time may be regular or irregular, fixed or random, and the variables measured at each occasion may be few or many. We also consider ILD to include situations involving continuous-time measurement of recurrent events, provided that the period of measurement is long enough for a large number of these events to be potentially observed for each subject.

For a brief moment, we were tempted to think that ILD were no different from any other kind of longitudinal data, except that there was more of it. After all, with the high speed and large memory capacity of modern computers, the same models and software that could handle five or ten occasions per subject a decade ago can easily handle hundreds of occasions today. Once we became involved in a study involving ILD, however, that myth was quickly dispelled. The major differences between ILD and other kinds of longitudinal data are not tied to the dimensionality of the database. Rather, the features that make ILD unique and worthy of special consideration pertain to the scientific motivations for collecting them, the unusual nature of the hypotheses they are intended to address, and the complex features of the data that need to be revealed and parameterized by statistical models.

Sources of Data

Some of the oldest ILD from human participants came from diary studies. Individuals were asked to complete by paper and pencil, usually at the end of the day, a log of their experiences or actions. For analytic purposes, diary data were often aggregated into means, totals, or other coarse summaries. In many cases, the high frequency of measurement was not central to the scientific questions being addressed; the primary reason for obtaining daily reports was to reduce bias and variance in the measurement process by having participants respond while the experiences were still relatively fresh in their minds.

Recent technological developments have made the collection of diary data more convenient for respondents and researchers alike. Subjects are now given small electronic devices (e.g., palmtop computers) which prompt them at various times throughout the day to ask questions and record responses. Among psychologists, the frequent recording of thoughts, feelings, or actions by electronic means has been referred to as the experience-sampling method (Csikszentmihalyi & Larson, 1987; ESM) and ecological momentary assessment (Stone & Shiffman, 2002; EMA). Other techniques, as covered by Nusser, Intille, and Maitra (this volume), involve automatic sensing of physical behaviors (e.g., number of steps taken) or bodily states (e.g., blood pressure or glucose levels), by using ambulatory devices that passively monitor participants in their natural environments outside of a laboratory. Frequently, ambulatory telemetric measurement devices and technology-enabled questionnaires have been employed in these studies. Of course, devices measuring actions and states may also be used in laboratories and clinics. Audio or video recordings of individuals as they interact with their physical environment or with one another, which are subsequently reviewed and coded by researchers, can also generate high volumes of ILD. Although most of the data examples used in this book involve human participants, ILD are also compiled from administrative records regarding institutions, organizational units, localities, retail outlets, and so on. Any of the methods in this volume could be applied to data collected in these settings as well.

Recurring Themes in ILD Modeling

As the chapters of this book took shape, common themes began to emerge that helped us to distinguish the modeling of ILD from other kinds of longitudinal analyses. The first theme is *the complexity and variety of individual trajectories and the need to move beyond simple time-graded effects*. With shorter series, patterns of average growth or change over time may be reasonably described by incorporating the effects of time through a linear or quadratic trend. Individual variation in trends, if necessary, can be accommodated by allowing intercepts, slopes, and so on, to randomly vary from one subject to another. With ILD, however, describing temporal change by conventional polynomials is rarely appropriate. Many waves of measurement produce complicated trajectories that are difficult to describe by simple parametric curves. Moreover, we often find that the shapes of these trajectories, even after smoothing over time, may vary wildly from one subject to another (Walls et al., forthcoming). With such strong variation, the relevance of a population-average time trend becomes questionable, because in comparison to the individual curves the average may be highly atypical. If the period of data collection represents a very narrow slice of the participants' lifespan, a pattern of average growth may be irrelevant or undetectable; long-term trends may be swamped by short-term variation. Sometimes it is reasonable to completely remove any absolute measure of time from the mean structure and view each subject's data as a realization of an autocorrelated process that is stable or stationary.

Another common theme, which is closely related to the previous one, is *the need to rethink the role of time as a covariate.* With ILD, there may be no obvious way to align the trajectories so that the time variable has an equivalent meaning across subjects; this problem is called curve registration (Ramsay & Li, 1998). If the study involves an intervention, the start of the intervention may provide a natural anchor point; in some cases, it could be the time of a major life event (e.g., death of a spouse). In other situations, no natural origin exists, and time may need to be characterized in some other way. Statisticians have grown accustomed to think of time as an unbroken continuum, and standard models tend to characterize effects of the passage of time as accumulating at a uniform rate.

With high intensities of measurement, however, we may need to recognize that time is heterogeneous. Morning is different from evening; Wednesday is different from Sunday; this Tuesday may be very different from next Tuesday. Effects may be cyclic, periodic, or vary randomly over the discrete units of time by which biological and social processes are organized. With ILD, descriptors of time, such as time of day or day of week, are often more influential than time itself, and the data analyst may need to introduce these features in new and creative ways. Most importantly, as we elaborate upon in our next theme, with these additional considerations also arises the need to decide whether an assumed temporal pattern holds for any one, any subset of, or all subjects.

A third theme of ILD analyses is that *effects of interest are often found in the covariance structure.* Most books and articles on longitudinal data assume that the parameters of greatest concern are the effects of covariates on the mean of a response variable, either in a population-average or subject-specific sense. Longitudinal analysis is usually presented as an extension of classical regression, and inferences about regression coefficients are seen as the primary goal. Although these coefficients are important, the most interesting features of ILD may lie elsewhere. With many waves of measurement, we may find that subjects vary not only in their means but also in their variances and covariances. Some individuals are stable, whereas others are erratic. Some may show strong positive relationships among measured variables, whereas others may show weak or negative relationships. If we seek to understand and explain this variation in the covariance structure, we can no longer treat that structure as a nuisance but must model it carefully and systematically.

A fourth emergent theme is *a focus on relationships that change over time.* In traditional longitudinal analyses, the effects of time-varying covariates are often taken to be fixed. That is, with a small number of occasions, one would typically estimate a single coefficient that reflects an average association between the time-varying covariate and the response. With a moderate number of occasions, we may discern that this association actually varies from one subject to another and include variance components to account for that variation. With intensive longitudinal measurement, however, we may have the opportunity to discover that the association not only varies among individuals, but also within individuals over time. Trends in association parameters are often complex, not easily described by simple time-by-covariate interactions, and we may need to consider models with nonparametrically time-varying coefficients.

A fifth theme we have identified is *an interest in autodependence and regulatory mechanisms.* Traditional longitudinal models focus on how a response varies over time, on the relationship between a time-varying response and covariate, or on a hazard rate that varies in relation to covariates. Many analyses of ILD, however, involve issues of an autodependent or self-regulatory nature. How does a high level of a response at one occasion, or a change in response from one occasion to the next, influence the distribution of the response at later occasions? Does the occurrence of an event temporarily elevate or depress the probability of additional arrivals later in time? Does self-regulation lead to oscillatory behavior? Questions like these prompt us to move away from basic regression into the realm of multiple-subject time series, dynamical systems, point process models, and control process.

Reactivity

Introducing monitoring devices and surveillance systems into research with human subjects immediately raises the specter of reactivity. Reactivity, also called reactive arrangements, means that individuals may change their behavior simply because they know they are being watched. More generally, it refers to any issue of external invalidity arising from novelty and surveillance effects in experiments and observational studies (Campbell & Stanley, 1963). Reactivity comes into play whenever a testing situation is not simply a passive recording of behavior but itself becomes a stimulus to change (Campbell & Stanley, 1963, p. 9). One of the earliest and most famous examples of reactivity is the Hawthorne effect, in which the productivity of factory workers was seen to improve regardless of the experimental condition applied (Franke & Kaul, 1978).

Diaries and electronic devices that collect ILD can produce new forms of reactivity. We might observe a priming effect, an initial rise or fall in the mean response, as participants gain familiarity with a device and gradually settle into a routine. Some researchers have noted decreasing variation and increasing rates of nonresponse over time. Subjects may tire when repeatedly prompted for what they perceive to be the same information and may respond inertly, settle on a favored score, or fail to respond altogether. Time and place also matter. When subjects are asked to respond in different contexts–homes, offices, play places, cars, and so on–and at different times of day, the nature of the measurement process could drastically change. Quite naturally, one might question the validity and reliability of self-report data in the initial stages of a study as participants grow accustomed to the device. Moreover, as the study progresses, one could imagine that individuals' responses may no longer be highly correlated with their instantaneous true scores; after subjects have adjusted to the protocol, their answers may become essentially random or hover within a relatively limited range.

In many cases, not much can be done to measure reactivity or adjust for its effects within a single study. An awareness of these issues, however–knowing, for example, that diminished within-person variation over time could be an artifact of measurement–may influence the way we interpret the results of our analyses and inform the design of future studies. Empirical evidence in experiments and observational studies indicate that reactivity tends to diminish as devices become easier to use (Hufford et al., 2002).

Scope of This Volume

Topic Areas Represented

In compiling this volume on ILD, it seemed logical to begin with existing techniques for longitudinal analysis. The initial chapters focus on multilevel models–which in various places are called hierarchical linear models, random-effects models, or mixed-effects models–and on marginal modeling through generalized estimating equations. These models can be fitted entirely with existing statistical software, but special considerations arise when applying them to ILD.

Later chapters introduce a variety of less well-known but highly useful methodological tools from item response theory, functional data analysis, time series, state-space modeling, analysis of dynamical systems through stochastic differential equations, engineering control systems, and point process models. Real data examples are drawn from psychology, studies of smoking and alcohol use, brain imaging, and traffic engineering; potential future applications to numerous other kinds of data are also mentioned.

Synopsis of the Chapters

In the first chapter, Walls, Jung, and Schwartz briefly discuss how ILD arise in the behavioral sciences. They review the multilevel linear model, both in terms of multiple stages of regression and as a single equation with multiple random effects, describing its properties and discussing its application to ILD in situations when calendar or clock time may not be highly relevant. Using data from a diary study of stress among Indian adolescents, they account for within-subject correlations by fitting two- and three-level models and demonstrate the use of transformed residuals and semivariograms to check assumptions about the covariance structure.

To complement that chapter, Schafer reviews a popular alternative to multilevel regression: marginal modeling of longitudinal data with generalized estimating equations (GEE). GEE treats the within-subject covariances as a nuisance, focusing instead on robust estimation of population-average regression coefficients. These average coefficients are useful, but with intensive longitudinal data they rarely tell the whole story; interesting features of the covariance structure may remain hidden in a GEE analysis. With GEE, one should also consider whether the sample is large enough for the approximations to be accurate, and whether the working assumptions about the correlation structure will lead to estimates that are sufficiently precise. This chapter reviews the theoretical development of GEE in the context of ordinary least squares, generalized linear models, and quasi-likelihood; compares GEE to multilevel modeling; and reanalyzes data previously examined by Walls, Schwartz, and Jung to illustrate the strengths and limitations of GEE when applied to ILD.

Li, Root, and Shiffman introduce us to the world of functional data analysis (FDA). FDA, a term coined by Ramsay and Silverman (1997), is a creative modeling philosophy that moves away from simple parametric curves to describe high-intensity data by more general and flexible functional forms. The authors insert the idea of local polynomial regression–originally developed as a scatterplot smoother for exploratory analysis–into a multilevel regression model of ecological momentary assessments from a study of smoking cessation. In a fascinating analysis, they show how the urge to smoke and its relationships to mood vary as study participants prepare to quit smoking and actually try to quit.

Hedeker, Mermelstein, and Flay describe item response theory (IRT) models, also called latent-trait models, which are used extensively in educational testing and psychological measurement. Although an IRT model is essentially a multilevel regression model for intensive longitudinal measurements, IRT has rarely been applied to longitudinal data outside of education and psychology. In this chapter, the authors present a concise overview of IRT models, relate them to multilevel regression models, and show how software for multilevel modeling can be used to estimate the IRT parameters. Using ecological momentary assessments from a study of adolescents' early smoking experiences, they show how IRT can be used to address some key questions about smoking behavior.

Similarly, Fok and Ramsay bring more FDA-related ideas into multilevel regression. Their chapter focuses on longitudinal series that exhibit both cyclic and long-term trends. Measurements collected at high frequency over a long span of time often contain periods (a day, a week, or a year) over which the data tend to repeat themselves. The series may also show gradual, long-term drifts, and the cyclic effects may also change gradually over time. Fok and Ramsay present strategies for incorporating these trends into multilevel regression through a combination of Fourier basis functions and B-splines. They then apply the techniques to personality data to study patterns of expression of interpersonal behaviors and changes in these patterns over time.

In multilevel regression, the mean response is expressed as a function of time and covariates. Time-series analysis takes a different approach, characterizing the momentary distribution of the response in terms of responses at previous occasions. Nevertheless, the two approaches can be related, and Rovine and Walls show one way to integrate them. Defining a first-order autoregressive or AR(1) equation for each subject, they allow the AR(1) parameters to vary randomly from one subject to another. Although this model can be fitted with standard multilevel software, it is more general than the usual AR(1) model for longitudinal data which fixes the autoregressive coefficients at a single value. Applying this model to daily alcohol consumption reports from moderate to heavy drinkers, they investigate individual differences in the stability of the drinking process.

State-space modeling is a more general framework that includes classical Box–Jenkins time series, multilevel regression, regression with time-varying coefficients, and dynamic factor analysis. These models, which were originally developed in the field of control engineering for space flight, have slowly made their way into economics, environmental science, finance, and behavioral science. Ho, Shumway, and Ombao present a self-contained treatment of Gaussian state-space modeling for those with little prior knowledge of the subject. They show how many well-known classes of models are actually special cases of the Gaussian state-space model, and describe how to estimate and interpret parameters of the state equations and observation equations. Their chapter concludes with two interesting applications: functional magnetic resonance imaging (fMRI) of the human brain, and a comparative study of traffic flows through improved and unimproved sections of California highways.

Ramsay offers a readable introduction to basic concepts in the design of feedback loops for controlling the performance of a process that converts one or more inputs into one or more outputs. These ideas, which are the basic tools of control systems engineering, open new doors for the analysis of multivariate longitudinal data in the behavioral and social sciences. Ramsay first explains the notion of a differential equation to describe the dynamic characteristics of a system. He then reviews three types of control strategies–proportional control, integrative control, and derivative control–which correspond to using information from the current state of the output, from past information, and from the current rate of change, respectively. The advantages and disadvantages of each type of control are explained, along with strategies for estimating the parameters of the differential equations from repeated observations of the process.

Differential equations are also key to the chapter by Boker and Laurenceau. They explain how dynamical systems models can describe self-regulation, a process by which a subject maintains equilibrium with respect to a variable by responding to information about temporal changes in that variable's state. Differential equations can also be coupled to allow regulation in one part of a system to influence the regulation of another part of a system. Boker and Laurenceau illustrate these ideas with an intriguing analysis of intimacy between husbands and wives. They show how the self-regulation of feelings of intimacy by each member influences the self-regulation of feelings of intimacy in the other, reflecting dyadic interdependence within the married couples. The existing literature on stochastic differential equations can be highly technical and daunting to nonexperts. Boker and Laurenceau fit their models by a practical, commonsense approach: they extract empirical estimates of derivatives from longitudinal data and feed these estimates into standard multilevel regression software. This and other systems-oriented models hold great promise for the analysis of ILD with respect to describing intraindividual processes.

Up to this point, each of the chapters has assumed that the data to be analyzed are quantitative or qualitative measurements taken on individuals over many points in time. In many studies, however, subjects are monitored continuously over time, and events or behaviors of interest are recorded as they occur. In these situations, the responses of interest are the moments in time when the events arrive, and the distribution of events through time can be modeled as a point process. Point process models are an important tool in environmental science, but in those contexts the processes tend to be spatial (distributed over two-dimensional space) or spatio-temporal (distributed over space and time). Modeling occurrences over a single dimension, time, can be considerably simpler, and because of this simplicity, a rich variety of models is available. Rathbun, Shiffman, and Gwaltney introduce point process models in the context of a smoking study which recorded for each individual the times at which cigarettes were lit. Beginning with simple Poisson processes, they show how individuals' rates of smoking tend to vary with time of day and day of the week. Effects of time-varying covariates on smoking rates are then incorporated into a modulated Poisson process. Point process models can potentially be applied to any kind of event history data–for example, the timings of riots, ethnic confrontations or labor protests, or the times at which firms are founded or go bankrupt.

The final chapter, by Nusser, Intille, and Maitra, looks ahead to the future of intensive longitudinal measurement. The development of tiny, low-cost digital sensors in combination with mobile and wireless technologies and global positioning systems will soon allow data collection on human behaviors and bodily states on a massive scale. This chapter poses many challenges to statisticians to develop practical and creative approaches to analyze the huge amounts of scientific data that will be generated by these studies in the years ahead.

Topics Not Covered

Although we have endeavored to include as many promising approaches as we could, we were forced by limitations of space, issues of timing, and the limited readiness of material in certain areas to omit some important topics. One obvious omission is event-history models for recurrent events and for processes involving multiple spells or states. The point process models described by Rathbun et al. are related to event-history models, but they focus on the arrival of only one kind of event. Another omission is the rapidly expanding body of techniques for causal inference, especially in the presence of time-varying confounding; some references to that literature are found in the chapter by Schafer. We have also chosen not to address techniques of variable reduction over many occasions of measurement. This topic is briefly mentioned, however, in the discussion of dynamic factor models by Ho et al. Finally, we do not actively consider the increasingly vast and important psychometric literature on inferential issues for single subject analyses.

Acknowledgments

We are deeply indebted to many people without whose help this volume would not have been possible. First, we thank those researchers who participated in a working group that met at Penn State University and other places between 2002 and 2004. The presentations and informal discussions that took place at these meetings helped us to identify interesting and useful approaches for analyzing ILD; they also helped us to decide which topics this volume should cover. These working-group meetings also led to fruitful, cross-disciplinary collaborations between statisticians, methodologists, and researchers who collect and analyze ILD.

Next, we whole-heartedly thank the authors. In addition to writing their chapters, they all made revisions at our request and in response to extensive reviews. These revisions were at times substantial. We asked many authors to provide more explanation so that relatively unexposed readers could understand the material more easily. In some cases, this required the development of additional examples and extensive expansion and reorganization of background material.

Third, we thank the many reviewers of chapters for our volume, who included, but were not limited to, the following people. Two anonymous reviewers of the entire manuscript contributed at the request of Oxford University Press. Nearly all of the chapter authors devoted time to reviewing one or more chapters contributed by other authors. In addition, the following reviewers provided assistance with one or more chapters: Naomi Altman, Daniel Bauer, Paul De Boeck, Ellen Hamaker, Robin Henderson, Peter Hovmand, Eric Loken, Han Oud, Christina Röcke, Anders Skrondal, and Alexander von Eye.

Fourth, we thank our staff at the Methodology Center, in particular Tina Meyers and Brenda Yorks, for their administrative support, and David Lemmon and Bob Lambert for developing our web page and handling other technical needs.

Fifth, we thank the many substantive researchers who contributed their data and time in helping our authors develop example applications of their models. They are either listed as coauthors or acknowledged in the text of the chapters.

Sixth, we thank the staff at Oxford University Press, in particular, our editors Maura Roessner and Joan Bossert.

We also express our appreciation to the many people who gave us input on the possibilities for or assisted in some way in the development of our volume, including David Almeida, Steve Armeli, Niall Bolger, Jim Bovaird, Linda Collins, Tamlin Connor, Ron Dahl, Bob Eberlein, John Eltinge, Kathleen Etz, Brian Flaherty, David Ford, Peter Hovmand, Jianhua Huang, Richard Jones, Frank Lawrence, Reed Larson, Dan Mroscek, Susan Murphy, Robert Nix, Wayne Osgood, Steve Raudenbush, Daniel Rivera, Judy Singer, David Walden, Alexander von Eye, Lourens Waldorp, and John Willett.

Finally, we gratefully acknowledge the support that we received from the National Institute on Drug Abuse through the center grant 2-P50-DA10075 which made this project possible.

The Editors

References

Campbell, D.T., & Stanley, J.C. (1963). *Experimental and Quasi-experimental Designs for Research.* Chicago: Rand McNally.

Csikszentmihalyi, M., & Larson, R. (1987). Validity and reliability of the experience-sampling method. *Journal of Nervous and Mental Disease, 175*, 526–536.

Franke, R.H., & Kaul, J.D. (1978). The Hawthorne experiments: First statistical interpretation. *American Sociological Review, 43*, 623–643.

Hufford, M.R., Shields, A.L., Shiffman, S., Paty, J., & Balabanis, M. (2002). Reactivity to ecological momentary assessment: An example using undergraduate problem drinkers. *Psychology of Addictive Behaviors, 16*, 205–211.

Ramsay, J.O., & Li, X. (1998). Curve registration. *Journal of the Royal Statistical Society, Series B*, *60*, 351–363.
Ramsay, J.O., & Silverman, B.W. (1997). *Functional Data Analysis.* New York: Springer.
Stone, A.A., & Shiffman, S. (2002). Capturing momentary, self-report data: A proposal for reporting guidelines. *Annals of Behavioral Medicine*, *24*, 236–243.
Stone, A.A., Shiffman, S., Atienza, A., & Nebelling, L. (Eds.) (forthcoming). *The Science of Real-Time Data Capture.* New York: Oxford University Press.
Walls, T.A., Höppner, B.B., & Goodwin, M.S. (forthcoming). Statistical issues in intensive longitudinal data analysis. In A. Stone, S. Shiffman, A. Atienza, & L. Nebelling. *The Science of Real-Time Data Capture.* New York: Oxford University Press.

Models for Intensive Longitudinal Data

1

Multilevel Models for Intensive Longitudinal Data

Theodore A. Walls, Hyekyung Jung, and Joseph E. Schwartz

1.1 Behavioral Scientific Motivations for Collecting Intensive Longitudinal Data

Psychologists increasingly employ diary-based methods to collect data reflecting study participants' experiences as they occur in daily life. This approach began with straightforward paper and pencil diaries that participants completed on fixed schedules (Hurlburt, 1979; Larson & Csikszentmihalyi, 1978; Sorokin & Berger, 1939; Wessman & Ricks, 1966). Over time, protocols have become more extensive, often including preprinted diaries with survey questions covering a diversity of experiential and contextual topics. The primary objective of these data collection strategies is to achieve greater ecological validity in the measurement of psychological states than can be realized through either infrequent administration of self-report surveys or via research conducted in decontextualized (e.g., laboratory) settings. The protocols also elicit information on underlying psychological processes that can only be revealed through the analysis of intensive longitudinal measurements.

The integration of technological devices into these data collection strategies has afforded considerable opportunities to researchers who employ diary-based methods. The approach now includes the use of outbound calling centers, pagers, interactive telephone-based systems, handheld computers, cell phones, and many more devices. By using these devices, study designers are able to randomize polling times, to order questions variably across occasions, to branch or create layers of questions, and to enable respondents to easily record recurrent events. These advantages have led to the emergence of a specific and rapidly expanding literature on the approach, found in social and personality psychology under the heading of the experience sampling method (ESM; Csikszentmihalyi & Larson, 1987; de Vries, 1992) and in the substance abuse and health sciences domain under the heading of ecological momentary assessment (EMA; Shiffman et al., 1994).

In the last decade, these studies have come to utilize extensive and sophisticated protocols to measure aspects of intimate relationships, moods, substance use, stress, physiological functioning, symptoms of illness, and several other psychological constructs. The nature of these protocols is reflected throughout this volume and in at least two forthcoming compilations (Csikszentmihalyi et al., in press; Stone et al., 2005). Over the last few years, the approach has seen broad application across the health and social sciences as a useful approach to the study of regulatory and process-level information inherent to a variety of phenomena. In general, research aims have been to produce descriptive statistics about samples, to assess within-person processes and their relations, and to investigate individual differences in processes of interest.

These studies uniquely contribute to our understanding of behavior and other phenomena by providing temporal information on proximal states and contexts, and by providing data on ephemeral events that are difficult to recall (e.g., affective states or behavior events) or that are subject to other forms of reporting bias. Representative studies using diary-based methods can be found across domains of psychological inquiry (Feldman Barrett et al., 1998; Kahneman et al., 1993; Stone & Shiffman, 2002).

We refer to the data resulting from these studies as intensive longitudinal data (ILD). The nature of these data is outlined in the introduction to this volume, but the main characteristic is the presence of long, intensively measured, independent series or sets of series that potentially reflect complex patterns of change over time. Diary-based studies are only one source of ILD. For example, studies using event calendars also produce ILD, such as monthly reports for periods of one to five years (Caspi et al., 1996; Freedman et al., 1988; Treloar et al., 1970). Studies in a variety of other disciplines, from operations management to physiology, have produced ILD. Many forthcoming studies will produce databases with similar characteristics, particularly with respect to having an intensive number of occasions of measurement that far exceeds what has traditionally been found in most areas of social science (cf. Nusser et al., chapter 11, this volume). Potentially complex patterns of change in person-specific series may be of interest in these studies.

The simplest approach to the analysis of these data is to simply average within-person scores for use in between-person analyses, such as group difference tests, or using traditional regression techniques. Recently, multilevel modeling has become the most predominant statistical approach for the analysis of these data (Laird & Ware, 1982; Moskowitz & Hershberger, 2002; Schwartz & Stone, 1998; see Bolger et al., 2003, for discussion of multilevel analysis of diary data). The multilevel model is also flexible in that quantities resulting from a range of other statistical modeling frameworks can be incorporated, as demonstrated in many other chapters in this volume. Finally, the model lends itself to the analysis of data that have been collected on both contextual influences and mechanisms potentially operating within the unit of analysis of interest (e.g., person, dyad, etc.).

Because the multilevel model is likely to have continued general applicability for the analysis of these data and many new ways of deploying the model continue to emerge, a preliminary goal of this chapter is to review the model. The central goal of this chapter, however, is to provide a thoughtful exposition of important considerations in conducting a multilevel model of intensive longitudinal data. A concomitant goal is to provide an applied demonstration of the model. We pursue these goals in three complementary sections.

First, we describe the multilevel model for longitudinal data. Because scientists routinely struggle to compare and contrast the features and variations of the model across scholarly publications tied to various software packages, we show how the two-level linear model and a linear mixed model version are mathematically equivalent. We note recent progress toward rectifying problems driven by software-specific nomenclature in applied methodological works such as Fitzmaurice et al. (2004), Snijders and Bosker (1999), and Willett and Singer (2003), and we aim to complement these efforts.

Second, an application of the model to an intensive longitudinal database is provided. We use ESM data from Indian schoolchildren regarding their perceived control beliefs in order to elucidate the kinds of special steps in data exploration, model specification, and diagnosis that may need to be made in multilevel modeling of ILD.

Third, recent innovations have utilized the multilevel model to handle specific intensive longitudinal data situations. We organize this work into four areas in which extensions have been undertaken and highlight some examples. We hope that this will enable practitioners to consider those extensions that may be most appropriate for their scientific questions and also provide a framework by which methodologists may consider new extensions. Discussion of special extensions and model limitations is particularly important at the outset of this volume (see Introduction), inasmuch as many of the new approaches included in the volume incorporate multilevel strategies. We conclude with a synopsis of our discussion and a few thoughts on future prospects in applied multilevel modeling of intensive longitudinal data.

1.2 Overview of Multilevel Models

The multilevel model originated in the scholarship of Laird and Ware (1982), who proposed a two-level random effects model for the analysis of longitudinal data. This model offered the ability to partition distinct variance components at two levels. Over the last twenty years, this model has become widely known as the linear mixed model or, informally, as the Laird and Ware model. It has seen extensive application in the social sciences and, particularly, in psychological and educational research. More extensive conceptual and pedagogical discourse on multilevel modeling can be found in Goldstein (2003), Hox (2002), Kreft and de Leeuw (1998), Raudenbush and Bryk (2002), and Snijders and Bosker (1999).

More complete historical reviews of the evolution of the multilevel model are contained in Hüttner and van den Eeden (1995) and Longford (1993).

The central idea of multilevel modeling is that a hierarchical structure in the data is accounted for via the use of random effects at various levels in the hierarchy. As the name suggests, the *multilevel* model is often used to analyze data from a hierarchically structured population, where lower-level observations are nested within higher level(s). That is, the data are collected from units that are nested within groups or clusters, such as students in classes, classes in schools, and so on. Similarly, repeated measurements over time can be nested within persons, who play the same role as groups in the hierarchical presentation. Explanatory variables measured at various levels of the hierarchy are combined into a single regression model. Because longitudinal databases include a series of repeated measures for each person, these series are specified at a separate within-person level. We shall refer to this first level as the *measurement occasions* level and to the second level as the *person* level. The applied literature on multilevel modeling includes extensive coverage of these model implementation possibilities (see, in particular, Fitzmaurice et al., 2004; Raudenbush & Bryk, 2002; Snijders & Bosker, 1999).

Multilevel models have a variety of names in the statistical literature, including random effects models (Laird & Ware, 1982), general mixed linear models (Goldstein, 1986), random coefficient models (de Leeuw & Kreft, 1986; Longford, 1993), variance component models (Longford, 1989), and hierarchical linear models (Bryk & Raudenbush, 1992; Raudenbush & Bryk, 1986, 2002). Similarly, the term latent growth curve (LGC) model refers to the use of multilevel models to account for between-person variation in temporal pattern(s), such as differences in the average rate of change. The model is a multivariate case of the multilevel model, in which time is treated as a primary predictor variable in order to capture essential characteristics of the shape of the trajectory. For example, random intercepts and slopes can be modeled on the predictor variables, as well as the response. For longitudinal data with fixed occasions in which only the response is modeled, comparable implementations of LGC produce the same estimates as multilevel models (Hox, 2000; Rovine & Molenaar, 2001). Ways in which multilevel modeling can be extended to discrete response data will be briefly described in section 1.4.2.

1.2.1 The Two-Level Linear Model

In this and the following section, generic representations of the multilevel model, as derived from both the seminal and applied statistical literature on multilevel models, are provided. Although Laird and Ware's seminal work expressed the model as a single linear combination, subsequent treatments often provide separate equations for each level of the model. We review these treatments and explain how they aggregate upwardly into the linear mixed model expressed as a single linear combination (cf. Raudenbush & Bryk, 2002; Rovine & Walls, chapter 6, this volume; Singer, 1998).

For simplicity, consider a model with only two levels. Extensions to additional levels are important and conceptually straightforward. Level 1 represents relations among variables at the measurement occasions level and level 2 represents the influence of variables at the person level (Bryk et al., 1996). A single observation appears for person i on occasion t,

$$\begin{aligned} y_{it} &= \pi_{i0} + \pi_{i1} z_{i1t} + \pi_{i2} z_{i2t} + \cdots + \pi_{iJ-1} z_{iJ-1t} + e_{it} \\ &= \pi_{i0} + \sum_{j=1}^{J-1} \pi_{ij} z_{ijt} + e_{it}, \end{aligned} \tag{1.1}$$

$$\begin{aligned} \pi_{ij} &= \beta_{j0} + \beta_{j1} w_{i1} + \beta_{j2} w_{i2} + \cdots + \beta_{jK-1} w_{iK-1} + b_{ij} \\ &= \beta_{j0} + \sum_{k=1}^{K-1} \beta_{jk} w_{ik} + b_{ij}. \end{aligned} \tag{1.2}$$

In equation (1.1), i indexes persons (level 2 units) and t indexes the measurement occasions (level 1 units). Note that the number of measurement occasions can vary across persons, so t goes from 1 to n_i for each person i. The term z_{ijt} is an explanatory variable at the measurement occasions level. Generally, time enters into the model as an explanatory variable at the measurement occasions level. The actual times at which measurements occur serve as time-varying covariates and are usually encoded into the z's in some fashion as linear terms, quadratic terms, and so on. The e_{it} are the residuals from the prediction of response, y_{it}, from the set of z_{ijt} assessed for person i. In general, the level 1 errors, e_{it}, are assumed to be independent, which implies that the responses of person i are independent. However, in multilevel models for longitudinal data with closely spaced observations, this assumption could be implausible; longitudinal data are ordered in time so observations that are close in time often are more similar than observations that are further apart (i.e., serially correlated). In addition, e_{it} are assumed to have a normal distribution with a mean zero and constant variance σ^2, that is, $N(0, \sigma^2)$.

In equation (1.2), the level 1 coefficients, that is, intercepts and/or slopes, can be assumed to vary randomly across persons and, therefore, can be estimated as random effects. Each of the level 1 coefficients, π_{ij}, becomes an outcome variable in the level 2 model. These coefficients are regressed upon level 2 covariates (e.g., sex), w_{ik}, producing a level 2 coefficient, β_{jk}, and person-dependent deviations, b_{ij}. Note that the w's in the level 2 model for π_{ij} are not necessarily different from the w's in the model for $\pi_{ij'}$ because they are chosen from a common pool of person-level predictors. In practice, they are the same for all level 1 coefficients, π_{ij} $(j = 0, \ldots, J-1)$.

The random effects, b_{ij}, are deviations between people's individual level 1 parameters and their respective expected values. In the longitudinal case, the b_{ij} reflect unexplained variance on the time dimension. The b_{ij} are assumed to be independent across persons and to have a normal distribution with mean zero and constant variance implying homoscedasticity. In addition, the residuals,

e_{it}, are assumed to be independent of b_{ij} of random effects; the measurement occasions (level 1 units) are assumed to be conditionally independent given the person-level (level 2) random effects and the covariates.

By substituting equation (1.2) into (1.1), the level 1 and level 2 models can be combined together into a *composite* model,

$$
\begin{aligned}
y_{it} &= \beta_{00} + \sum_{j=1}^{J-1} \beta_{j0} z_{ijt} + \sum_{k=1}^{K-1} \beta_{0k} w_{ik} + \sum_{j=1}^{J-1} \sum_{k=1}^{K-1} \beta_{jk} w_{ik} z_{ijt} \\
&\quad + b_{i0} + \sum_{j=1}^{J-1} b_{ij} z_{ijt} + e_{it} \\
&= \sum_{j=0}^{J-1} \sum_{k=0}^{K-1} \beta_{jk} w_{ik} z_{ijt} + \sum_{j=0}^{J-1} b_{ij} z_{ijt} + e_{it}.
\end{aligned}
\tag{1.3}
$$

The composite model, as a mathematically equivalent model specification to the level 1/level 2 specification, is more parsimonious and is, in fact, the model being fitted to the data in many multilevel statistical software programs.

In equation (1.3), $\sum_{j=1}^{J-1} \beta_{j0} z_{ijt}$ represents the main effects of measurement occasions level (or within-person) predictors, $\sum_{k=1}^{K-1} \beta_{0k} w_{ik}$ represents the main effects of person-level (or between-person) predictors, and $\sum_{j=1}^{J-1} \sum_{k=1}^{K-1} \beta_{jk} w_{ik} z_{ijt}$ represents an interaction effect of cross-level predictors, which will be discussed further in section 1.2.2. In equation (1.3), by letting w_{i0} represent 1 and letting z_{i0t} represent 1, the response y_{it} is predicted from the explanatory variables, w_i and z_{it}, which reside at the measurement occasions level and person level, respectively.

In the composite model, all of the level 1 coefficients are assumed to randomly vary from one person to another, but many of them are actually fixed to the same value. In practice, many of the level 2 regressions except the one for π_{i0} are extremely simple—just a constant plus an error, or a constant plus nothing. In our presentation, in order to allow the intercept only to vary randomly, the b_{i0} are allowed to vary randomly and $b_{ij} (j \neq 0)$ is set to zero. More detailed explanation on components in the composite model is given by Willett and Singer (2003).

It is possible to construct two or more levels above the measurement occasions level in at least two situations: when persons are grouped into some units, or, as we demonstrate later, when time can be subdivided into meaningful discrete periods. In the former case, three-level models will have repeated observations at the first level, persons at the second level, and groups at the third level. The latter case occurs when there are natural divisions of time in which it is meaningful to think of time differently, such as by breaking time into subunits (i.e., day of week, weekday/weekend). Other plausible higher-level models in the latter situation will have repeated observations on hierarchically structured groupings of time units, such as time of day at the first level, day of week at the second level, and persons at the third level. This approach has been undertaken rarely, if at all, in the past; however, it is a natural choice in the case of ILD because time periods have distinct meaning. Because of the great potential relevance of this strategy, we develop a full example in section 1.3.3.

1.2.2 The Laird–Ware Linear Mixed Model

The two-level linear models organize the systematic and random variation at each level of the model. By contrast, Laird and Ware's (1982) model integrates the systematic and random variation in one linear prediction equation. The model in this representation is commonly encountered in statistics and is increasingly utilized in the social and behavioral sciences. In Laird and Ware's model, for person i,

$$\boldsymbol{Y}_i = \boldsymbol{X}_i\boldsymbol{\beta} + \boldsymbol{Z}_i\boldsymbol{b}_i + \boldsymbol{e}_i, \tag{1.4}$$

for $i = 1, \ldots, n_i$. $\boldsymbol{Y}_i$ represents a vector of response values, consisting of all the multiple assessments of person i made at a different occasion t. The known design matrices $\boldsymbol{X}_i$ and $\boldsymbol{Z}_i$ are, respectively, linking $\boldsymbol{\beta}$ and $\boldsymbol{b}_i$ to $\boldsymbol{Y}_i$. The second design matrix, $\boldsymbol{Z}_i$ is made up of time-varying explanatory variables. $\boldsymbol{\beta}$ and $\boldsymbol{b}_i$ indicate vectors of unknown fixed effects and unknown random effects. The $\boldsymbol{e}_i$ indicates a vector of unobservable random errors.

In equation (1.4), the first design matrix, $\boldsymbol{X}_i$, includes three kinds of terms: explanatory variable(s) at the measurement occasions level, explanatory variable(s) at the person level, and interaction term(s) between these variables. The second design matrix, $\boldsymbol{Z}_i$, only includes explanatory variables at the measurement occasions level. The interaction term(s) between variables measured at the measurement occasions (first) level and at the person (second) level is referred to as a *cross-level interaction*, because it involves explanatory variables from different levels. A cross-level interaction exists when the effects of time-varying level 1 explanatory variables differ by the values of time-invariant level 2 explanatory variables. Consider an example in which test performance over many occasions is the response and alertness is a predictor of interest. If measurement occasions at different times of day are defined as the level 1 covariate and a person's alertness during a day is defined as the level 2 covariate, we might hypothesize that alertness influences the slope of test performance according to the time of day the test is taken. In this case, the interaction between the alertness variable and the slope of alertness over measurement occasions would be a cross-level interaction.

The $\boldsymbol{e}_i$ are assumed to be normally distributed with mean vector zero and variance–covariance (henceforth, covariance) matrix $\sigma^2\boldsymbol{V}_i$, where $\boldsymbol{V}_i$ may have different correlation structures. This topic is covered by Schafer (chapter 2, this volume). The most common correlation structure—especially for short series—has been to assume that $\boldsymbol{V}_i = \boldsymbol{I}$, where $\boldsymbol{I}$ is an identity matrix. In other words, we assume that the correlation that exists among observations within a person i comes only from $\boldsymbol{b}_i$, and not from any other sources.

The random coefficients, $\boldsymbol{b}_i$, also have a random distribution. They are normally distributed random variables with mean vector zero and with covariance matrix $\boldsymbol{\psi}$, which is a general $(J \times J)$ covariance matrix. These variables are assumed to be independent of the $\boldsymbol{X}_i$'s and of the random errors, $\boldsymbol{e}_i$. Given the vector of random effects, $\boldsymbol{b}_i$, $\boldsymbol{Y}_i$ are defined to be independent and normally distributed with mean vector $\boldsymbol{X}_i\boldsymbol{\beta} + \boldsymbol{Z}_i\boldsymbol{b}_i$ and with covariance matrix $\sigma^2\boldsymbol{V}_i$. Therefore, the marginal distribution of $\boldsymbol{Y}_i$ is independent and normal with mean vector $\boldsymbol{X}_i\boldsymbol{\beta}$ and covariance matrix $\boldsymbol{\Sigma}_i = \boldsymbol{Z}_i\boldsymbol{\psi}\boldsymbol{Z}_i' + \sigma^2\boldsymbol{V}_i$. The assumption that $\boldsymbol{b}_i \sim N(0, \boldsymbol{\psi})$ may be unattractive if the persons are of a few different types and the types behave quite differently. In light of this, Nagin (1999) replaces the normal distribution with a discrete distribution with mass at a number of points, where the points are predetermined and estimated in the multivariate space. For example, we may reasonably assume that $\boldsymbol{b}_i$ has multinomial distribution if we have few types of persons. Muthén and Muthén (2000) extend the model to a low-dimensional finite mixture of multivariate normals. Both approaches are potentially useful in situations where the population may be overrepresented by a small number of groups and it seems appropriate to represent the population with a small number of prototypical curves.

We provide the detailed matrix form of the Laird–Ware linear mixed model for easy review:

$$\underbrace{\begin{pmatrix} y_{i1} \\ \vdots \\ y_{in_i} \end{pmatrix}}_{\boldsymbol{Y}_i} = \underbrace{\begin{pmatrix} 1 & x_{i11} & \cdots & x_{iQ-11} \\ \vdots & \vdots & \ddots & \vdots \\ 1 & x_{i1n_i} & \cdots & x_{iQ-1n_i} \end{pmatrix} \begin{pmatrix} \beta_0 \\ \vdots \\ \beta_{Q-1} \end{pmatrix}}_{\boldsymbol{X}_i\boldsymbol{\beta}}$$

$$+ \underbrace{\begin{pmatrix} 1 & z_{i11} & \cdots & z_{iJ-11} \\ \vdots & \vdots & \ddots & \vdots \\ 1 & z_{i1n_i} & \cdots & z_{iJ-1n_i} \end{pmatrix} \begin{pmatrix} b_{i0} \\ \vdots \\ b_{iJ-1} \end{pmatrix}}_{\boldsymbol{Z}_i\boldsymbol{b}_i} + \underbrace{\begin{pmatrix} e_{i1} \\ \vdots \\ e_{in_i} \end{pmatrix}}_{\boldsymbol{e}_i}$$

From this representation, we can easily obtain the equation for the single response of person i on occasion t,

$$\begin{aligned} y_{it} &= \beta_0 + \beta_1 x_{i1t} + \beta_2 x_{i2t} + \cdots + \beta_{Q-1} x_{iQ-1t} \\ &\quad + b_{i0} + b_{i1} z_{i1t} + b_{i2} z_{i2t} + \cdots + b_{iJ-1} z_{iJ-1t} + e_{it} \\ &= \sum_{q=0}^{Q-1} \beta_q x_{iqt} + \sum_{j=0}^{J-1} b_{ij} z_{ijt} + e_{it}. \end{aligned} \tag{1.5}$$

Of note, $\boldsymbol{e}_i$'s are generally assumed to be independent, that is, each person's distribution over the measurement occasions is independent of the other's. From the view of the two-level regression model, the Laird–Ware model combines the lower and the higher levels into one linear combination reflecting the "fixed effects" and "random effects" components. The fixed effects part of the model contains the regression coefficients (β's). The random effects part of the model contains the measurement occasions error term ($\boldsymbol{e}_i = \{e_{it}\}$) and the person-level error terms ($\boldsymbol{b}_i = \{b_{ij}\}$) and carries subscripts for persons because the error terms represent the deviations of each person's regression coefficients from their overall mean over occasions.

Even though the two-level linear model introduced in section 1.2.1 is remarkably similar to the Laird–Ware linear mixed model, Fitzmaurice et al. (2004) point out a notable difference between them. The regression version places some unnecessary constraints on the choice of the design matrix for the fixed effects, which are not needed in the linear mixed model expression. Further explanation is given by Fitzmaurice et al. (2004).

1.2.3 Estimation

The parameters to be estimated in multilevel models are generally divided into two parts: a fixed part and a random part. The parameters of the fixed part are regression coefficients, β's, whereas those of the random part are variance components ($\sigma^2 \boldsymbol{V}_i$, $\boldsymbol{\psi}$). The most widely used statistical method for the estimation of parameters in multilevel modeling is *maximum likelihood*(ML), also called *full information maximum likelihood* (FIML) (Hox, 2002; Kreft & de Leeuw, 1998; Snijders & Bosker, 1999). This produces estimates for both the fixed parameters and random parameters that maximize the probability of recovering the observed variance–covariance matrix (cf. Eliason, 1993). *Restricted maximum likelihood* (REML) is a close relative of maximum likelihood estimation.

An alternative approach to FIML and REML is the generalized estimation equation (GEE), which provides a general and unified approach for analyzing correlated responses that can be either discrete or continuous (Fitzmaurice et al., 2004). The topic of discrete responses is covered by Schafer (chapter 2, this volume).

In full maximum likelihood estimation, we create the log-likelihood function based on $\boldsymbol{Y}_i \sim N(\boldsymbol{X}_i\boldsymbol{\beta}, \Sigma_i(\boldsymbol{\alpha}))$, where $\Sigma_i(\boldsymbol{\alpha}) = (\boldsymbol{Z}_i{}'\boldsymbol{\psi}\boldsymbol{Z}_i + \sigma^2\boldsymbol{V}_i)$ and $\boldsymbol{\alpha}$ contains the nonredundant elements of $\boldsymbol{\psi}$ and free parameters of $\boldsymbol{V}_i$.

The log-likelihood function is

$$\ell = \sum_i \left\{ -\frac{1}{2}\log|\Sigma_i(\boldsymbol{\alpha})| - \frac{1}{2}(\boldsymbol{Y}_i - \boldsymbol{X}_i\boldsymbol{\beta})'\Sigma_i(\boldsymbol{\alpha})(\boldsymbol{Y}_i - \boldsymbol{X}_i\boldsymbol{\beta}) \right\},$$

and maximize the joint likelihood, ℓ, with respect to $\boldsymbol{\beta}$ and $\boldsymbol{\alpha}$.

The FIML estimator has useful properties such as consistency, efficiency, and asymptotic normality. However, the FIML method does not adjust for the loss of the degrees of freedom involved in estimating the fixed parameters, so it produces biased estimates of the random parameters ($\boldsymbol{\alpha}$) in small samples.

Similarly, restricted maximum likelihood applies maximum likelihood only to residuals ($\boldsymbol{e}_i$, $\boldsymbol{b}_i$) so only the random parameters are included in the likelihood function. The REML estimator for $\boldsymbol{\alpha}$ can be obtained by removing fixed effects,

$\boldsymbol{\beta}$, from the likelihood function through integrating,

$$\ell^* = \log \int \exp(\ell) d\boldsymbol{\beta},$$

and then by maximizing the REML likelihood function, ℓ^*, with respect to $\boldsymbol{\alpha}$. Because the REML method removes the fixed effects and then estimates the variance components from the likelihood function, which reduces the degrees of freedom, it produces less biased estimates for the random components (Hox, 2002; Snijders & Bosker, 1999). In general, the FIML and REML methods produce similar estimates except when the number of covariates is large.

Except in trivial cases (e.g., $\Sigma_i = \sigma^2 \boldsymbol{I}$), neither the FIML nor the REML estimates can be expressed in closed form for the maximizer of the likelihood and must be computed by iterative procedures such as expectation-maximization algorithms (Jennrich & Schluchter, 1986; Laird & Ware, 1982; Laird et al., 1987; Liu & Rubin, 1995), Newton–Raphson algorithms, and Fisher scoring procedures (Jennrich & Schluchter, 1986; Lindstrom & Bates, 1988). Software packages developed for estimating the two-level regression models and linear mixed models using either FIML or REML include HLM (Bryk et al., 1996), MLn (Multilevel Models Project, 1996), PROC MIXED in SAS (Littell et al., 1996), S-PLUS (MathSoft, 1997), MIXREG (Hedeker & Gibbons, 1993), and Stata (StataCorp, 1997).

FIML and REML give estimates of $(\boldsymbol{\alpha}, \sigma^2)$ and then those estimates are plugged into Generalized Least Squares (GLS) to obtain the estimates of fixed effects, $\boldsymbol{\beta}$. From equation (1.4), the marginal distribution of the $\boldsymbol{Y}_i$ is independent and normally distributed with mean vector $\boldsymbol{X}_i\boldsymbol{\beta}$ and with covariance matrix $\Sigma_i = \boldsymbol{Z}_i\boldsymbol{D}\boldsymbol{Z}_i' + \sigma^2\boldsymbol{V}_i$. In this case,

$$\hat{\beta} = \left(\sum_i \mathbf{X}_i' \Sigma_i^{-1} \mathbf{X}_i\right)^{-1} \sum_i \mathbf{X}_i' \Sigma_i^{-1} \mathbf{Y}_i, \tag{1.6}$$

where $\Sigma_i = \boldsymbol{Z}_i\boldsymbol{D}\boldsymbol{Z}_i' + \sigma^2\boldsymbol{V}_i$ and the covariance matrix Σ_i can be estimated by REML or FIML.

Once the β's and covariance parameters $(\boldsymbol{\alpha}, \sigma^2)$ have been estimated, we can estimate the person-specific random coefficients by empirical Bayes (EB). In the EB method, from Bayes's theorem, $\boldsymbol{b}_i$'s are independent and normally distributed given $(\boldsymbol{Y}, \boldsymbol{\beta}, \sigma^2, \boldsymbol{\psi})$. Then, their posterior moments given $(\boldsymbol{Y}, \boldsymbol{\beta}, \sigma^2, \boldsymbol{\psi})$ are

$$E(\boldsymbol{b}_i | \boldsymbol{Y}, \boldsymbol{\beta}, \sigma^2, \boldsymbol{\psi}) = (\boldsymbol{\psi}^{-1} + \mathbf{Z}_i' \boldsymbol{V}_i^{-1} \mathbf{Z}_i)^{-1} \mathbf{Z}_i' \boldsymbol{V}_i^{-1} (\boldsymbol{Y}_i - \mathbf{X}_i \boldsymbol{\beta}), \tag{1.7}$$

$$Var(\boldsymbol{b}_i | \boldsymbol{Y}, \boldsymbol{\beta}, \sigma^2, \boldsymbol{\psi}) = \sigma^2 (\boldsymbol{\psi}^{-1} + \mathbf{Z}_i' \boldsymbol{V}_i^{-1} \mathbf{Z}_i)^{-1}. \tag{1.8}$$

Substituting estimates for $(\boldsymbol{\beta}, \sigma^2, \boldsymbol{\psi})$ into equations (1.7) and (1.8), we obtain

$$\hat{b}_i^{EB} = (\hat{\psi}^{-1} + \mathbf{Z}_i' \hat{V}_i^{-1} \mathbf{Z}_i)^{-1} \mathbf{Z}_i' \hat{V}_i^{-1} (\boldsymbol{Y}_i - \mathbf{X}_i \hat{\beta}). \tag{1.9}$$

The predicted response profile is

$$\begin{aligned}
\hat{Y}_i &= \boldsymbol{X}_i\hat{\beta} + \boldsymbol{Z}_i\hat{b}_i \\
&= \boldsymbol{X}_i\hat{\beta} + \boldsymbol{Z}_i(\hat{\psi}^{-1} + \boldsymbol{Z}_i'\hat{V}_i^{-1}\boldsymbol{Z}_i)^{-1}\boldsymbol{Z}_i'\hat{V}_i^{-1}(\boldsymbol{Y}_i - \boldsymbol{X}_i\hat{\beta}) \\
&= [\boldsymbol{I}_n - \boldsymbol{Z}_i(\hat{\psi}^{-1} + \boldsymbol{Z}_i'\hat{V}_i^{-1}\boldsymbol{Z}_i)^{-1}\boldsymbol{Z}_i'\hat{V}_i^{-1}]\boldsymbol{X}_i\hat{\beta} + \boldsymbol{Z}_i(\hat{\psi}^{-1} + \boldsymbol{Z}_i'\hat{V}_i^{-1}\boldsymbol{Z}_i)^{-1}\boldsymbol{Z}_i'\hat{V}_i^{-1}\boldsymbol{Y}_i \\
&= \boldsymbol{\lambda}_i\boldsymbol{X}_i\hat{\beta} + (\boldsymbol{I}_n - \boldsymbol{\lambda}_i)\boldsymbol{Y}_i, \qquad (1.10)
\end{aligned}$$

where $\boldsymbol{\lambda}_i = \boldsymbol{I}_n - \boldsymbol{Z}_i(\hat{\psi}^{-1} + \boldsymbol{Z}_i'\hat{V}_i^{-1}\boldsymbol{Z}_i)^{-1}\boldsymbol{Z}_i'\hat{V}_i^{-1}$.

From equation (1.10), we see that the predicted response profile is the weighted sum of predicted value ($\boldsymbol{X}_i\hat{\beta}$) and observed value ($\boldsymbol{Y}_i$). Equation (1.10) shows how the EB causes shrinkage of the predicted response profile toward the population mean. This is why EB is often referred to as a shrinkage estimate. Further discussion of the EB method is given by Carlin and Louis (2000). The EB estimates can be obtained from software such as MLwiN (Goldstein et al., 1998; Rasbash & Woodhouse, 1995) or PROC MIXED in SAS. As an alternative to EB, fully Bayesian estimates can be calculated. For example, the software program BUGS (Spiegelhalter et al., 1996) uses the Gibbs sampler, which is a simulation-based procedure for fully Bayesian estimates.

1.3 Applying Multilevel Modeling to Intensive Longitudinal Data

In this section, an analysis of an intensive longitudinal database is reported. The focus is on the decision-making process followed among the model-building steps, including some exploratory analyses, the model specification, and diagnostic tools utilizing residuals to check for the appropriateness of model assumptions. Specifically, we present a two-level model and a three-level model, each with covariates.

1.3.1 Control and Choice in Indian Schoolchildren

The data for this analysis were contributed by Verma et al. (2002). The researchers conducted a diary study of stress in Indian early adolescents using the experience-sampling technique and, in general, found that students spending more time doing homework experienced lower average emotional states and more internalizing problems. Conversely, students who spent less time doing homework and more time in leisure activities had higher academic anxiety, higher general mental health, and lower scholastic achievement. Participants ($n = 100$) were polled up to 9 times daily by a wrist watch and asked to complete a short paper and pencil survey covering activities and emotional states, such as stress and motivational state. Each participant contributed approximately 40 reports over as many as 7 days.

This analysis adopts a hypothesis unrelated to the researchers' original scientific goals. Specifically, perceived sense of control over an activity may be related positively to a sense of choice about involvement in the activity for some participants and related negatively for others. No psychological studies have considered the association of these two affective states over many successive moments. Therefore, this analysis attempts to characterize this association for individual persons and to explain variability in the association over intensive measurements. Two variables from the original motivational state measure were selected as markers of participants' views–namely, sense of control ("How were you feeling when you were signaled, in control?") and sense of choice ("How much choice did you have about what you were doing?"). The scale items were measured on 9-point ordinal scales with four labeled points "not at all/kind of/pretty much/very much." The change of this association over the study term is not of interest; rather, only the overall association for each individual, independent of time and across activities, is of interest. Of further interest is the extent to which covariates account for individual differences in this association. For this reason, sex differences, effects related to day of the week, and sources of extra day-to-day variation are investigated.

1.3.2 Exploratory Analyses

The distributions across persons and over time within persons were reviewed. Some participants reflected possible ceiling effects in responding to items over time; however, because this occurred infrequently, normality of responses was assumed. Participant curves were registered by day of the week. A loess curve of the response (control) over time for four representative individual trajectories is shown in figure 1.1. The individual trajectories reflected substantial variability over the group trajectory; most participants reflected a plot like the one in the lower left-hand corner. About 25% of the plots reflected other shapes; however, there was no theoretical expectation of an overall population trend in the population over the week. Hence, time was not modeled using polynomials and person-specific time-graded variability was transferred into the error covariance structure. In other studies in which the person-specific variability is deemed to be theoretically meaningful, this decision would be inappropriate.

In addition, correlation coefficients of control and choice were computed for all 100 persons. A histogram of these correlations is shown in figure 1.2. The mean correlation between control and choice was 0.0924, with a minimum of −0.379 and a maximum of 0.865. These correlations suggest that although the average relationship seems to be positive, there is substantial variation in the correlation of control and choice across persons.

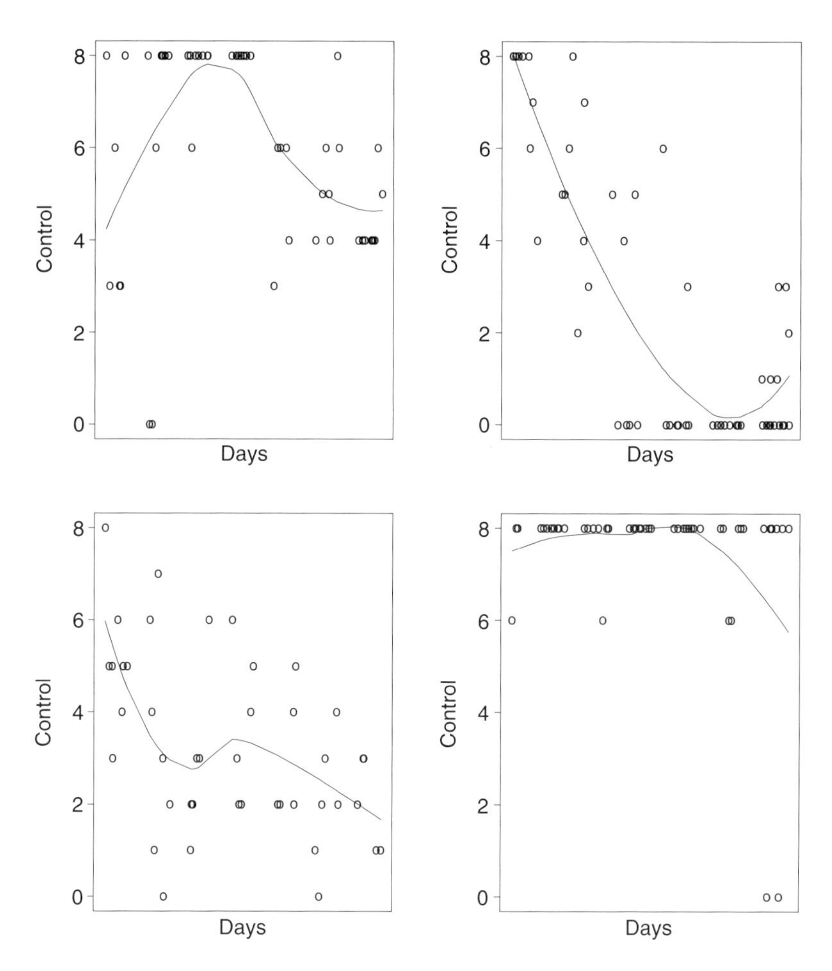

Figure 1.1 Loess curves for four randomly selected persons.

1.3.3 The Two-Level Model

In order to assess the time-varying influence of choice on control, as well as the possible explanatory role of day of the week and other covariates, two multilevel models were developed using SAS PROC MIXED. For brevity, a subset of the possible covariates was selected, namely, choice, day of the week, and sex. Models with richer substantive meaning would include covariates such as age, socioeconomic status, activity at time of polling, grade in school, and their interactions.

The first level of the model included the time-varying effect of choice on control and the second level included the effect of day of the week and sex on the intercept and slope coefficients of the first level. The choice variable was

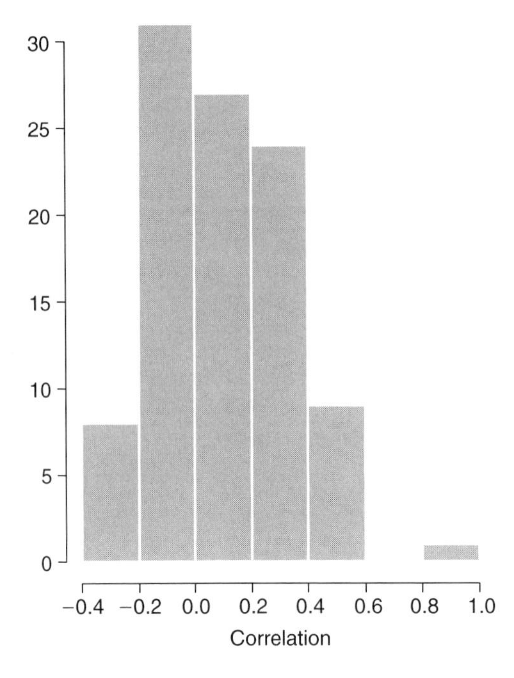

Figure 1.2 The histogram of the correlations between control and choice.

utilized as the main covariate of interest, instead of a variable reflecting an absolute measure of time (calendar or clock time). Some temporal influences were accounted for by including a day of the week effect. Within-day effects were not modeled because participants moved among classes differently during the school day, making contextual effects difficult to dissociate from within-day estimates of within-person states. These decisions make sense because, as mentioned above, the phenomenon of interest here is not longitudinal change per se, rather, the nature and explanation of the correlation of choice and control over all occasions. From figure 1.2, it is clear that this correlation varies by person. Preliminary analyses also showed some effect of day of the week. Therefore, we investigated models of the following form.

Let y_{it} denote the control measurement for person i on occasion t. The two-level model is

$$\text{Level 1: } y_{it} = \pi_{i0} + \pi_{i1} * choice_{it} + e_{it}.$$

$$\begin{aligned}\text{Level 2: } \pi_{i0} &= \beta_{00} + \beta_{01} * sex_i + \beta_{02} * Monday_i + \beta_{03} * Tuesday_i + \beta_{04} * Wednesday_i \\ &\quad + \beta_{05} * Thursday_i + \beta_{06} * Friday_i + \beta_{07} * Saturday_i + b_{i0}, \\ \pi_{i1} &= \beta_{10} + \beta_{11} * sex_i + \beta_{12} * Monday_i + \beta_{13} * Tuesday_i + \beta_{14} * Wednesday_i \\ &\quad + \beta_{15} * Thursday_i + \beta_{16} * Friday_i + \beta_{17} * Saturday_i + b_{i1}.\end{aligned}$$

In fitting this model, we found significant fixed effects for sex, day of the week, and choice on the intercept of control, but no significant effect of sex or day of the week on the random slope at the second level. That is, at the second level, estimates of the variability in random intercepts and slopes for choice were significant, but the selected cross-level interactions were not significant and were removed from the model. Expressing the model in composite form, the reduced model became

$$\begin{aligned} y_{it} = {} & \beta_{00} + \beta_{01} * sex_i + \beta_{02} * Monday_i + \beta_{03} * Tuesday_i + \beta_{04} * Wednesday_i \\ & + \beta_{05} * Thursday_i + \beta_{06} * Friday_i + \beta_{07} * Saturday_i + \beta_{10} * choice_{it} \\ & + b_{i0} + b_{i1} * choice_{it} + e_{it}, \end{aligned}$$

where e_{it} is normally distributed with mean zero and variance σ^2 and $\boldsymbol{b}_i = (b_{i0}, b_{i1})'$ is normally distributed with mean vector zero and covariance matrix,

$$\boldsymbol{\psi} = \begin{bmatrix} \psi_{00} & \psi_{01} \\ \psi_{10} & \psi_{11} \end{bmatrix},$$

and the e_{it} and $\boldsymbol{b}_i$ are mutually independent.

In table 1.1, the specific estimates for the dummy-coded day variables, other fixed effects, and variance components are reported. In table 1.2, the type 3 tests of fixed effects are shown. The estimate of ψ_{11} and its relatively small standard error indicated that substantial heterogeneity in the effect of choice on control still existed among persons. None of the selected covariates, however, explained this variation. We were only able to explain the heterogeneity of control intercepts.

Table 1.1 Estimates for two-level model with random intercept and slope

Variable	Day of week	Sex	Estimate	SE	Z
Intercept			4.6823	0.2935	15.95
Choice			0.0941	0.0196	4.81
Day of week	Mon.		0.2215	0.1211	1.83
Day of week	Tues.		0.0372	0.1188	0.31
Day of week	Wed.		−0.1414	0.1201	−1.18
Day of week	Thurs.		−0.2219	0.1176	−1.89
Day of week	Fri.		−0.2725	0.1180	−2.31
Day of week	Sat.		−0.3040	0.1187	−2.56
Day of week	Sun.		0	—	—
Sex		Male	−0.9850	0.3901	−2.52
Sex		Female	0	—	—
Level 2 variance:					
ψ_{00}			3.7687	0.6298	
ψ_{01}			−0.05422	0.04390	
ψ_{11}			0.02037	0.005318	
Level 1 variance:					
σ^2			4.4830	0.09458	

Table 1.2 Tests of significance for fixed effects

Effect	Num. DF	Den. DF	F value	Pr> F
Choice	1	4585	23.13	< 0.0001
Day of week	6	558	5.16	< 0.0001
Sex	1	98	6.37	0.0132

This simple two-level model assumes that the within-person correlations $y_{it}, y_{it'}$ are due only to random intercepts and random choice slopes. It assumes that no additional autocorrelation is present. This assumption must be checked, a need that the empirical semivariogram addresses well.

We first briefly review how to transform residuals resulting from the multilevel model and semivariograms for pedagogical and illustrative purposes.

Analysis of residuals As in ordinary regression analysis, diagnostics for the fitted multilevel model are necessary to ensure that the assumptions underlying the model have been met. This is typically pursued through the examination of residuals. In multilevel models, there are residuals at each level. All model diagnostics used for the residuals of the ordinary regression model can also be applied to the residuals at each level in the multilevel model. However, instead of this stepwise analysis, we here examine the residuals obtained from the composite model, which are expressed in terms of the level 1 residual and the level 2 residuals. For example, the composite residuals in our two-level model have the form of $(b_{0i} + b_{1i} * choice_{it} + e_{it})$. They describe the differences between the observed response and the predicted mean response. Further discussion on analysis of residuals in multilevel models is given by Willett and Singer (2003) and by many standard textbooks on multilevel models.

To keep matters simple, composite residuals from the two-level model, r_{it}, are used for demonstration. This can be conceptually extended to higher-level models, r_{ijt} in the three-level model, and so on. To check the assumptions made on errors, ϵ_{it}, we usually use the residuals, which are the estimates of errors and are defined as

$$\epsilon_{it} = y_{it} - \boldsymbol{X}_i\boldsymbol{\beta},$$
$$\hat{\epsilon}_{it} = r_{it} = y_{it} - \boldsymbol{X}_i\hat{\beta}.$$

From our model specification, the responses of person i, $\boldsymbol{Y}_i$, have a normal distribution with mean vector $\boldsymbol{X}_i\boldsymbol{\beta}$ and covariance matrix Σ_i, that is, $\boldsymbol{Y}_i \sim N(0, \Sigma_i)$, where Σ_i consists of $\boldsymbol{Z}_i\boldsymbol{\psi}\boldsymbol{Z}_i' + \sigma^2\boldsymbol{I}$ in the two-level model, and $\boldsymbol{Z}_i\boldsymbol{\psi}\boldsymbol{Z}_i' + \boldsymbol{\xi} + \sigma^2\boldsymbol{I}$ in the three-level model, so the $\boldsymbol{\epsilon}_i$'s are normally distributed with mean vector zero and covariance matrix Σ_i. The residuals would approximate the distribution of errors if our assumed model is true so they can reasonably reflect the behaviors of unobserved errors.

One possible way to check the appropriateness of assumptions on errors is to plot residuals against the predicted values ($\boldsymbol{X}_i\hat{\beta}$), or time t. Under the assumed model, residuals would have a corresponding covariance structure. As a result, the plot of residuals against the predicted values or time would show some systematic patterns because of the correlation among residuals. For this reason, plotting the untransformed residuals is not a good strategy to check the appropriateness of assumptions. Instead, the standardized residuals should be used by transforming the residuals so that they are uncorrelated and have constant unit variance of 1. For this transformation of the residuals, we need matrix $\boldsymbol{L}_i$, where

$$Cov(\boldsymbol{L}_i^{-1}\boldsymbol{r}_i) = (\boldsymbol{L}_i)^{-1}Cov(\boldsymbol{r}_i)(\boldsymbol{L}_i')^{-1} = \boldsymbol{I}.$$

We can find lower triangular matrix $\boldsymbol{L}_i$ by using the Cholesky decomposition. If $\hat{\Sigma}_i$ is an approximate covariance matrix for residuals, that is, $Cov(\boldsymbol{r}_i) = \hat{\Sigma}_i$, the Cholesky decomposition of $\hat{\Sigma}_i$ creates lower triangular matrix $\boldsymbol{L}_i$ such that $\hat{\Sigma}_i = \boldsymbol{L}_i\boldsymbol{L}_i'$. By multiplying the inverse of lower triangular matrix $\boldsymbol{L}_i$, the residuals are standardized with mean vector zero and covariance matrix $\boldsymbol{I}$. Let $\boldsymbol{r}_i^*$ be the vector of standardized residuals,

$$\boldsymbol{r}_i^* = \boldsymbol{L}_i^{-1}\boldsymbol{r}_i = \boldsymbol{L}_i^{-1}(\boldsymbol{Y}_i - \boldsymbol{X}_i\hat{\beta}).$$

Under this transformation, if our model is correct, r_{it}^* are no longer correlated so we can use these standardized residuals for diagnostics much as we do residuals in any ordinary linear regression analysis.

Semivariograms In addition to the basic diagnostics that have been used in the ordinary regression analysis, in the fields of spatial and geostatistics the semivariogram is frequently used to depict the covariance structure of data collected over space. In the case of longitudinal data, the semivariogram reflects the covariance structure for the time-graded data points (Fitzmaurice et al., 2004; Verbeke et al., 1998; Verbeke & Molenberghs, 2000).

The definition of a semivariogram for time-related data is

$$\rho(h) = \frac{var(r_t - r_{t'})}{2}, \tag{1.11}$$

where $h = |t - t'|$ and r_t and $r_{t'}$ are residuals of measurements each at time t and at time t' for the same person. Since these residuals are standardized with mean 0 and variance 1, the formula for the semivariogram in equation (1.11) becomes

$$\rho(h) = \frac{var(r_t - r_{t'})}{2} = \frac{[\,var(r_t) + var(r_{t'}) - 2cov(r_t, r_{t'})]}{2}$$
$$= 1 - cov(r_t, r_{t'}).$$

If r_t and $r_{t'}$ are not correlated (i.e., $cov(r_t, r_{t'})$=0), the semivariogram for the standardized residuals will be equal to 1. To produce the semivariogram plot, we use an empirical semivariogram, which is the average of the squared difference between pairs of residuals, plotting h versus $\rho(h)$. If our assumed covariance structure is proper, the points in the semivariogram plot would deviate randomly around 1 and reflect no systematic variation over time. Otherwise, the semivariogram plot will show a trend over time which implies that the assumed covariance structure is not adequate. In this sense, the semivariogram plot can be used to determine the appropriateness of the assumed covariance structure.

We can construct a semivariogram using the transformed residuals described above. For example, suppose r^*_{it} is the transformed residual for person i at time t. Instead of $\frac{1}{2}E(r^*_{it} - r^*_{it'})^2$, we calculate the squared difference between r^*_{it} and $r^*_{it'}$, and divide by 2, where $t \neq t'$, and the elapsed time $|t - t'|$. By plotting $\frac{1}{2}(r^*_{it} - r^*_{it'})^2$ versus $|t - t'|$, an empirical semivariogram for person i can be drawn.

Figure 1.3 shows loess smoothed curves of the empirical semivariogram for all persons from the two-level model residuals. Clearly, the smoothed line of the residuals reveals a trend, indicating that the covariance structure may not be adequate. Figure 1.4 shows the semivariogram for some randomly chosen persons based on the two-level model. These plots demonstrate that several possible trends may exist over days for given individuals.

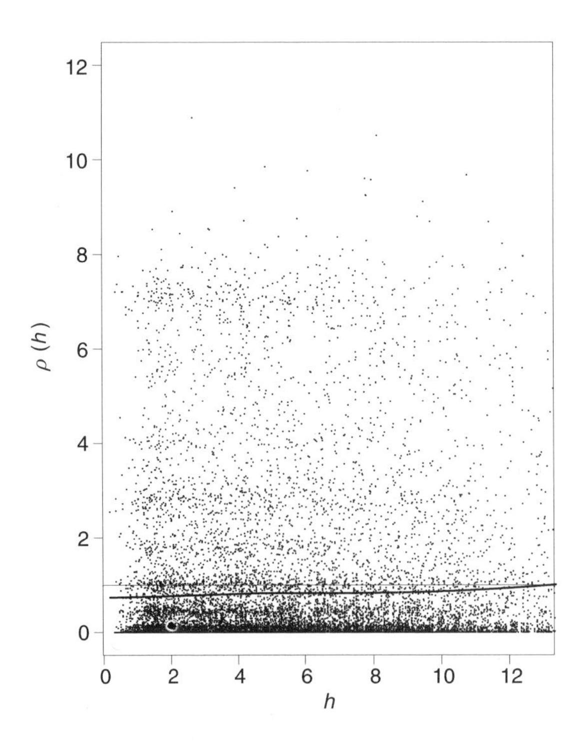

Figure 1.3 Loess smoothed curve of the empirical semivariogram for all persons from the two-level model.

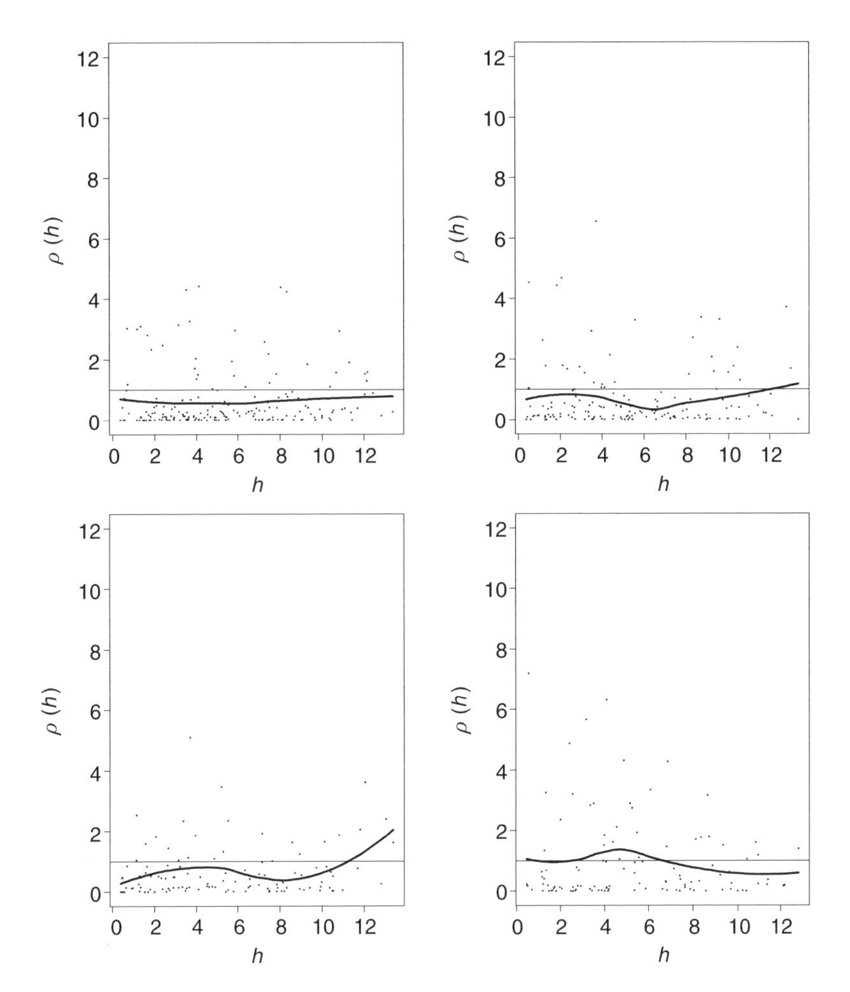

Figure 1.4 Loess smoothed curve of the empirical semivariogram for randomly chosen participants from the two-level model.

1.3.4 The Three-Level Model

Because of the notable ψ_{11} coefficient, there is a suspicion that additional day-to-day unexplained variation within individuals can arise. This variation could account for the apparent autocorrelation in the semivariogram. If this variation is not accounted for, it could lead to misleading inferences about some of the fixed effects in the model, notably, the day of the week terms.

The assumptions of the two-level model led to the following covariance structure, $\boldsymbol{Z}_i\boldsymbol{\psi}\boldsymbol{Z}_i' + \sigma^2\boldsymbol{I}$. This covariance structure does not accurately reflect the phenomenon, however, because there is likely to be some autocorrelation in the moment-to-moment reports of control. That said, the within-day moment-to-moment states cannot easily be separated from influences caused by varying classes among subjects within a day in this database. The autocorrelation that may exist from day to day may be more important. Because setting up an error covariance structure to vary by day would have proven difficult, and an autocorrelation was expected from day to day, the model was specified to account for the effect of measurements of within a day by adding another level to the model. Specifically, we included random effects of day in order to account for the correlation by day among the moment-to-moment reports of control and to see if days vary substantially for each person.

Hence the three-level model was specified as follows. The first level is the time-varying effect of choice on control. The intercepts for control and slopes for the association of choice and control for each person were specified as random effects. The second and third levels were created by nesting the day of measurement (day) within person (ID), that is, days were nested within person.

Letting y_{ijt} denote the control measurement for person i in day j on occasion t, the three-level model model was written as:

$$
\begin{aligned}
\text{Level 1: } y_{ijt} &= \pi_{ij0} + \pi_{ij1} * choice_{ijt} + e_{ijt}. \\
\text{Level 2: } \pi_{ij0} &= \delta_{i00} + \delta_{01} * Monday_i + \delta_{02} * Tuesday_i + \delta_{03} * Wednesday_i \\
&\quad + \delta_{04} * Thursday_i + \delta_{05} * Friday_i + \delta_{06} * Saturday_i + \eta_{ij0}, \\
\pi_{ij1} &= \delta_{i10} + \delta_{11} * Monday_i + \delta_{12} * Tuesday_i + \delta_{13} * Wednesday_i \\
&\quad + \delta_{14} * Thursday_i + \delta_{15} * Friday_i + \delta_{16} * Saturday_i + \eta_{ij1}. \\
\text{Level 3: } \delta_{i00} &= \beta_{00} + \beta_{01} * sex_i + b_{i0}, \\
\delta_{i10} &= \beta_{10} + \beta_{11} * sex_i + b_{i1},
\end{aligned}
$$

where e_{ijt} is normally distributed with mean zero and variance σ^2, $\boldsymbol{\eta}_{ij} = (\eta_{ij0}, \eta_{ij1})'$ is normally distributed with mean vector zero and covariance matrix

$$
\boldsymbol{\xi} = \begin{bmatrix} \xi_{00} & \xi_{01} \\ \xi_{10} & \xi_{11} \end{bmatrix},
$$

and $\boldsymbol{b}_i = (b_{i0}, b_{i1})'$ is normally distributed with mean vector zero and covariance matrix

$$
\boldsymbol{\psi} = \begin{bmatrix} \psi_{00} & \psi_{01} \\ \psi_{10} & \psi_{11} \end{bmatrix}.
$$

Fitting the three-level model with random intercepts and slopes, the estimated variance of the random slopes was close to zero. Therefore, the three-level model was refined by letting only the intercepts vary randomly at the second level while allowing both intercepts and slopes to vary at the third level. Moreover, the interaction terms among sex, day, and choice were not significant, so all cross-level interactions were excluded from the three-level model. (Note, however, from a

Table 1.3 Estimates for a more parsimonious three-level model

Variable	Day of week	Sex	Estimate	SE	Z
Intercept			4.7361	0.3074	15.40
Choice			0.08647	0.01646	5.25
Day of week	Mon.		0.2227	0.1869	1.19
Day of week	Tues.		0.0523	0.1836	0.28
Day of week	Wed.		−0.1631	0.1842	−0.89
Day of week	Thurs.		−0.2261	0.1822	−1.24
Day of week	Fri.		−0.2619	0.1827	−1.43
Day of week	Sat.		−0.3108	0.1840	−1.69
Day of week	Sun.		0	—	—
Sex		Male	−1.0138	0.3940	−2.57
Sex		Female	0	—	—
Level 3 variance:					
ψ_{00}			3.6026	0.6202	
ψ_{01}			−0.01989	0.03681	
ψ_{11}			0.01063	0.003689	
Level 2 variance:					
ξ			1.0023	0.09419	
Level 1 variance:					
σ^2			3.6722	0.0825	

substantive standpoint that exploratory analyses utilizing additional covariates that reflect time-varying aspects of the school or nonschool environment suggest possible explanations of this variation.) The final three-level model in composite form is

$$\begin{aligned} y_{ijt} = \beta_{00} &+ \beta_{01} * sex_i + \beta_{02} * Monday_i + \beta_{03} * Tuesday_i \\ &+ \beta_{04} * Wednesday_i + \beta_{05} * Thursday_i + \beta_{06} * Friday_i + \beta_{07} * Saturday_i \\ &+ \beta_{10} * choice_{ijt} + \eta_{ij} + b_{i0} + b_{i1} * choice_{ijt} + e_{ijt}. \end{aligned}$$

Results from this three-level model are shown in tables 1.3 and 1.4 and a plot of the estimates for each day is shown in figure 1.5.

Table 1.4 Tests of significance for fixed effects

Effect	Num. DF	Den. DF	F value	Pr>F
Choice	1	4585	27.59	< 0.0001
Day of week	6	558	2.25	0.0374
Sex	1	98	6.62	0.0116

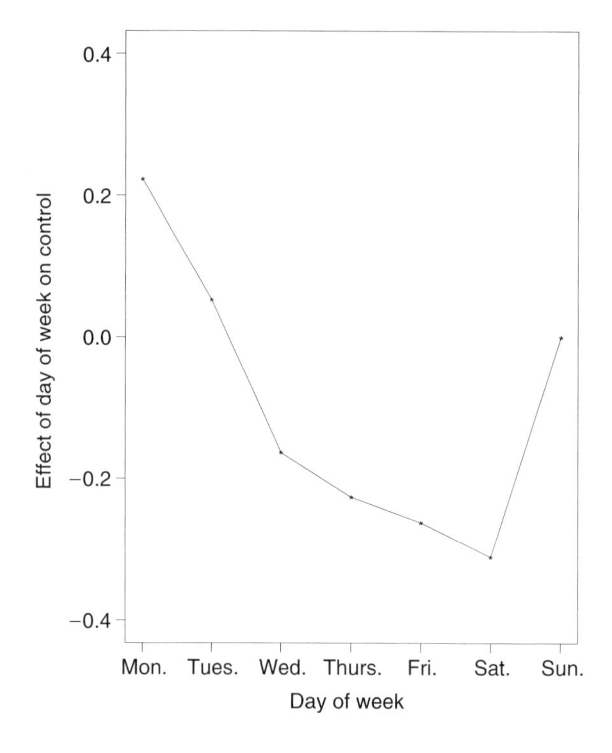

Figure 1.5 The effect of day of the week on control in the three-level model.

1.3.5 Interpretation of the Three-Level Model

The three-level model demonstrates how to take advantage of longitudinal data in the absence of anticipated time-graded effects and in service of a multivariate theory of individual functioning. The plot in figure 1.5 shows the day-by-day coefficients of control from the three-level model and demonstrates the need to adjust for time effects, where Sunday is an aberrant day, perhaps caused by having more persons from private schools with relaxed schedules contributing to scores on that day.

There are only minor differences between the two- and three-level models in terms of the interpretation of the estimates; however, the standard errors from the three-level model are wider and because the semivariogram indicates that the covariance structure accounts for the residual variation well, these error estimates are more realistic. Also, note that the level 2 variance component is highly significant, showing that there is substantial day-to-day variation. Table 1.4 provides the results of the three-level model.

If specific or prototypical individuals are of interest, it is possible to produce estimates for the effect of choice for each person. If ML or REML is used for estimation, we can obtain these specific effects using empirical Bayes rules. The empirical Bayes estimates and confidence intervals for the coefficient of choice

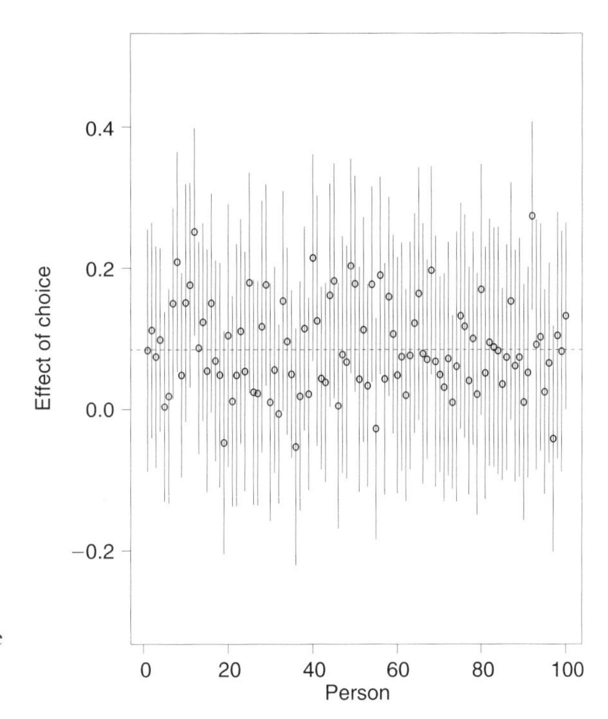

Figure 1.6 Empirical Bayes estimates of the effect of choice for 100 individuals with 95% confidence intervals.

across persons are shown in figure 1.6, where the dotted line represents the population average effect of choice on control. This figure shows the situation in which the population effect of choice does not explain that of some individuals at all. With other modeling methods for longitudinal data such as marginal models, the effect of choice on control on average would be clear, but how individuals may vary in this regard would not be clear (0.086 ± 0.04). The effect of choice on control for some individuals could be negative, even though the average for the population is positive. If so, this is a potentially important finding that would never be discovered if only marginal models were fitted.

Figures 1.7 and 1.8 show loess smoothed curves of the empirical semivariogram of the three-level model for all participants and for some randomly chosen participants. Note that all of the smoothed lines are closer to 1 and reflect less trending than seen in the corresponding plots from the two-level model.

Briefly, recall that figure 1.3 shows the semivariogram for the whole sample using the transformed residuals from the two-level model, estimated by a loess smoothed curve. In the semivariogram of the two-level model the points deviate around 0.7, which suggested that the two-level model was not appropriate to account for the correlation among measurements within days. For this reason, the three-level model was developed. Examination of the semivariogram for the

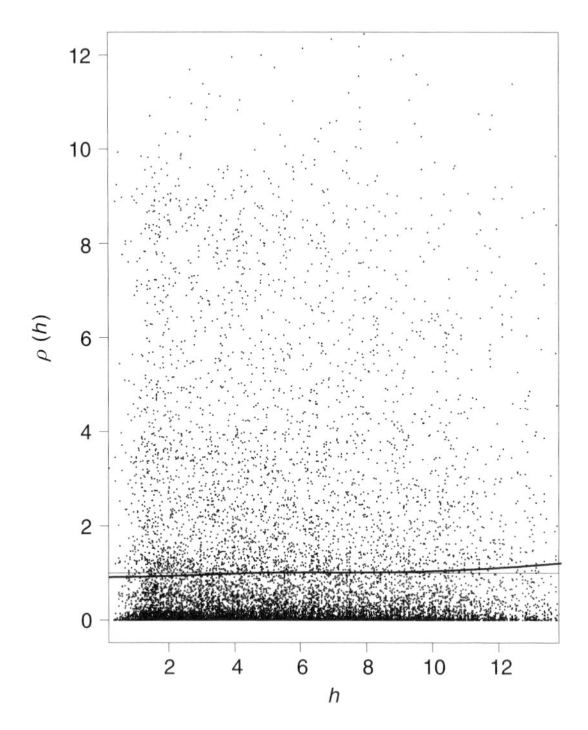

Figure 1.7 Loess smoothed curve of the empirical semivariogram for all persons from the three-level model.

three-level model indicated that the assumed covariance matrix, $\boldsymbol{Z\psi Z'} + \boldsymbol{\xi} + \sigma^2\boldsymbol{I}$, described the time-graded structure of these data rather well, as the points randomly deviate around 1 (see figure 1.7). Hence, there was no evidence to suggest that there is any correlation of the residuals over time beyond what the three-level model accounted for.

After fitting the final multilevel model to the data, it was necessary to check if the assumptions about the model were valid. The normal quantile plot of residuals and the plot of the transformed residuals against the predicted values provide diagnostics for checking the normality of errors, outliers, and constant error variance.

The normal quantile plot of the residuals for the three-level model (figure 1.9) shows that there is no significant departure from normality or outliers, suggesting that the assumption of normality is reasonable. This is especially remarkable given that the response is measured on an ordinal scale. The plot of the transformed residuals against the predicted values for the three-level model (figure 1.10) does not reveal any obvious patterns. That is, there is no fanning out of the graph from either side to the other. This suggests that the variances are well modeled. More dramatically, the plot of the absolute values of the transformed residuals against the predicted values (figure 1.11) reveals no systematic trends, which implies constant variance of the errors. There is a mild indication of heteroscedasticity

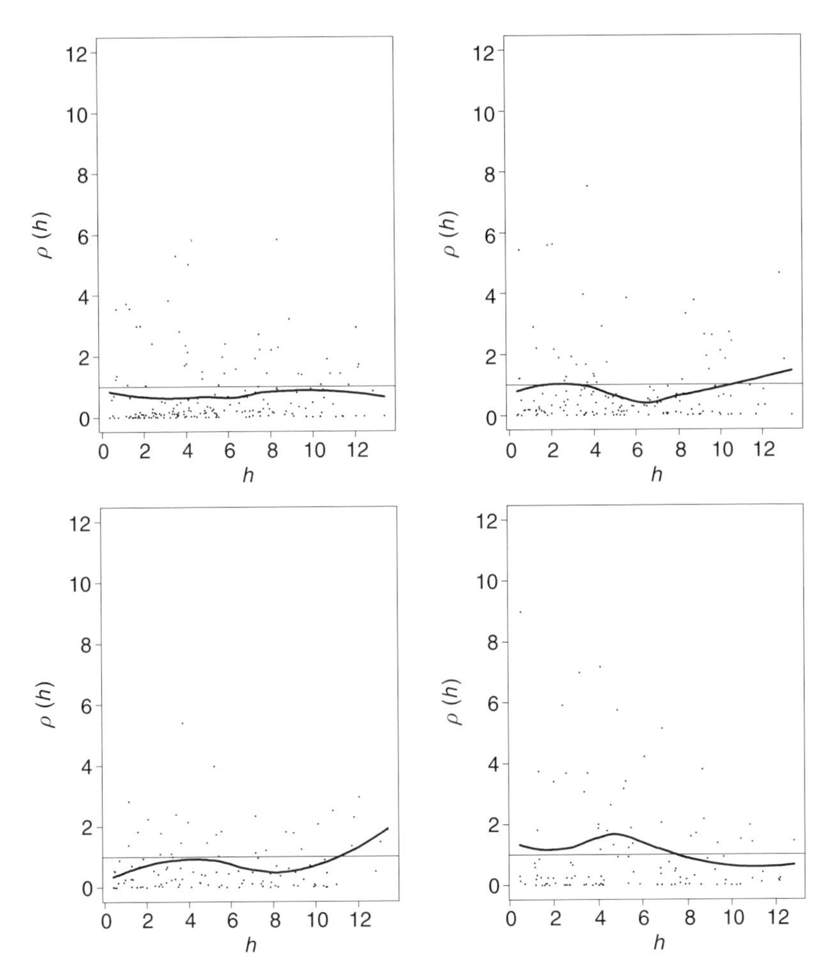

Figure 1.8 Loess smoothed curve of the empirical semivariogram for randomly chosen participants from the three-level model.

reflected in the downturn of the line; this is most likely due to a ceiling effect for several respondents on the response scale. More detailed explanation and examples of diagnostic residual analysis are available in Fitzmaurice et al. (2004).

1.4 Application: Control and Choice in Indian Schoolchildren

The purpose of this section is to stimulate ideas on how to apply multilevel models to ILD. Innovative strategies undertaken in this volume and elsewhere are highlighted.

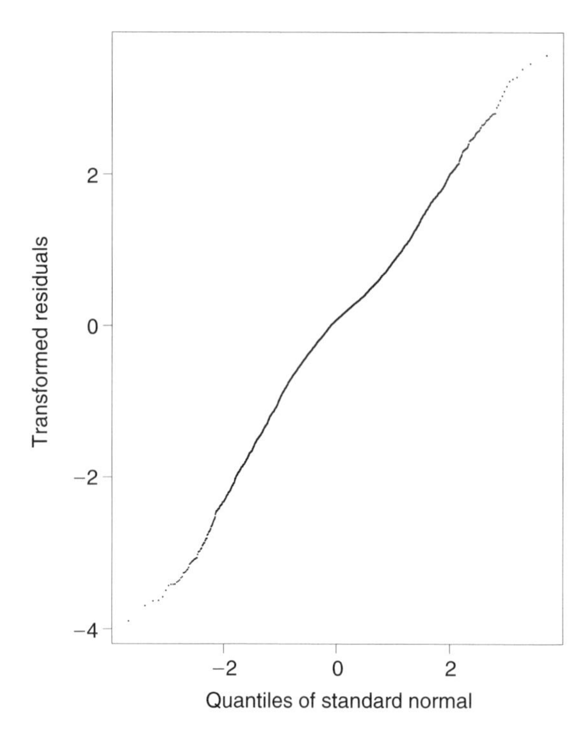

Figure 1.9 Normal quantile plot of the transformed residuals in the three-level model.

1.4.1 Using Transformed Variables in the Multilevel Model

In order to address the challenges described above, several chapters in this volume describe ways of developing alternative inputs to the multilevel model at the first level of the model and also as time-varying covariates. These quantities are typically derived values developed to characterize other properties of the person-specific series. For example, Rovine and Walls (chapter 6, this volume) demonstrate the incorporation of lagged variables reflecting a first-order autoregressive process. This approach is promising to better describe the regularity of individual person's series. Similarly, Boker and Laurenceau (chapter 9, this volume) describe the multilevel modeling of derivatives reflecting heterogeneity of rates and changes in rates within persons. Hedeker et al. (chapter 4, this volume), as described below, incorporate item parameters into the first level of a multilevel model. In general, the spirit of these new models is on diversification of the inputs used in the measurement occasions level of the model, to reflect processes of various types, regulatory, change or item parameter based.

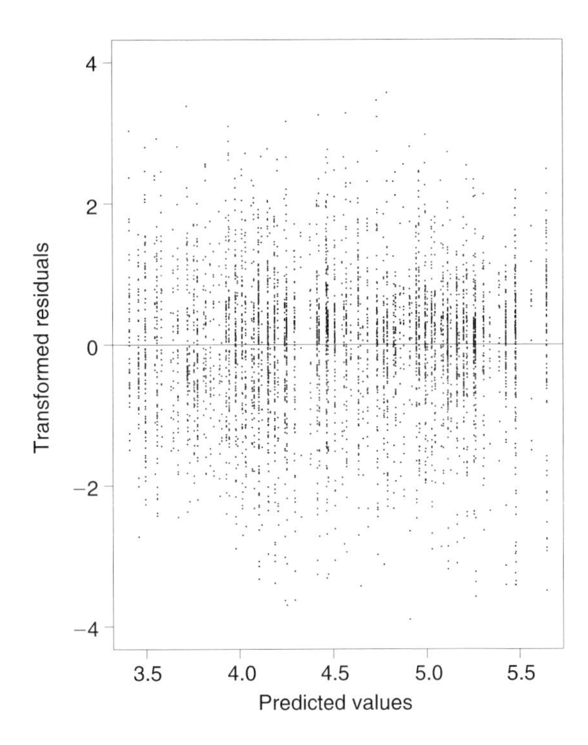

Figure 1.10 Plot of the transformed residuals against the predicted values in the three-level model.

1.4.2 Nonlinear Modeling Options

In general, multilevel models assume that the predictors are related to the response linearly, including polynomial or transformed variables that are linear in their parameters, so linear regression models are satisfactory approximations for multilevel modeling. However, as in the case of multilevel models for binary and count data where the relationship between the response and the predictors is not linear, nonlinear approaches are needed for occasions when an empirically indicated or a theoretically justified nonlinear model is more appropriate. This is also the case when data are not intensive, but it may be particularly necessary to account for nonlinear relations in the case of ILD. For example, in a nonlinear model for binary responses, we can utilize logistic regression for each person and allow one or more coefficients to randomly vary across persons (see Hedeker et al., chapter 4, this volume).

There are several algorithms available for maximum likelihood estimation of multilevel models for normally distributed outcomes. However, in the case of estimation of the nonlinear models, maximum likelihood procedures can be intractable, involving irreducibly high dimensional integrals over the random effects to obtain the unconditional distribution of the response. Some relevant

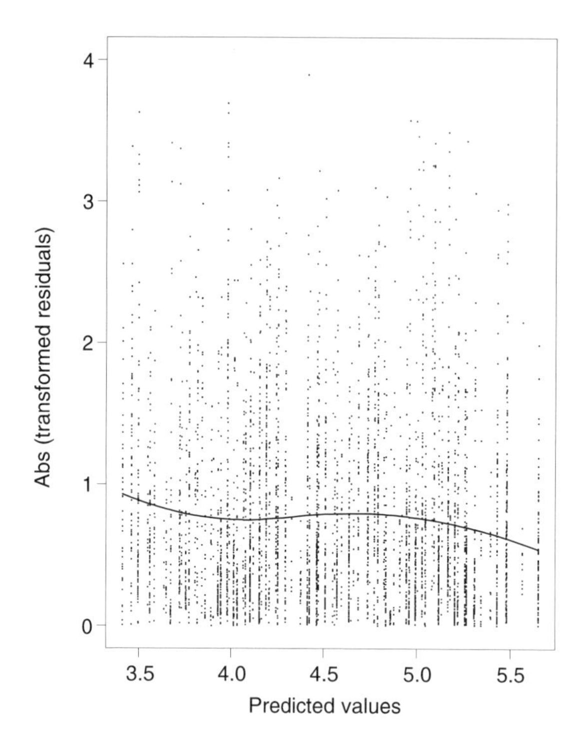

Figure 1.11 Plot of the absolute value of the transformed residuals against the predicted values in the three-level model.

scholarship in this area includes Breslow and Clayton (1993), Cudeck and du Toit (2002), and Rodrígues and Goldman (2001) .

1.4.3 Modeling in the Absence of Hypothesized Absolute Time Effects

In our analysis, there are no hypothesized influences of absolute (or calendar) time. That is, the variability over days is important to the description and explanation of within- and between-person effects, but we are simply using many occasions of measurement to enable the best estimates of an association of interest over the length of individual series. In a sense, time is a confound that we would not want to influence the model structure.

Another strategy for modeling time effects is a situation when the time trend is highly anticipated and can be modeled based on an a priori known smooth function. In this situation, the goal is to describe the function of time within individuals well. The following description of a way of fitting a sinusoidal curve model is a representative strategy in this situation.

As described earlier, much scholarship presents growth curve models in which the level 1 equation includes the linear effect of time (and possibly low-order polynomials), treated as random effects in order to model individual differences in the temporal pattern of the outcome variable. When only a linear term for time is included, individual differences in the average rate of change are modeled and one can test to determine whether these differences are related to other person-level characteristics (e.g., by sex).

While the inclusion of higher-order polynomials or splines (see Li et al., chapter 3, this volume; Ramsay & Silverman, 1997) offers one strategy for modeling nonlinear temporal patterns, this strategy does not describe complex patterns that one might *expect* temporal patterns to exhibit. In the presence of nonsystematic variation over time, these models have great appeal. At the same time, many functions, more complex ones than low-order polynomials, but short of those inherent to nonparametric estimation approaches, can be employed in the presence of strong theoretical suppositions about the series. For example, diurnal patterns repeat themselves every 24 hours, day-of-the-week effects repeat themselves across weeks, and seasonal patterns repeat themselves from one year to the next. In this section, we briefly describe how such patterns might be specified in a multilevel model.

Weekday versus weekend differences Suppose there is an outcome, y, that tends on average to be higher on weekends than weekdays, perhaps cheerfulness. The average level of cheerfulness varies from person to person, and the difference between weekends and weekdays may also vary among persons. Let us define $z_{it} = 1$ if assessment t for person i occurs on a weekend; otherwise $z_{it} = 0$.

In the level 1 equation,

$$y_{it} = \pi_{0i} + \pi_{1i}z_{1it} + e_{it}, \tag{1.12}$$

π_{0i} will be person i's average cheerfulness on weekdays and π_{1i} will be the difference between his/her average cheerfulness on weekends compared to weekdays. If π_{1i} is treated as a fixed effect (assumed to equal a constant, π_1, for all i), then the estimated variance of the π_{0i} corresponds to the between-person variance in overall average cheerfulness. However, if π_{1i} is treated as a random effect, then the estimated variance of the π_{0i} corresponds to the between-person variance in average weekday cheerfulness and the variance of the π_{1i} corresponds to the between-person variability in the weekend versus weekday difference in average cheerfulness. If data on y are collected over several weeks, this model implies that the weekend versus weekday effect is consistent from week to week (i.e., periodicity is built into this model). In addition, one could add a level 2 predictor, for example, sex, and its interaction with z to explain the first level. In addition to weekend versus weekday effects, there may be other day-of-the-week effects. For example, people may be more cheerful on Fridays than on other weekdays, least cheerful on Mondays, more cheerful on Saturdays than Sundays, and so on. By choosing a reference day, or days, of the week, and defining multiple dummy-coded indicator variables (different $z_{1it}, z_{2it}, \ldots$), one could simultaneously incorporate each of these hypothesized effects into a model. This same type of specification could be used to model seasonal effects (differences between spring, summer, autumn, and winter) or time-of-day effects (morning, afternoon, evening).

A potential problem with the above "indicator variable" approach to modeling temporal patterns is that it assumes a step function, with instantaneous shifts in the outcome variable as one crosses the boundary between adjacent time periods (e.g., from Friday night to Saturday morning, or from 11:59 A.M. to noon and from 5:59 P.M. to 6:00 P.M.). It also raises questions about where the boundaries should be specified, for example, should the weekend begin on Friday night at midnight, or on Friday afternoon at 5 P.M.? An alternative is to specify a smooth curve that nonetheless has the property of periodicity referred to above.

Specifying a sinusoidal function Perhaps the simplest smooth curves to exhibit periodicity are the sine and cosine functions. The problem with either one of these alone is that each predefines where the nadir and peak of the curve will occur. However, in combination, one can specify a sinusoidal (sinelike) function in which the nadir and peak are determined empirically. The one feature of this model that must be specified a priori is the periodic rate, the time interval after which the function begins repeating itself. For example, diurnal patterns repeat themselves every 24 hours. To specify a model with a sinusoidal diurnal pattern, one must define two level 1 variables. If T_{it} is the time of day that y_{it} was assessed, measured in minutes past midnight, then we can define $z_{1it} = \sin(360T_{it}/1440) = \sin(T_{it}/4)$ and $z_{2it} = \cos(T_{it}/4)$.

In the level 1 equation,

$$y_{it} = \pi_{0i} + \pi_{1i}z_{1it} + \pi_{2i}z_{2it} + e_{it}, \tag{1.13}$$

a separate sinusoidal function is implicitly fitted to each person's data. Together, π_{1i} and π_{2i} indicate the amplitude of the sinusoidal function (half the difference between the predicted values of y_{it} at the peak and nadir of the curve for person i is $\sqrt{\pi_{1i}^2 + \pi_{2i}^2}$ and the time of day at which the peak $= 540(\pi_{2i} < 0) + \sin(\pi_{2i}) \arcsin(\pi_{2i}/\sqrt{\pi_{1i}^2 + \pi_{2i}^2}$ minutes past midnight and the nadir (12 hours later/earlier) occur). Applications of this kind of model appear in Jacob and colleagues (1999) and Schwartz (2005).

1.4.4 Modeling in the Presence of Hypothesized Absolute Time Effects

In the presence of hypothesized absolute or calendar time effects, we can undertake a number of strategies. Simple methods for encoding time (e.g., linear, quadratic, etc.) are often inadequate. Polynomials are unlikely to describe well either the population average or the individual person trends. Sometimes the effects of time can be coded meaningfully as parameter estimates, that is, when the effects are not of specific interest, as in our example. Refer to the earlier individual participant figures of loess curves as examples, had we wanted to account for time-graded variability in the individual lines.

Instead, a rich set of flexible methods for fitting curves is needed, such as those found in Ramsay and Silverman (1997). There are two families of such models: splines (global approximation) and local polynomial fitting techniques (local approximation). In addition, some techniques covered in later chapters of this volume take advantage of nonparametric estimation in that they do not yield estimates that are intended for interpretation. Rather, the estimate of interest is the fitted curve itself, hence the heavy reliance on graphical displays for data interpretation. Fok and Ramsay (chapter 5, this volume) undertake both of these strategies in the same model, by combining trends that are modeled as splines with periodic effects. Moreover the family of generalized additive models (GAM) exploits the advantages of splines (Hastie & Tibshirani, 1990). Finally, for an application of local polynomial fitting functions to ILD, see Li et al. (chapter 3, this volume).

1.5 Summary

In general, the multilevel modeling framework is useful for practitioners in need of analyzing intensive longitudinal data from any source when a unit or person-specific trajectory is anticipated and variability in response intercepts or slopes is anticipated. Two example models using diary data were presented, key literature and resources were described, and some potentially useful extensions were outlined. Scientists may be able to utilize the two- and three-model specification procedure and concomitant residual analysis process provided herein for similar questions.

We also questioned the degree to which multilevel models are appropriate without careful implementation to characterize long series when clear theoretical expectations about the error structure cannot be made. Unfortunately, it is often the case that theoretical expectations are insufficient to inform modeling choices about the error structure. In addition, the nature of scientific questions under study may suggest the need to deploy the model in new ways, using sophisticated model specification strategies for the measurement occasions level and/or by modifying the nature of the inputs at the first level beyond raw data.

A new practical distinction involving whether intensive longitudinal studies bear theory that relies on absolute (calendar or clock) time or not in relation to the study of the phenomenon of interest was highlighted in this chapter. To reprise, if a long-term trend is expected, then careful consideration of appropriate error covariance structures is needed in estimating time-graded effects at the first level. This may be facilitated by use of semivariograms as we have shown. It may also be necessary to account for discontinuities in measurements over time, perhaps through piecewise regression along the time trend (not discussed herein, but see Neter et al., 1996, for an overview). However, if absolute time is not believed to be meaningful, as in our case, there may still be time-graded aspects of the model to account for, such as day of the week. Our models were thus analogous to multilevel models that consider clusters, but we are using a clustering conception to organize effects of time into levels in an attempt to control for their influence. In an alternative approach, Schafer (chapter 2, this volume) demonstrates the utility of marginal models in the absence of hypothesized absolute time effects, using the same databases described in this chapter.

References

Bolger, N., Davis, A., & Rafaeli, E. (2003). Diary methods: Capturing life as it is lived. *Annual Review of Psychology, 54*, 579–616.

Breslow, N.E., & Clayton, D.G. (1993). Approximate inference in generalized linear mixed models. *Journal of the American Statistical Association, 88*, 9–25.

Bryk, A.S., & Raudenbush, S.W. (1992). *Hierarchical Linear Models, Applications and Data Analysis Methods*. Newbury Park, CA: Sage.

Bryk, A.S., Raudenbush, S.W., & Congdon, R.T. (1996). *HLM: Hierarchical Linear and Nonlinear Modeling with the HLM/2L and HLM/3L Programs*. Chicago: Scientific Software International, Inc.

Carlin, B., & Louis, T. (2000). *Bayes and Empirical Bayes Methods for Data Analysis* (2nd ed.). Boca Raton, FL: Chapman & Hall.

Caspi, A., Moffit, T.E., Thornton, A., Freedman, D., Amell, J.W., & Harrington, H. (1996). The life history calendar: A research and clinical assessment method for collecting retrospective event history data. *International Journal of Methods in Psychiatric Research, 6*, 101–114.

Csikszentmihalyi, M., Hektner, J., & Schmidt, J. (in press). *Measuring the Quality of Everyday Life: The ESM Handbook*. Thousand Oaks, CA: Sage.

Csikszentmihalyi, M., & Larson, R.W. (1987). Validity and reliability of the experience sampling method. *Journal of Nervous and Mental Disease, 175*, 526–536.

Cudeck, R., & du Toit, S.H.C. (2002). Nonlinear multilevel models for repeated measures data. In S.P. Reise & N. Duan (Eds.), *Multilevel Modeling Methodological Advances, Issues, and Applications* (pp. 1–24). Mahwah, NJ: Lawrence Erlbaum.

de Leeuw, J., & Kreft, I.G.G. (1986). Random coefficient models for multilevel analysis. *Journal of Educational Statistics, 11*, 57–85.

de Vries, M.W. (Ed.). (1992). *The Experience of Psychopathology: Investigating Mental Disorders in Their Natural Settings*. Cambridge: Cambridge University Press.

Eliason, S.R. (1993). *Maximum Likelihood Estimation: Logic and Practice*. Newbury Park, CA: Sage.

Feldman Barrett, L., Robin, L., Pietromonaco, P.R., & Eyssell, K. M. (1998). Are women the "more emotional sex": Evidence from emotional experiences in social context. *Cognition and Emotion, 12*, 555–578.

Fitzmaurice, G.M., Laird, N.M., & Ware, J.H. (2004). *Applied Longitudinal Analysis*. Hoboken, NJ: John Wiley.

Freedman, D., Thornton, A., Camburn, D., Alwin, D., & Young-DeMarco, L. (1988). The life history calendar: A technique for collecting retrospective data. *Sociological Methodology, 18*, 37–68.

Goldstein, H. (1986). Multilevel mixed linear model analysis using iterative generalized least squares. *Biometrika, 73*, 43–56.

Goldstein, H. (2003). *Multilevel Statistical Models* (3rd ed.). New York: Oxford University Press.

Goldstein, H., Rasbash, J., Plewis, I., Draper, D., Browne, W., Yang, M., Woodhouse, G., & Healy, M. (1998). *A User's Guide to MLwiN*. London: Multilevel Models Projects, Institute of Education, University of London.

Hastie, T., & Tibshirani, R.J. (1990). *Generalized Additive Models.* New York: Chapman & Hall.

Hedeker, D., & Gibbons, R.D. (1993). A random-effects ordinal regression model for multilevel analysis. *Biometrics, 50*, 933–944.

Hox, J.J. (2000). Multilevel analysis of grouped and longitudinal data. In T.D. Little, K.U. Schnabel, & J. Baumert (Eds.), *Modeling Longitudinal and Multilevel Data* (pp. 15–32). Mahwah, NJ: Lawrence Erlbaum.

Hox, J.J. (2002). *Multilevel Analysis: Techniques and Applications*. Mahwah, NJ: Lawrence Erlbaum.

Hurlburt, R.T. (1979). Random sampling of cognitions and behavior. *Journal of Research in Personality, 13*, 103–111.

Hüttner, H.J.M., & van den Eeden, P. (1995). *The Multilevel Design: A Guide with an Annotated Bibliography, 1980–1993*. Westport, CT: Greenwood Press.

Jacob, R.G., Thayer, J.F., Manuck, S.B., Muldoon, M.F., Tamres, L.K., Williams, D.M., Ding, Y., & Gatsonis, C. (1999). Ambulatory blood pressure responses and the circumplex model of mood: A 4 day study. *Psychosomatic Medicine, 61*, 319–333.

Jennrich, R.I., & Schluchter, M.D. (1986). Unbalanced repeated-measures models with structured covariance matrices. *Biometrics, 42*, 805–820.

Kahneman, D., Fredrickson, B.L., Schreiber, C.A., & Redelmeier, D.A. (1993). When more pain is preferred to less: Adding a better end. *Psychological Science, 4*, 401–405.

Kreft, I.G.G., & de Leeuw, J. (1998). *Introducing Multilevel Modeling.* Thousand Oaks, CA: Sage.

Laird, N.M., Lange, N., & Stram, D. (1987). Maximum likelihood computations with repeated measures: Application of the EM algorithm. *Journal of the American Statistical Association, 82*, 97–105.

Laird, N.M., & Ware, H. (1982). Random-effects models for longitudinal data. *Biometrics, 38*, 963–974.

Larson, R.W., & Csikszentmihalyi, M. (1978). Experiential correlates of time alone in adolescence. *Journal of Personality, 46*, 677–693.

Lindstrom, M.J., & Bates, D.M. (1988). Newton-Raphson and EM algorithms for linear mixed-effects models for replicated measures data. *Journal of the American Statistical Association, 83*, 1014–1022.

Littell, R.C., Milliken, G.A., Stroup, W.W., & Wolfinger, R.D. (1996). *SAS System for Mixed Models*. Cary, NC: SAS Institute.

Liu, C., & Rubin, D.B. (1995). Application of the ECME algorithm and the Gibbs sampler to general linear mixed models. Proceedings of the 17th International Biometric Conference, 97–107.

Longford, N.T. (1989). A Fisher scoring algorithm for variance component analysis of data with multilevel structure. In R.D. Bock (Ed.), *Multilevel Analysis of Educational Data.* San Diego, CA: Academic Press.

Longford, N.T. (1993). *Random Coefficient Models*. New York: Oxford University Press.

MathSoft (1997). SPLUS 4.0. Cambridge, MA: MathSoft Inc.

Moskowitz, D.S., & Hershberger, S.L. (Eds.) (2002). *Modeling Intraindividual Variability with Repeated Measures Data: Method and Applications*. Mahwah, NJ: Erlbaum.

Multilevel Models Project (1996). Multilevel modeling applications: A guide for users of MLn (ed. Woodhouse, G.). London: Institute of Education, University of London.

Muthén, B.O., & Muthén, L.K. (2000). Integrating person-centered and variable-centered analysis: Growth mixture modeling with latent trajectory classes. *Alcoholism: Clinical and Experimental Research, 24*, 882–891.

Nagin, D.S. (1999). Analyzing developmental trajectories: A semi-parametric, group-based approach. *Psychological Methods, 4*, 139–157.

Neter, J., Kutner, M.H., Nachtsheim, C.J., & Wasserman, W. (1996). *Applied Linear Statistical Models* (4th ed.). Boston: McGraw-Hill.

Ramsay, J.O., & Silverman, B.W. (1997). *Functional Data Analysis*. New York: Springer.

Rasbash, J., & Woodhouse, G. (1995). *MLn Command Reference: Version 1.0a*. London: Institute of Education, University of London, Multilevel Models Project.

Raudenbush, S.W., & Bryk, A.S. (1986). A hierarchical model for studying school effects. *Sociology of Education, 59*, 1–17.

Raudenbush, S.W., & Bryk, A.S. (2002). *Hierarchical Linear Models: Applications and Data Analysis Methods*. Thousand Oaks, CA: Sage.

Rodrígues, G., & Goldman, N. (2001). Improved estimation procedures for multilevel models with binary response: a case-study. *Journal of the Royal Statistical Society, Series A, 164*, 339–355.

Rovine, M.J., & Molenaar, P.C.M. (2001). A structural equations modeling approach to the general linear mixed model. In L. Collins & A. Sayer (Eds.), *New Methods for the Analysis of Change* (pp. 65–96). Washington, DC: American Psychological Association.

Schwartz, J. (2005). The analysis of real-time data: A practical guide. In A. Stone, S. Shiffman, A. Atienza, & L. Nebeling. (Eds.), *The Science of Real-Time Data Capture: Self-Reports in Health Research*. New York: Oxford University Press (forthcoming).

Schwartz, J.E., & Stone, A.A. (1998). Strategies for analyzing ecological momentary assessment data. *Health Psychology, 17*, 6–16.

Shiffman, S., Fischer, L.A., Paty, J.A., Gnys, M., Hickcox, M., & Kassel, J.D. (1994). Drinking and smoking: A field study of their association. *Annals of Behavioral Medicine, 16*, 203–209.

Singer, J.D. (1998). Using SAS PROC MIXED to fit multilevel models, hierarchical models, and individual growth curve models. *Journal of Educational and Behavioral Statistics, 24*, 323–355.

Snidjers, T.A.B., & Bosker, R.L. (1999). *Multilevel Analysis: An Introduction to Basic and Advanced Multilevel Modeling*. Thousand Oaks, CA: Sage.

Sorokin, P.A., & Berger, C.F. (1939). *Time-Budgets of Human Behavior*. Oxford, UK: Harvard University Press.

Spiegelhalter, D., Thomas, A., Best, N., & Gilks, W. (1996). *BUGS 0.5: Bayesian Inference Using Gibbs Sampling-Manual* (version ii). Cambridge, UK: Medical Research Council Biostatistics Unit.

StataCorp (1997). *Stata Statistical Software: Release 5.0*. College Station, TX: Stata.

Stone, A.A., & Shiffman, S. (2002). Capturing momentary, self-report data: A proposal for reporting guidelines. *Annals of Behavioral Medicine, 24*, 236–243.

Stone, A., Shiffman, S., Atienza, A., & Nebeling, L. (Eds.) (2005). *The Science of Real-Time Data Capture: Self-Reports in Health Research*. New York: Oxford University Press (forthcoming).

Treloar, A.E., Boynton, R.E., Behn, B.G., & Brown, B.W. (1970). Variation of the human menstrual cycle through reproductive life. *International Journal of Fertility 12*, 77–126.

Verbeke, G., Lesaffre, E., & Brant, L.J. (1998). The detection of residual serial correlation in linear mixed models. *Statistics in Medicine, 17*, 1391–1402.

Verbeke, G., & Molenberghs, G. (2000). *Linear Mixed Models for Longitudinal Data*. New York: Springer.

Verma, S., Sharma, D., & Larson, R.W. (2002). School stress in India: Effects on time and daily emotions. *International Journal of Behavioral Development, 26*, 500–508.
Wessman, A.E., & Ricks, D.F. (1966). *Mood and Personality*. Oxford, UK: Holt, Rinehart & Winston.
Willett, J.B., & Singer, J.D. (2003). *Applied Longitudinal Data Analysis*. New York: Oxford University Press.

2

Marginal Modeling of Intensive Longitudinal Data by Generalized Estimating Equations

Joseph L. Schafer

The growth of semiparametric regression modeling through generalized estimating equations (GEE) (Liang & Zeger, 1986) is one of the most influential recent developments in statistical practice. GEE methods are attractive both from a theoretical and a practical standpoint; they are flexible, easy to use, and make relatively weak assumptions about the distribution of the response of interest. They are closely related to multilevel models and are commonly regarded as robust cousins of the linear mixed model described by Walls et al. (chapter 1, this volume) and generalized linear mixed models described by Hedeker et al. (chapter 4, this volume).

Because of longstanding tensions between different schools of statistical thought–in this case, those who favor full parametric models versus those who eschew distributional assumptions wherever possible–some who handle longitudinal data may rely either on multilevel models or GEE but not both. This is unfortunate. Rather than regarding the two as competitors, we see them as complementary. In certain cases, parameter estimates obtained from GEE will be nearly or precisely identical to those obtained from multilevel models. In fact, one could argue that the two methods, if applied well, should yield essentially the same inferences for any parameters they share in common. Yet GEE is sufficiently different from multilevel modeling in terms of its assumptions, how it is understood and applied, and its statistical properties that it warrants separate treatment in this volume.

Many good references on GEE are available (Diggle et al., 2002; Hardin & Hilbe, 2003). As with multilevel models, however, this literature tends to emphasize short longitudinal series (e.g., fewer than ten occasions). Applications to intensive longitudinal data with many occasions and potentially high correlations are less common, perhaps because the number of subjects N in such studies is often too small for the large-sample approximations to be accurate.

In the remainder of this chapter, we review the motivation, application, interpretation, and properties of GEE regression, pointing out similarities and differences between GEE and multilevel models. We apply GEE to the data regarding stress in Indian adolescents previously analyzed by Walls et al. (chapter 1, this volume) and compare our results. We then carry out a small simulation experiment to get a rough idea of how well GEE may be performing relative to multilevel models for this particular dataset.

2.1 What Is GEE Regression?

This section traces the development of GEE in the context of ordinary least squares, generalized linear models, and quasi-likelihood. Readers less interested in technical details are still encouraged to skim this section to become familiar with the terminology and properties of GEE and its relationships to other types of regression.

2.1.1 Ordinary Least Squares with a Heteroscedastic Response

To understand the nature of GEE, let us first begin with classical linear regression. Suppose that a numeric response y_i is recorded for subjects $i = 1, \ldots, N$ whom we regard as independent. Suppose that $y_i = \beta_1 x_{i1} + \cdots + \beta_p x_{ip} + \epsilon_i$, where the x_{ij}'s are known (typically $x_{i1} \equiv 1$) and the disturbance ϵ_i is normally distributed with constant variance. That is, we assume y_i is normal with mean

$$E(y_i) = \mu_i = x_i^T \beta \tag{2.1}$$

and variance

$$Var(y_i) = \sigma_i^2 = \sigma^2, \tag{2.2}$$

where $x_i = (x_{i1}, \ldots, x_{ip})^T$ and $\beta = (\beta_1, \ldots, \beta_p)^T$. Under these conditions, the unknown coefficients are optimally estimated by ordinary least squares (OLS). Taking $y = (y_1, \ldots, y_N)^T$, and assuming that $X = (x_1, \ldots, x_N)^T$ has full rank, the OLS estimate of β is

$$\hat{\beta} = (X^T X)^{-1} X^T y = \left(\sum_{i=1}^{N} x_i x_i^T \right)^{-1} \left(\sum_{i=1}^{N} x_i y_i \right),$$

and the usual unbiased estimate of σ^2 is

$$\hat{\sigma}^2 = \frac{1}{N-p} (y - X\hat{\beta})^T (y - X\hat{\beta}) = \frac{1}{N-p} \sum_{i=1}^{N} (y_i - x_i^T \hat{\beta})^2.$$

Inferences about β proceed from the fact that

$$\hat{\beta} - \beta \sim N(0, \sigma^2 (X^T X)^{-1}), \tag{2.3}$$

and standard errors for the estimated coefficients are obtained from the diagonal elements of

$$\widehat{Var}(\hat{\beta}) = \hat{\sigma}^2(X^TX)^{-1}. \tag{2.4}$$

Texts on classical regression do not strongly emphasize the assumption of normality, because (2.3) is approximately true for nonnormal y_i provided that (2.1) and (2.2) still hold and N is large. Gross violations of (2.1) are difficult to contemplate, because if this mean structure is misspecified, then the meaning of the β_j's becomes dubious. For example, if important covariates have been omitted from x_i, then, depending on the omitted variables' relationships to x_i, β_j may or may not have its usual interpretation (Clogg et al., 1992).

Relaxing the assumption of normality but supposing that the mean structure is still plausible, what can we say about departures from homoscedasticity? If the variances σ_i^2 are not constant, then $\hat{\beta}$ is still unbiased but no longer efficient; alternatives to OLS will produce estimates that are more precise. If the σ_i^2's were known up to some constant of proportionality, as in $\sigma_i^2 \propto w_i^{-1}$, then the optimal strategy would be a weighted least-squares fit using w_i's as weights. The existence of a better alternative, however, does not automatically render the OLS estimate useless, especially when σ_i^2's vary in ways that are unknown. If N is sufficiently large, $\hat{\beta}$ may still be good enough to estimate the effects of interest with reasonable accuracy. In those applications, the real danger is not that $\hat{\beta}$ performs poorly, but that standard errors derived from (2.4) may not accurately reflect the true precision, distorting the behavior of tests and confidence intervals and leading us to incorrect conclusions.

Remarkably, there is a simple alternative to (2.4) which protects us against danger from residual variances σ_i^2 that vary in unspecified ways. The actual error in the OLS estimate can be written as

$$\begin{aligned}\hat{\beta} - \beta &= (X^TX)^{-1}X^T(y - X\beta) \\ &= \left(\frac{1}{N}X^TX\right)^{-1}\left[\frac{1}{N}\sum_{i=1}^{N} x_i(y_i - x_i^T\beta)\right].\end{aligned} \tag{2.5}$$

The second term on the right-hand side of (2.5) is the average of independent but nonidentically distributed random vectors; the expectation of this average is $(0, \ldots, 0)^T$, and its true covariance matrix is

$$\frac{1}{N^2}\sum_{i=1}^{N} \sigma_i^2 x_i x_i^T = \frac{1}{N^2} X^T \Sigma X, \tag{2.6}$$

where $\Sigma = \text{Diag}(\sigma_i^2)$ is the $N \times N$ matrix with $\sigma_1^2, \ldots, \sigma_N^2$ on the diagonal and zeros elsewhere. As N grows, it becomes reasonable to replace the unknown σ_i^2's in (2.6) by the rough empirical estimates $\tilde{\sigma}_i^2 = (y_i - x_i^T\hat{\beta})^2$, so that Σ is replaced by $\tilde{\Sigma} = \text{Diag}(\tilde{\sigma}_i^2)$. Notice that the $\tilde{\sigma}_i^2$'s do not individually converge to their respective σ_i^2's, and the dimensions of $\tilde{\Sigma}$ and Σ are both increasing. But the size of $X^T\Sigma X$

remains fixed ($p \times p$), and in the limit the difference between $N^{-1}X^T\tilde{\Sigma}X$ and $N^{-1}X^T\Sigma X$ becomes small. Substituting $\tilde{\Sigma}$ for Σ in (2.6), it follows from (2.5) that the covariance matrix of $\hat{\beta}$ can be estimated by

$$\widetilde{Var}(\hat{\beta}) = \left(X^TX\right)^{-1}\left(X^T\tilde{\Sigma}X\right)\left(X^TX\right)^{-1}. \tag{2.7}$$

This method, which was first proposed by the econometrician White (1980), is for obvious reasons called a sandwich estimate. A similar method was presented by Huber (1967) in the context of maximum likelihood estimation under a misspecified model. The "bread" of the sandwich, $(X^TX)^{-1}$, is an estimate of $Var(\hat{\beta})$ under the usually implausible assumption that $\sigma_i^2 = 1$ for all i; if that were true, then $(X^T\tilde{\Sigma}X)(X^TX)^{-1}$ would converge to the identity matrix. With moderate-sized samples, (2.7) tends to slightly understate the true variance of $\hat{\beta}$, and in practice it is often multiplied by the adjustment factor $N/(N-p)$.

2.1.2 Generalized Linear Models

Pairing the OLS with standard errors derived from the sandwich (2.7) is one special case of the GEE methodology developed by Liang and Zeger (1986). GEE extends this idea in two important ways: first, it replaces OLS with a more general estimation procedure motivated by linear exponential families; second, it moves from a single response per subject to a longitudinal response. Let us first consider the extension to exponential families.

Exponential-family regression models—commonly known as generalized linear models (GLIMs)—were introduced by Nelder and Wedderburn (1972) and popularized by McCullagh and Nelder (1989). GLIMs represent a unified treatment of linear models, logistic models for binary or binomial responses, loglinear models for counts, and many other kinds of regression. GLIM is also the name of the computer program that first implemented this class of models in its full generality (Francis et al., 1993). Procedures for fitting GLIMs are now available in S-PLUS, Stata, SAS PROC GENMOD, and many other commercial packages. Among these, we find PROC GENMOD to be especially useful because it also implements the procedures for GEE described later in this section.

Several versions of the GLIM have been proposed, but the most common one supposes that the probability distribution of the response y_i can be written in the form

$$L_i = \exp\left\{\frac{y_i\theta_i - b(\theta_i)}{\phi w_i^{-1}} + c(y_i, \phi)\right\}, \tag{2.8}$$

where b and c are known functions, θ_i is a parameter related to the mean $\mu_i = E(y_i)$, w_i is a known weight, and ϕ is a dispersion parameter. For our purposes, it does not really matter if the response is discrete or continuous, or if its support (i.e., its range of possible values) is the entire real line $(-\infty, \infty)$, the positive reals $(0, \infty)$, the nonnegative integers $\{0, 1, 2, \ldots\}$, or some other fixed set. The reason it does not matter is that, in the GLIM framework, L_i is used not as a probability distribution or density per se but as a likelihood function. In practice we create a GLIM not by specifying the exact form of L_i, but by a sequence of choices that determines the mean function and the variance structure.

To set up the mean function, we must decide which covariates to include in the covariate vector $x_i = (x_{i1}, \ldots, x_{ip})^T$ and how they relate to μ_i. Rather than requiring the relationship to be linear as in (2.1), we now assume more generally that

$$\eta_i = g(\mu_i) = x_i^T \beta, \tag{2.9}$$

where g is a monotonically increasing or decreasing function known as the *link*.[1] Considerations in choosing the link function are theoretical, practical, and empirical. One issue is to prevent μ_i from straying out of its allowable range. If y_i is a binary indicator or proportion, then μ_i must lie between 0 and 1, which suggests that a logit link, $\eta_i = \log(\mu_i/(1 - \mu_i))$, may be appropriate. If y_i is a frequency or count, then μ_i must be positive, in which case $\eta_i = \log \mu_i$ becomes attractive. The link function heavily influences the meaning of β, so we want to choose g for which the coefficients are readily interpretable. With a logit link, for example, an exponentiated coefficient $\exp(\beta_j)$ is an odds ratio. Finally, the link function should be chosen to fit the data well. Diagnostics for assessing the fit of GLIMs are described by McCullagh and Nelder (1989).

The variance structure describes how the variance of y_i is related to μ_i. In the homoscedastic normal model (2.2), the variance is a constant. Other response distributions posit specific relationships between the variance and the mean. With a Poisson model, for example, the mean and variance are constrained to be equal. In general, the variance of the response in a GLIM is taken to be

$$\text{Var}(y_i) = \Sigma_i = \phi w_i^{-1} v(\mu_i), \tag{2.10}$$

where v is the so-called variance function. The dispersion parameter ϕ is either fixed or estimated, and the weight w_i, if present, is often a measure of sample size associated with case i. Variance structures corresponding to the most common exponential-family models are shown in table 2.1.

The unknown coefficients in a GLIM are typically estimated by an iterative procedure known as Fisher scoring (McCullagh & Nelder, 1989); the resulting values can be interpreted as maximum-likelihood (ML) estimates for β given

Table 2.1 Variance structures for the most common types of GLIMs

Type of response	Variance structure
Homoscedastic normal, $y_i \sim N(\mu_i, \sigma^2)$	$\phi = \sigma^2$, $w_i = 1$, $v(\mu_i) = 1$
Heteroscedastic normal, $y_i \sim N(\mu_i, \sigma^2/\kappa_i)$	$\phi = \sigma^2$, $w_i = \kappa_i$, $v(\mu_i) = 1$
Poisson count	$\phi = 1$, $w_i = 1$, $v(\mu_i) = \mu_i$
Binary (0/1)	$\phi = 1$, $w_i = 1$, $v(\mu_i) = \mu_i(1 - \mu_i)$
Binomial proportion from n_i trials	$\phi = 1$, $w_i = n_i$, $v(\mu_i) = \mu_i(1 - \mu_i)$

a fixed value for the dispersion parameter.[2] Fisher scoring finds the vector $\hat{\beta} = (\hat{\beta}_1, \ldots, \hat{\beta}_p)^T$ at which the first derivatives of the log-likelihood function $\sum_{i=1}^{N} \log L_i$ with respect to the elements of β are simultaneously zero. One can show that

$$\frac{\partial \log L_i}{\partial \beta_j} = d_{ij} \Sigma_i^{-1}(y_i - \mu_i), \tag{2.11}$$

where $d_{ij} = \partial\mu_i/\partial\beta_j = (\partial\mu_i/\partial\eta_i)x_{ij}$; note that $\partial\mu_i/\partial\eta_i = 1/g'(\mu_i)$ is determined by the link. Collecting the derivatives (2.11) for $j = 1, \ldots, p$ into a $p \times 1$ vector, we obtain

$$U_i = D_i^T \Sigma_i^{-1}(y_i - \mu_i),$$

where $D_i = (d_{i1}, \ldots, d_{ip}) = \partial\mu_i/\partial\beta^T = (\partial\mu_i/\partial\eta_i)x_i^T$. The estimate $\hat{\beta}$ is defined as the solution to the simultaneous equations

$$\sum_{i=1}^{N} U_i = (0, \ldots, 0)^T. \tag{2.12}$$

The U_i's are commonly known as score vectors, and (2.12) is called the score equations. In each iteration of Fisher scoring, we first evaluate η_i, μ_i, Σ_i, and D_i at the current estimate $\hat{\beta}^{cur}$; the updated estimate is then

$$\begin{aligned} \hat{\beta}^{new} &= \hat{\beta}^{cur} + \left(\sum_{i=1}^{N} D_i^T \Sigma_i^{-1} D_i\right)^{-1} \left(\sum_{i=1}^{N} U_i\right) \\ &= (X^T W X)^{-1} X^T W z, \end{aligned} \tag{2.13}$$

where $W = \text{Diag}[\,\Sigma_i^{-1}(\partial\mu_i/\partial\eta_i)^2\,]$, $z_i = \eta_i + (\partial\eta_i/\partial\mu_i)(y_i - \mu_i)$, and $z = (z_1, \ldots, z_N)^T$. Upon convergence, standard errors for the elements of $\hat{\beta}$ may be obtained from

$$\widehat{Var}(\hat{\beta}) = (X^T W X)^{-1}. \tag{2.14}$$

It is easy to see that the dispersion parameter cancels out of this estimation procedure, because $D_i^T \Sigma_i^{-1} D_i$ and U_i are both proportional to ϕ^{-1}. The estimate $\hat{\beta}$, therefore, does not depend at all on the assumed value for ϕ, but the estimated covariance matrix (2.14) does. When ϕ is not regarded as known, it is often estimated by $\hat{\phi} = (N - p)^{-1} \sum_{i=1}^{N} e_i^2$, where $e_i = (y_i - \mu_i)/\sqrt{\Sigma_i}$ is the Pearson residual.

2.1.3 Quasi-likelihood and Robust Estimation

Notice that Fisher scoring was described not in terms of the exponential family model (2.8), but only in terms of the mean function μ_i and variance Σ_i. In practice, there is no need to restrict ourselves to the variance structures implied by the traditional exponential families (e.g., those shown in table 2.1) if they do not agree with the data. If y_i is a count, for example, the data may show clear evidence that $\Sigma_i > \mu_i$; in that case, we are free to replace the dispersion parameter $\phi = 1$ implied by the Poisson distribution with a larger value, even though this modification takes us outside the realm of familiar parametric models. If we choose a variance structure that does not correspond to a traditional exponential family, the U_i's are no longer the derivatives of a true log-likelihood but of a quasi-likelihood (Wedderburn, 1974); in that case, the U_i's are called estimating functions, and the equations (2.12) are called estimating equations.

When the responses y_i depart from familiar parametric models, there are theoretical reasons why it is still appropriate and desirable to use the Fisher scoring procedure and its resulting estimate $\hat{\beta}$. These reasons, which are closely related to why OLS is sometimes appropriate for heteroscedastic responses, pertain to the behavior of $\hat{\beta}$ when the variances $\Sigma_1, \ldots, \Sigma_N$ have been misspecified. Gourieroux et al. (1984) proved that when the mean function is correct—that is, when $\mu_i = g^{-1}(x_i^T \beta)$ for some β—but the variance structure or other aspects of the assumed density are incorrect, an ML estimate derived under an incorrect parametric model is consistent for the true β if and only if the assumed density belongs to a linear exponential family; see also Gourieroux and Montfort (1993). A nice overview of this and other results for misspecified regression models is given by Arminger (1995).

The practical implication of this result is that, if little is known about the variance or higher moments of y_i, it is often reasonable to take an educated guess at what the form of Σ_i might be and proceed with Fisher scoring (2.13) as if that guess were true. If N is large relative to p, we are protected against biases in $\hat{\beta}$ in the event that the guess is incorrect. Note that this protection does not extend to the estimated covariance matrix; standard errors based on (2.14) could seriously mislead. Just as in section 2.1.1, however, the estimated covariance matrix can be repaired by another application of the sandwich.

To see how this works, we can apply a trick that exploits the asymptotic equivalence of $\hat{\beta}$ and the estimate that we would get by performing only a single iteration of the Fisher scoring procedure (2.13). That is, as $N \to \infty$, a single iteration of scoring from any starting value produces a one-step estimate that is only negligibly different from the final convergent solution $\hat{\beta}$ (McCullagh & Nelder, 1989). If we imagine computing this one-step estimate beginning at the unknown true value β, we see that the estimation error is approximately

$$\hat{\beta} - \beta \approx \left(\frac{1}{N} \sum_{i=1}^{N} D_i^T \Sigma_i^{-1} D_i \right)^{-1} \left(\frac{1}{N} \sum_{i=1}^{N} U_i \right). \tag{2.15}$$

The second term, $N^{-1} \sum_i U_i$, is the average of N independent but nonidentically distributed random vectors with mean zero. If N is sufficiently large, it is reasonable to approximate the covariance matrix of $N^{-1}\sum_i U_i$ by

$$\frac{1}{N^2}\sum_i U_i U_i^T = \frac{1}{N^2}\sum_i D_i^T \Sigma_i^{-1}\tilde{\Sigma}\Sigma_i^{-1}D_i,$$

where $\tilde{\Sigma}$ is the $N \times N$ matrix with squared empirical residuals $(y_i - g^{-1}(x_i^T\hat{\beta}))^2$, $i = 1,\ldots,N$, on the diagonal and zeros elsewhere. Combining this empirical covariance matrix with the linear approximation (2.15), the new estimated covariance matrix for $\hat{\beta}$ is

$$\widetilde{Var}(\hat{\beta}) = \left(\sum_{i=1}^{N} D_i^T \Sigma_i^{-1} D_i\right)^{-1}\left(\sum_{i=1}^{N} U_i U_i^T\right)\left(\sum_{i=1}^{N} D_i^T \Sigma_i^{-1} D_i\right)^{-1}. \qquad (2.16)$$

In the statistical literature, (2.16) is called a robust, empirical, or sandwich estimate, whereas (2.14) is called parametric, model-based, or naïve. In our opinion, the term naïve is unnecessarily pejorative. Parametric assumptions should not be equated with naïveté, as all statistical procedures rely on models to varying degrees. If the assumed form for Σ_i is reasonably accurate, then the conventional standard errors from (2.14) are actually preferable to those from the sandwich (2.16), as the former can be considerably more precise, especially when N is only moderately large. Moreover, the parametric form for Σ_i need not be proposed blindly; this assumption can be checked, just as we can and should check the adequacy of the proposed mean function $\mu_i = g^{-1}(x_i^T\beta)$.

2.1.4 The GEE Procedure

The GEE method of Liang and Zeger (1986) is a conceptually and notationally straightforward generalization of quasi-likelihood regression to longitudinal responses. Let y_{it} denote the response for subject i at occasion t, and let $y_i = (y_{i1},\ldots,y_{in_i})^T$ denote the $n_i \times 1$ vector of responses for subject i. Let $x_{it} = (x_{it1},\ldots,x_{itp})^T$ be a $p \times 1$ vector of covariates describing subject i at occasion t. This vector will typically contain a constant term, $x_{it1} \equiv 1$; it may also include characteristics of the subject that do not change (e.g., a dummy indicator for sex) and time-varying covariates as well. Letting $\mu_{it} = E(y_{it})$, we shall suppose that $g(\mu_{it}) = x_{it}^T\beta$ for some link function g, so that $\mu_{it} = g^{-1}(x_{it}^T\beta)$. In the GEE literature, it has become customary to write the mean of the entire y_i vector as an $(n_i \times 1)$-dimensional function of β,

$$E(y_i) = \mu_i(\beta),$$

where the covariates x_{it} for all occasions $t = 1,\ldots,n_i$ have been subsumed into this function.

To come up with a procedure for estimating β, we shall also need to make some working assumptions about the variance of each y_{it} and the covariances between y_{it} and $y_{it'}$ for $t \neq t'$. Let us momentarily suppose that the $n_i \times n_i$

covariance matrices

$$Var(y_i) = \Sigma_i$$

for $i = 1, \ldots, N$ were completely known. The optimal estimate $\hat{\beta}$ could then be computed by a simple generalization of the scoring procedure (2.13),

$$\hat{\beta}^{new} = \hat{\beta}^{cur} + \left(\sum_{i=1}^{N} D_i^T \Sigma_i^{-1} D_i \right)^{-1} \left(\sum_{i=1}^{N} U_i \right), \tag{2.17}$$

where $D_i = \partial \mu_i / \partial \beta^T$ is now the $n_i \times p$ matrix whose (t,j)th element is $\partial \mu_{it} / \partial \beta_j$, and the estimating function or quasi-score $U_i = D_i^T \Sigma_i^{-1} (y_i - \mu_i)$ is still $p \times 1$. Upon convergence, standard errors for the elements of $\hat{\beta}$ are typically obtained from the sandwich

$$\widetilde{Var}(\hat{\beta}) = \left(\sum_{i=1}^{N} D_i^T \Sigma_i^{-1} D_i \right)^{-1} \left(\sum_{i=1}^{N} U_i U_i^T \right) \left(\sum_{i=1}^{N} D_i^T \Sigma_i^{-1} D_i \right)^{-1}.$$

If Σ_i were an accurate representation of $Var(\Sigma_i)$, then the "bread" of the sandwich, $(\sum_i D_i^T \Sigma_i^{-1} D_i)^{-1}$, would also provide consistent standard errors.

In most applications of GEE, the user does not specify the working covariance matrices $\Sigma_1, \ldots, \Sigma_N$ entirely, but defines them as functions of the mean vector μ_i and one or more unknown parameters. The most common way to do this is to decompose each matrix as

$$\Sigma_i = \phi A_i^{1/2} R_i(\alpha) A_i^{1/2}, \tag{2.18}$$

where ϕ is a positive dispersion factor, A_i is an $n_i \times n_i$ matrix with elements $v(\mu_{it})$, $t = 1, \ldots, n_i$ on the diagonal and zeros elsewhere, v is a variance function, and $R(\alpha)$ is a patterned correlation matrix that may depend on one or more unknown parameters α. The variance function, which accounts for heteroscedastic dependence of $Var(y_{ij})$ upon μ_{ij}, is usually chosen from the same group of functions applied to GLIMs. If y_{ij} were a frequency or count, for example, then we would probably take $v(\mu_{ij}) = \mu_{ij}$ to mimic the behavior of a Poisson variate; the distribution is not required to be Poisson, however, as we can freely estimate the value of ϕ rather than fixing it at $\phi = 1$.

Dependence among the responses from the same subject at different occasions is captured in the correlation matrix $R_i(\alpha)$. If these correlations were thought to be weak, one could take R_i to be the $n_i \times n_i$ identity matrix, which imposes a working assumption of independence; in that case, the coefficients $\hat{\beta}$ would be identical to those estimated by a GLIM that ignores the grouping of observations within subjects. Another common choice is

$$R_i(\alpha) = \begin{bmatrix} 1 & \rho & \rho & \cdots \\ & 1 & \rho & \cdots \\ & & 1 & \cdots \\ & & & \ddots \end{bmatrix} \tag{2.19}$$

with $\alpha = \rho$, which is called an exchangeable or compound symmetry assumption. This is the pattern implied by a multilevel linear model with a single random coefficient, a random intercept, for each subject. The exchangeable form assumes that every pair of observations for a subject is correlated to the same degree, no matter how far apart the measurement occasions are. If the correlations are thought to diminish over time, an alternative is

$$R_i(\alpha) = \begin{bmatrix} 1 & \rho^{|t_{i1}-t_{i2}|} & \rho^{|t_{i1}-t_{i3}|} & \cdots \\ & 1 & \rho^{|t_{i2}-t_{i3}|} & \cdots \\ & & 1 & \cdots \\ & & & \ddots \end{bmatrix},$$

where $t_{i1}, t_{i2}, t_{i3}, \ldots$ are the measurement times for subject i; this structure is called first-order autoregressive or AR(1) (Jones, 1993; Rovine & Walls, chapter 6, this volume). Another popular alternative, which is called "m-dependent" in the documentation for PROC GENMOD, imposes a banded structure for R_i, setting the correlation between y_{it} and $y_{i,t+j}$ to ρ_j for $j = 1, 2, \ldots, m$ and zero for $j > m$. In some settings, it may even be practical to allow R_i to have no particular structure and estimate the correlation between y_{it} and $y_{it'}$ independently for every pair (t, t'); this approach is practical when subjects are measured at a small number of common occasions.

When the working covariance matrix is decomposed as in (2.18), the unknown parameters in Σ_i are estimated at each cycle of the quasi-scoring procedure (2.17). Each time a new estimate of β is produced, we compute the fitted means $\mu_i(\beta)$ and the Pearson residuals

$$e_{it} = \frac{y_{it} - \mu_{it}}{\sqrt{v(\mu_{it})}},$$

and then reestimate ϕ and α from the e_{it}'s by an appropriate method; estimation formulas for some popular working covariance structures are given in the documentation accompanying PROC GENMOD. Liang and Zeger (1986) point out that ϕ and α may be estimated by any method that is "$\sqrt{N}$-consistent," although the meaning of this consistency is dubious when the form of Σ_i is misspecified. In reality, as long as the sequence of estimates for (ϕ, α) stabilizes to something as $N \to \infty$, the desirable theoretical properties of the GEE estimator $\hat{\beta}$ and sandwich-based standard errors are maintained.

It is easy to criticize (2.18) on the grounds that it is motivated by analogy to linear models and may not accurately represent the joint distribution of correlated nonnormal responses. The arbitrary pairing of a variance function with a patterned correlation matrix may in fact produce a model that does not correspond to any joint distribution for $y_i = (y_{i1}, \ldots, y_{in_i})^T$. Some proponents of GEE would argue that, despite this conceptual difficulty, the working covariance matrix need only be a rough approximation to the true covariance structure, and the statistical theory of estimating equations stands on its own without recourse to joint models. At any rate, the decomposition (2.18) is not central to the GEE methodology, and we are free to replace it with other forms more suitable to the data and scientific questions at hand. For example, Thall and Vail (1990) describe parametric forms for $\Sigma_i(\alpha)$ consistent with a Poisson model with log-linear random effects. Carey et al. (1993) present models for correlated binary observations that capture the association between y_{it} and $y_{it'}$ in terms of log-odds ratios.

Finally, it deserves mention that the theory of estimating equations goes well beyond the modeling of mean structures. If variances or covariances are of key interest, we can parameterize these aspects of the distribution of y_i and obtain estimates that are robust to departures from working assumptions about the higher moments. This generalization, which is called GEE-2, is described by Liang et al. (1992). The sandwich method for variance estimation pioneered by Huber (1967) is also quite general and can be applied to many kinds of ML estimates. Many of the software packages for multilevel modeling mentioned by Walls et al. (chapter 1, this volume) now provide sandwich-based standard errors upon request along with the usual model-based standard errors. With large samples, the sandwich helps to protect us against misspecification in parts of the model that are not parameterized.

2.2 Practical Considerations in the Application of GEE

Having covered the theoretical development of the GEE methodology, we now turn to practical issues of how to use it, especially in applications to intensive longitudinal data.

2.2.1 Choosing a Covariance Structure

Because the large-sample properties of the GEE estimate $\hat{\beta}$ and the sandwich depend only on the correctness of the mean function $\mu_i(\beta)$, it is tempting to adopt a carefree attitude in selecting the form of the working covariance matrices. This temptation should be resisted. If $\Sigma_1, \ldots, \Sigma_N$ have been correctly specified up to the unknown parameters ϕ and α, then the resulting $\hat{\beta}$ is not only asymptotically unbiased but efficient. For reasons of efficiency, therefore, we should like the working covariance structure to be reasonably accurate. Liang and Zeger (1986) suggest that the working assumption of independence ($R_i = I$) is often adequate and nearly optimal for longitudinal data when the number of observations per subject n_i is small and the within-subject correlations are weak. An independence structure may also be appropriate for applications to clustered data (e.g., children nested within schools) which may have large n_i's but rarely have large intracluster correlations. Intensive longitudinal data (ILD), however, tend to have both large n_i's and large correlations; in these settings, a working assumption of independence can produce estimates that are grossly inefficient. Dramatic evidence of this inefficiency is found in the results of our data analysis and simulation which we present in the next section.

Given that independence may be a poor choice, it is also clear that the popular working correlation patterns available in PROC GENMOD and elsewhere may not be particularly well suited to ILD either. These patterns are motivated by studies in which the subjects are measured at a common set of occasions, often equally spaced, and they tend to assume that the correlation between y_{it} and $y_{it'}$ diminishes as a function of the simple difference or time interval between t and t' alone. With ILD, however, individuals may be observed at irregular or random intervals over weeks or days with multiple observations per day. The presence of natural day-to-day variation may cause a pair of adjacent observations, y_{it} and $y_{i,t+1}$, to be highly correlated if they fall on the same day but weakly correlated if they came from different days. One could imagine R_i having a block pattern with blocks corresponding to days and different correlations in the diagonal and off-diagonal blocks. It may even be reasonable to propose a structure like the Kronecker product of two AR(1) correlation matrices, each with its own autoregressive coefficient, reflecting different rates of decay within a day and across days. (If the number of observations and their spacing vary from day to day, however, the pattern cannot be represented exactly by a Kronecker product.) Specialized covariance structures like these are unlikely to become available in the near future as standard options in GEE software; implementing them, although conceptually straightforward, may require a nontrivial amount of programming on the part of a data analyst.

Another aspect of covariance structures relevant to ILD is that the dependence between y_{it} and $y_{it'}$ is likely to be affected not only by the relative proximity of the occasions t and t', but by relationships between y_{it} and other time-varying covariates that may vary across individuals. In the data example used by Walls et al. (chapter 1, this volume), to which we shall return in the next section, the questions of interest focus on the relationship between students' sense of control and their sense of choice. The moderately large numbers of responses ($n_i \approx 40$) allow us to estimate a random effect of choice on control for each subject, which induces a strong intrasubject correlation. In studies where n_i is even larger, it becomes possible to address hypotheses about coefficients that vary over time as well as across subjects (Li et al., chapter 3, this volume). Indeed, the scientific motivations for collecting ILD are often related to the study of complex relationships among time-varying variables, rather than the more traditional study of change in the levels of a response over time. The dependences generated by these processes tend to be considerably more complicated than the simple, time-dependent patterned correlation matrices mentioned in section 2.1.4.

2.2.2 On the Relationship Between GEE and Multilevel Models

Multilevel models estimate relationships between the mean response and covariates for specific individuals; GEE estimates the relationship between mean response and covariates for the population of individuals. In some cases the two are the same; in other cases they can be radically different.

To clarify, consider the two-level linear model reviewed by Walls et al. (chapter 1, this volume); in the familiar notation of Laird and Ware (1982), the model is

$$y_i = X_i\beta + Z_ib_i + \epsilon_i, \tag{2.20}$$

where y_i is the vector of responses for individual i; X_i and Z_i are matrices of covariates; β is a vector of fixed coefficients common to all individuals; $b_i \sim N(0, \psi)$ is a vector of random coefficients specific to individual i; and $\epsilon_i \sim N(0, \sigma^2 V_i)$ is a vector of residuals independent of b_i. Averaging over the distribution of the unobserved b_i, this model implies that y_i is normally distributed in the population with a mean function $\mu_i(\beta) = X_i\beta$ and a patterned covariance matrix $\Sigma_i = Z_i\psi Z_i^T + \sigma^2 V_i$. With a large number of individuals, we could fit a model in PROC GENMOD using the same mean function $\mu_i(\beta) = X_i\beta$ and virtually any working covariance structure, and we would obtain an estimate for β similar to what we would find if we fit the multilevel model (2.20) in PROC MIXED. In fact, in the special case of a random-intercepts model with $Z_i = (1, 1, \ldots, 1)^T$ and $V_i = I$, the correlation matrix would have the exchangeable form (2.19), and GEE under the exchangeable working assumption would be essentially identical to fitting the multilevel model.[3]

Now consider the interpretation of a single coefficient β_j within the vector β. Suppose the jth column of X_i represents a variable that is a fixed property of individual i–for example, a dummy indicator for sex. In that case, β_j should be viewed as the difference in mean response between two groups of individuals in the population, groups that have a one-unit difference in the value for that covariate (e.g., males versus females) but identical values for all other covariates.

If the covariate in question is time-varying, then β_j still represents the change in mean response associated with a unit change in that covariate *in the population of all individuals.* If the covariate has not been placed into a column of Z_i, then β_j is also the increase in $E(y_{it})$ associated with a unit increase in the covariate *for every individual.* This is not to say, however, that the value of the response for an individual at any occasion would be expected to rise by β_j units if his or her covariate were suddenly increased by one unit. Even putting aside any suggestion of causality, the coefficient of a time-varying covariate may simply reflect differences among the subjects rather than changes for any subject. If the average value of the covariate varies considerably from one individual to another, then β_j reflects a combination of two associations: the tendency of individuals with higher average values of that covariate to have higher (or lower) average values of the response, and the tendency for an individual who experiences an increase in the covariate from one occasion to another to also experience an increase (or decrease) in the response over the same interval of time. In other words, β_j represents an overall relationship between the covariate and the response due to both between-subject differences and to within-subject differences. Further discussion of this point will be given shortly.

A very different situation arises, however, for a time-varying covariate that appears both in the jth column of X_i and in the kth column of Z_i. In that case, a one-unit increase in the value of the covariate for subject i is accompanied by an increase in mean response of $\beta_j + b_{ik}$, where b_{ik} denotes the kth element of b_i. Because the b_{ik} has a mean of zero, the average of these regression coefficients $\beta_j + b_{ik}$ in the population is still β_j, the same parameter estimated by a GEE model with mean function $\mu_i(\beta) = X_i\beta$. A large variance for the level 2 disturbance b_{ik}, however, would indicate substantial heterogeneity among individuals with respect to their relationships between the response and the covariate in question. This would naturally raise doubts about the usefulness of β_j as a sole summary measure of the relationship. If the true value of β_j is not substantially different from zero, a GEE analysis might lead us to believe that the covariate is essentially unrelated to the response; a multilevel analysis, however, could reveal that relationships are indeed present at the individual level, and that these relationships vary. Individuals whose relationships are positive may be qualitatively very different from those whose relationships are negative, and these differences might be explainable by other individual characteristics.

This discussion reveals one of the greatest potential weaknesses of GEE methodology in situations involving ILD. GEE, as it is commonly applied today, tends to parameterize the within-subject covariance structure only in terms of crude time-graded effects, and not in terms of any other time-varying covariates. One of the emergent themes of the new field of ILD, as described in the introduction to this volume, is that time-graded effects on the response may be among the least interesting aspects of the data, whereas relationships among individuals as they vary and evolve over time are often key. On the other hand, a population-average coefficient β_j is still a meaningful, albeit incomplete, summary measure of a relationship between a response and a time-varying covariate, and it deserves to be estimated well. GEE may allow us to make quick and reliable inferences about β_j's without having to invest a great deal of effort into building an accurate and elaborate model for Σ_i.

2.2.3 On the Interpretation of Regression Coefficients

As we have just noted, a fixed coefficient β_j for a time-varying covariate represents a combination of two types of covariation: the tendency for individuals who differ in their mean values of the covariate to exhibit different mean levels of response, and the tendency for temporal changes in the covariate for any individual to be associated with temporal changes in the response. In studies that collect ILD, the latter may often be more relevant. For multilevel analyses, some advocate removing of the between-subject portion by centering the covariate values for each subject at his or her individual mean (Raudenbush & Bryk, 2002; Singer & Willett, 2003). Centering could be used in GEE analyses as well, and the interpretation of the coefficient would be the same as in the multilevel setting. A stronger remedy, the so-called conditional or fixed-effects analysis, removes the effects of all stable characteristics of subjects simultaneously whether or not they have been measured (Green, 1990). In a linear model, a conditional or fixed-effects analysis is implemented by including $N - 1$ dummy indicators to distinguish among the N subjects; this requires the removal of all non-time-varying covariates to avoid redundancy.

The discussion in section 2.2.2 on the equivalent meaning of β_j in a multilevel model and a GEE analysis is based on the premise that the mean response is a linear function of the covariates, that is, that we are using the identity link $g(\mu_{it}) = \mu_{it}$. Nonlinear link functions, which are easily applied in GEE and are becoming increasingly available in multilevel models, introduce a discrepancy in the meaning of β_j. Consider a GEE analysis of repeated binary observations using a logit link,

$$g(\mu_{it}) = \log\left(\frac{\mu_{it}}{1 - \mu_{it}}\right) = x_{it}^T \beta.$$

The coefficient β_j is the additive increase in log-odds of "success" ($y_{it} = 1$) across the population associated with a one-unit increase in x_{itj}, holding all other elements of x_{it} constant. The exponentiated coefficient $\exp(\beta_j)$ is the multiplicative effect on the odds of success in the population. The multilevel analog to this model would be

$$g(\mu_{it}) = \log\left(\frac{\mu_{it}}{1 - \mu_{it}}\right) = x_{it}^T \beta + z_{it}^T b_i,$$

where $b_i \sim N(0, \psi)$ is a random effect for subject i. The latter is an example of a generalized linear mixed model (GLMM); a brief overview of GLMMs is given by Hedeker et al. (chapter 4, this volume). In the GLMM, β_j is the increase in log-odds of success for any individual associated with a one-unit increase in x_{itj} (assuming that this covariate has not also been included in z_{it}). Averaging across individuals, β_j is also the increment in log-odds for the population, because the mean value of b_i in the population is zero. But $\exp(\beta_j)$ is not the average multiplicative effect on the odds in the population, because the average of an antilog is not the same as the antilog of an average. The nonlinearity of the link function requires us to make a distinction between the meaning of β_j in the marginal and multilevel analyses. In the former, it is called a population-average effect; in the latter, it is said to be subject-specific. A very lucid and thorough discussion on the differences between population-average and subject-specific effects is given by Fitzmaurice et al. (2004).

It is also crucial to distinguish the coefficients of a regression model from causal effects. With a linear model, it is tempting to view β_j as the estimated increase in y_{it} that would be seen if we could suddenly intervene in the life of subject i and change his or her value of x_{itj} to $x_{itj} + 1$ without altering any other element of x_{it}. This thinking implicitly compares outcomes under two different scenarios–the one that actually occurred, in which x_{itj} took its realized value, and a counterfactual reality in which x_{itj} would actually have been $x_{itj} + 1$. Analyses based on the notion of counterfactual outcomes were introduced by Neyman (1923, reprinted with discussion in 1990) and Rubin (1978) and have spawned a revolution in the way statisticians now think about causality. Any serious consideration of β_j as a causal effect must take into account the mechanism by which the values of x_{itj} came to be realized. If that mechanism is not confounded with the distribution of the response (e.g., as in a randomized experiment), then the causal interpretation will hold; in observational studies where this assignment is not randomized, strong assumptions and careful analyses are required. The tendency of a novice to think of all regression coefficients as causal effects ought to be strongly discouraged; in general, it is unwise to try to make causal inferences for more than one covariate at a time. Methods for causal inference go far beyond the scope of this chapter. A summary of how modern statisticians view causality is given in the chapters of the newly edited volume by Gelman and Meng (2004); a gentler introduction for social scientists is given by Winship and Sobel (2004). For a good discussion of special issues in causal inference pertaining to time-varying covariates, refer to Fitzmaurice et al. (2004).

2.2.4 How Large Is Large?

GEE regression is a large-sample procedure; the consistency of $\hat{\beta}$ and sandwich-based standard errors are theoretically attained as the number of occasions per subject n_i remain fixed and the number of subjects N goes to infinity. Inferences from a particular GEE analysis rest on the belief that the actual N is large enough for the approximations to be accurate. Multilevel modeling also rests upon theoretical results about the optimality of maximum likelihood (ML) and related techniques as $N \to \infty$ with n_i fixed. The situation is a little different, though, because the multilevel model is assumed to be true whereas the marginal model is assumed to be partly misspecified. It is generally recognized that a larger N may be required for GEE than for ML under their respective assumptions. If N is only moderately large, we may be willing to sacrifice some of the theoretical robustness properties of GEE in favor of a multilevel analysis which, although it makes stronger distributional assumptions, leads to more accurate and efficient inferences if the model is approximately true.

The existing literature on GEE reveals some vague guidelines about how large N should be. Because simulations by Liang and Zeger (1986) demonstrated reasonable accuracy for small n_i and $N = 30$, some data analysts have apparently come to rely on $N \geq 30$ as a rule-of-thumb. Kauermann and Carroll (2001) investigated the accuracy of standard errors from the sandwich and found them to be considerably less precise than their model-based counterparts; they also propose a degrees-of-freedom correction to improve the coverage of sandwich-based confidence intervals. The sample size needed to guarantee accuracy will undoubtedly vary from one application to another; it will depend on the number of covariates, the number and spacing of observations over time, and the structure and complexity of the working covariance matrix. Fitzmaurice et al. (2004) point out that the asymptotic arguments supporting the use of the sandwich are based on replication of the covariate matrices X_i. These considerations can be especially troubling for studies involving ILD, where the expense and intensity of the data collection protocols may permit only small numbers of participants. Large numbers of occasions are not necessarily harmful, because evidence about parameters of simple working covariance structures such as AR(1) may tend to accumulate as n_i grows. With more elaborate working assumptions, however, the combination of large n_i and small N could spell disaster.

If little is known about the behavior of GEE in a particular setting, it may be feasible to carry out a simulation experiment to assess performance in a population designed to mimic essential features of the given data. This approach will be used in the next section.

2.2.5 Impact of Missing Values

Algorithms and software for both multilevel modeling and GEE are designed to handle unbalanced data in which the number of occasions n_i and the occasions themselves vary by subject. In both paradigms, it has become commonplace to ignore the missingness and analyze the incomplete data as if they were complete but unbalanced. That is, if some responses y_{it} for some subjects are not recorded for reasons beyond the investigators' control, most analysts will simply ignore the missed occasions and analyze the remaining ones if they were the only occasions that were intended for data collection in the first place. This practice is not necessarily bad. It does, however, make certain implicit assumptions about the processes that produced missing values, and these assumptions are different for multilevel models and for GEE.

Rubin (1976) demonstrated that ignoring the mechanism of missingness is appropriate in a likelihood-based analysis when the missing values are missing at random (MAR). MAR is a technical condition under which the probabilities of missingness depend only on quantities that are observed and present in the parametric model. Statistical procedures that do not rely on a likelihood function, which include GEE methods, are generally appropriate only when the missing values are missing completely at random (MCAR). MCAR, which is stronger and often less plausible than MAR, implies that the probabilities of missingness do not depend on any observed or missing data.

Based on Rubin's (1976) results, one could say that standard approaches in multilevel modeling require the missing data to be MAR, whereas GEE requires the stronger assumption of MCAR. In reality, however, the situation with GEE is more subtle. Liang and Zeger (1986) pointed out that if the working assumptions are correct, GEE becomes asymptotically equivalent to ML under a full parametric model, which makes the estimated coefficients $\hat{\beta}$ appropriate under MAR. Even in that setting, however, standard errors based on the sandwich would still require MCAR because they are not based on a likelihood function. Even the model-based standard errors (from the bread of the sandwich) may require MCAR, because they may use expected derivatives of the score vectors rather than the actual derivatives (Kenward & Molenberghs, 1998).

In practice, this means that a data analyst may need to pay greater attention to missing values when using GEE than when fitting multilevel models. Biases in coefficients due to departures from MCAR can be corrected by weights (Robins et al., 1994, 1995), which requires the analyst to build a parametric model for the missingness mechanism. Simple weighted estimators can be inefficient. More recent work by Scharfstein et al. (1999) shows how to increase efficiency by introducing additional information to predict missing values in a manner reminiscent of imputation. At present, these new techniques have been applied only in relatively simple longitudinal settings (Davidian et al., 2005). More discussion on the role of missing data in multilevel and GEE analyses of longitudinal data is given by Verbeke and Molenberghs (2000) and Fitzmaurice et al. (2004).

2.3 Application: Reanalysis of the Control and Choice Data Using GEE

2.3.1 Brief Recapitulation of Results from Multilevel Analyses

Walls, Jung, and Schwartz (chapter 1, this volume; hereafter, WJS) analyzed a set of intensive longitudinal measurements provided by Verma et al. (2002). In this study, $N = 100$ adolescents in India were prompted several times per day to answer questions related to stress. WJS's analyses focused on the relationship between two measures: an individual's feeling of control over his or her situation and the perceived degree of choice over the activities in which he or she was engaged. Treating control as a response variable and choice as a predictor, WJS constructed multilevel models in which the effect of choice on control was allowed to randomly vary by subject. Average trajectories of the response over calendar time were essentially flat, indicating no overall growth or developmental trends with respect to time. Some modest differences in mean levels of the response were found with respect to day of the week, and WJS also found a significant difference between boys and girls.

Regarding the covariance structure, WJS found substantial variation in the response from one day to the next. A random perturbation to the intercept for each subject-day was found to have high variance, but a random perturbation to the slope with respect to choice was unnecessary. The final WJS model had three levels: the first or lowest level corresponded to repeated observations within a day, the second level to days, and the third level to subjects. Fixed effects were included for choice, day of the week, and sex. Intercepts were allowed to vary by subject and day, and the choice-slopes were allowed to vary by subject. Analysis of Cholesky-transformed residuals revealed no strong evidence of heteroscedasticity or nonnormality, except for a mild ceiling effect in the ordinal scale. Semivariograms revealed no apparent residual dependences between pairs of observations within a day or across days. Taken together, these diagnostics suggested that the three-level model fitted the data quite well.

2.3.2 Results from GEE

After fitting the three-level model, WJS computed an empirical Bayes estimate of each individual's random coefficient for choice. These estimates indicated that, although the average relationship between choice and control in the population is positive, individuals do vary; some exhibit a strong positive relationship, others show little or no relationship, and for some the relationship is apparently negative. This subject-level variation is part of the covariance structure which remains hidden in a standard GEE analysis. GEE treats the covariance structure as a nuisance and focuses on the population-average effects on the mean. Nevertheless, if the same covariates (choice, sex, day of the week) are used in a GEE analysis, and if the GEE methodology works as it should, then the GEE-based coefficient estimates should resemble the estimated fixed effects from the multilevel model, because the population parameters in these two approaches are in fact the same.

Estimated coefficients and sandwich-based standard errors from three GEE analyses are shown in table 2.2, along with the results obtained by WJS using restricted maximum-likelihood (REML) under the three-level model. All three GEE analyses use an identity link and a constant variance function; they differ only in the choice of the working correlation structure. The three correlation structures tried were independent, exchangeable, and first-order autoregressive. The working assumption of independence is clearly unrealistic; the strong subject-level variation reported by WJS indicates that observations within a subject are highly related. The exchangeable and AR(1) models do allow correlation, but neither seems realistic. The exchangeable model fails to reflect the fact that observations from the same day are more strongly related than observations from different days. The AR(1) model allows the autocorrelation to die down smoothly as time passes, and may not account for a sudden drop as a subject transitions from one day to the next. Despite the apparent misspecification of the working correlation structure, GEE is still supposed to provide estimates and standard errors that are reasonably unbiased, although perhaps not efficient.

Examining table 2.2, we see that the coefficients and standard errors for sex are extremely close under all four methods. Sex is a fixed characteristic of the subject, and all four models treat the subjects as independent. Effects for days of the week are more volatile; standard errors under the four methods are fairly close, but the estimated coefficients from REML fall one or two standard errors or more away from the GEE estimates. Perhaps this should have been expected, because the three-level model makes assumptions different from those of GEE about variability with respect to day. The coefficient of greatest interest to us, the average effect of choice, is relatively close for the exchangeable, AR(1), and three-level models. Under the independence working assumption, however, the estimate is much larger, and its standard error has more than doubled. This inflation of the standard error suggests that the independence working assumption produces an estimate that is much less efficient than those produced by other models. This seems to agree with a notion that is becoming common wisdom

Table 2.2 Estimated coefficients (with standard errors in parentheses) from marginal models with working assumptions of independence (GEE-IND), exchangeable (GEE-EX), and first-order autoregression (GEE-AR), along with a restricted maximum-likelihood fit under a three-level model (REML)

Term	GEE-IND	GEE-EX	GEE-AR	REML
Intercept	4.552	4.922	4.796	4.736
	(0.361)	(0.296)	(0.306)	(0.307)
Choice	0.184	0.099	0.104	0.086
	(0.050)	(0.019)	(0.019)	(0.016)
Day 2	−0.188	−0.188	−0.092	0.223
	(0.180)	(0.169)	(0.170)	(0.187)
Day 3	−0.503	−0.411	−0.290	0.052
	(0.203)	(0.196)	(0.201)	(0.186)
Day 4	−0.466	−0.478	−0.439	−0.163
	(0.236)	(0.210)	(0.215)	(0.184)
Day 5	−0.514	−0.490	−0.372	−0.226
	(0.208)	(0.194)	(0.212)	(0.182)
Day 6	−0.630	−0.544	−0.503	−0.262
	(0.229)	(0.224)	(0.217)	(0.183)
Day 7	−0.297	−0.234	−0.143	−0.311
	(0.185)	(0.172)	(0.176)	(0.184)
Sex	−0.950	−1.002	−0.927	−1.014
	(0.401)	(0.393)	(0.400)	(0.394)

among users of GEE: if measurements within a subject are highly interrelated, estimates will be much more efficient if that correlation is somehow taken into account in the estimation procedure. The exact form of the working correlation is not especially important, as long as some within-subject correlation is allowed.

2.3.3 A Simulation Experiment

Results from the three GEE analyses seem consistent with those from the multilevel model and with what we know about the theoretical properties of GEE in large samples. Yet an important question remains. Relatively little is known about the performance of GEE in studies involving modest N, large n_i, and high within-subject correlation. Is this sample of $N = 100$ individuals truly large enough for GEE to work as advertised? Or should the fully parametric multilevel analysis, which makes stronger assumptions, be preferred for a sample of this size?

Table 2.3 Simulated performance of estimated coefficients for choice (actual value is 0.08647): average of the estimates, raw bias, relative bias, standard deviation, and root mean squared error

	Avg.	Bias	Rel. bias	SD	RMSE
REML	0.08648	0.00001	0.0	0.01666	0.01665
ML	0.08646	−0.00001	0.0	0.01667	0.01666
GEE-IND	0.08389	−0.00258	−3.0	0.04127	0.04133
GEE-EX	0.08611	−0.00036	−0.4	0.01789	0.01788
GEE-AR	0.08593	−0.00054	−0.6	0.02015	0.02015

To address these questions, we ran a simulation to see how GEE performs for data resembling this sample. We drew new response vectors from normal distributions, $y_i \sim N(X_i\beta, \Sigma_i)$, $i = 1, \ldots, 100$, where the design matrices $X_1, \ldots, X_{100}$ were set equal to those of the individuals in our sample. The columns of each design matrix included a constant, the measurement for choice, dummy indicators for day of the week and sex, exactly as shown in table 2.2. The elements of β were fixed at the REML estimates reported by WJS for their three-level model, and the covariance matrices Σ_i were set equal to their estimates under the three-level model. In other words, we supposed that WJS's three-level model were true, and that the parameters of the model were identical to the REML estimates. After drawing a new sample $y_1, \ldots, y_{100}$, we fitted the three-level model to the simulated data by ML and REML, and we also fitted marginal models using independence, exchangeable, and AR(1) working correlations. The entire procedure was repeated 1,000 times.

For simplicity, we shall focus attention on the main parameter of interest, the population-average coefficient for choice. The performance of the estimated coefficients for choice over the 1,000 repetitions is summarized in table 2.3. For each method, the table reports the average of the 1,000 estimates, the raw bias (the difference between the average and the actual value), the relative bias (the raw bias as a percentage of the actual value), the standard deviation of the estimates, and the root mean squared error (the square root of the average squared distance between the estimate and the actual value). The actual population value for this parameter is 0.08647, and on average each of the estimation methods was able to reproduce this value very closely; biases in the estimates are practically insignificant. The fully parametric ML and REML methods are clearly the most precise, followed by the GEE method under exchangeable and AR(1) working assumptions. GEE under the implausible independence working assumption is essentially unbiased, as it should be, but compared to the other methods it is grossly inefficient. Inefficiency can be a major concern when the number of subjects is not large. Based on these results, we would strongly discourage the use of the independence working assumptions for typical analyses involving ILD. The exchangeable and AR(1) assumptions, although unrealistic in this case, do

Table 2.4 Simulated performance of standard errors for the coefficient of choice: variance of the estimates, average of the squared standard errors, coverage of nominal 95% intervals, and average interval width

	Var.	Avg.(SE^2)	Coverage	Avg. width
REML	0.000277	0.000269	94.5	0.06413
ML	0.000278	0.000266	94.5	0.06375
GEE-IND	0.001703	0.001626	94.4	0.15710
GEE-EX	0.000320	0.000305	94.1	0.06818
GEE-AR	0.000406	0.000385	93.5	0.07672

rather well and produce estimates that are only slightly less efficient than ML and REML.

Although the behavior of the estimated coefficients is important, it is equally crucial to assess the accuracy of standard errors. Standard errors that tend to be too small or too variable can adversely affect the coverage of confidence intervals and impair the performance of hypothesis tests. The performance of these standard errors is summarized in table 2.4. The first column in this table is the sample variance among the 1,000 coefficients; it estimates the actual variance. The second column is the average of the 1,000 squared standard errors; if the standard errors are accurate, then the second column should closely match the first. In this respect, all five methods look good. The model-based standard errors from REML and ML do accurately describe the variability of those estimation methods, and the sandwich-based standard errors from GEE accurately reflect the larger variances of the less efficient GEE estimates. The third column reports the percentage of nominal 95% intervals (computed as the estimated coefficient plus or minus two standard errors) that captured the true parameter value of 0.08647. The simulated coverage rates are all slightly below 95%, but none are drastically lower. Accuracy in the coverage rate of confidence intervals translates into accuracy of type I error rates in hypothesis testing. Finally, the last column reports the average width of the nominal 95% intervals. It is here that an important difference appears; intervals from GEE with an independence working assumption are on average more than twice as wide as the intervals from the other methods. Wider intervals lead to reduced power and higher rates of type II error. In this setting, statistical inferences based on GEE-IND are not in any sense invalid; the estimates have low bias and the standard errors accurately measure the uncertainty. They do not, however, make efficient use of the data in estimating the population-average coefficient for choice.

2.3.4 Conclusions

This example suggests that marginal modeling with GEE can be useful for estimating population-average mean effects from ILD if the number of subjects is sufficiently large. Considerations of efficiency should prompt us to use working assumptions that allow correlations within subjects, but the actual form of these correlations may not need to be especially accurate.

Because our simulation computed REML and ML estimates under the true model, we had no opportunity to see whether any of the GEE methods performed better than a misspecified parametric analysis. WJS built their final model carefully and found no major problems in any of the diagnostic plots, so we have no reason to believe that the departures from that model are severe. GEE does protect a data analyst from the harmful effects of certain types of misspecification if the sample is large enough, but a careful analysis should involve model checking and adjustment to fix gross problems anyway.

One clear advantage of multilevel modeling in this example is that it provides an estimate of the choice coefficient for each subject, allowing us to see how variable these coefficients really are. Based on a GEE analyis, all we know is that, on average across individuals, the effect of choice on control is about 0.086 ± 0.04; we have no idea how individuals may vary in this regard. The fact that the coefficients for some individuals are nearly zero or negative is a potentially important finding that deserves further study; this variation could perhaps be explained by additional subject-level covariates. The random day-to-day variation in feeling of control is also substantial and could perhaps be partly explained by additional characteristics of the subject or the day. These interesting effects are part of the covariance structure, which a standard GEE analysis regards as a nuisance.

ACKNOWLEDGMENTS

This research was supported by the National Institute on Drug Abuse, 2-P50-DA10075. Thanks to Suman Verma for providing data from the Indian stress study, and to Hyekyung Jung for help in running the analyses and simulations reported in this chapter.

NOTES

1. In some cases, the mean structure is taken to be $\eta_i = g(\mu_i) = o_i + x_i^T\beta$, where o_i is an offset, a covariate whose coefficient is assumed to be 1. The most common use of an offset arises in a loglinear model when y_i is an event count and o_i is the logarithm of some measure of exposure to which μ_i is thought to be approximately proportional. For example, y_i could be the number of highway fatalities reported last year in state i, and o_i could be the state's log-population. Discussion of loglinear models with offset terms is given by Agresti (2002).

2. The practice of treating ϕ as fixed in the estimation procedure is sometimes called limited-information maximum likelihood (LIML), as opposed to full-information maximum likelihood (FIML) which maximizes the likelihood function with respect to ϕ as well as β. More discussion on LIML versus FIML is given by Hardin and Hilbe (2003).

3. Minor differences would arise merely because the GEE software's $\sqrt{N}$-consistent method for estimating the dispersion parameter and intrasubject correlation differs slightly from the multilevel modeling technique of maximum likelihood or restricted maximum likelihood.

References

Agresti, A. (2002). *Categorical Data Analysis* (2nd ed.). New York: John Wiley.

Arminger, G. (1995). Specification and estimation of mean structures: Regression models. In G. Arminger, C.C. Clogg, & M.E. Sobel (Eds.), *Handbook of Statistical Modeling for the Social and Behavioral Sciences*. New York: Plenum Press.

Carey, V.C., Zeger, S.L., & Diggle, P.J. (1993). Modelling multivariate binary data with alternating logistic regressions. *Biometrika, 80*, 517–526.

Clogg, C.C., Petkova, E., & Shihadeh, E.S. (1992). Statistical methods for analyzing collapsibility in regression models. *Journal of Educational Statistics, 17*, 51–74.

Davidian, M., Tsiatis, A.A., & Leon, S. (2005). Semiparametric estimation of treatment effect in a pretest-posttest study with missing data. *Statistical Science, 20*, in press.

Diggle, P.J., Heagerty, P., Liang, K.Y., & Zeger, S.L. (2002). *The Analysis of Longitudinal Data* (2nd ed.). Oxford: Oxford University Press.

Fitzmaurice, G.M., Laird, N.M., & Ware, J.H. (2004). *Applied Longitudinal Analysis*. New York: John Wiley.

Francis, B., Green, N., & Payne, C. (1993). *The GLIM System: Generalised Linear Interactive Modelling, Release 4 Manual*. Oxford: Clarendon Press.

Gelman, A., & Meng, X.L. (Eds.) (2004). *Applied Bayesian Modeling and Causal Inference from Incomplete-Data Perspectives*. New York: John Wiley.

Gourieroux, C., & Montfort, A. (1993). Pseudo-likelihood methods. In G.S. Maddala, C.R. Rao, & H.D. Vinod (Eds.), *Handbook of Statistics*, Vol. 11. Amsterdam: Elsevier.

Gourieroux C., Monfort, A., & Trognon, A. (1984). Pseudo-maximum likelihood methods: Theory. *Econometrica, 52*, 681–700.

Green, W.H. (1990). *Econometric Analysis*. New York: Macmillan.

Hardin, J.W., & Hilbe, J.M. (2003). *Generalized Linear Models*. New York: Chapman & Hall/CRC Press.

Huber, P.J. (1967). The behavior of maximum likelihood estimates under non-standard conditions. In *Fifth Berkeley Symposium in Mathematical Statistics and Probability* (pp. 221–233). Berkeley: University of California Press.

Jones, R.H. (1993). *Longitudinal Data with Serial Correlation: A State-Space Approach*. London: Chapman & Hall.

Kauermann, G., & Carroll, R.J. (2001). A note on the efficiency of sandwich covariance matrix estimation. *Journal of the American Statistical Association, 96*, 1387–1396.

Kenward, M.G., & Molenberghs, G. (1998). Likelihood based frequentist inference when data are missing at random. *Statistical Science, 13*, 236–247.

Laird, N.M., & Ware, J.H. (1982). Random-effects models for longitudinal data. *Biometrics, 38*, 963–974.

Liang, K.Y., & Zeger, S.L. (1986). Longitudinal data analysis using generalized linear models. *Biometrika, 73*, 13–22.

Liang, K.Y., Zeger, S.L., & Qaqish, B. (1992). Multivariate regression analyses for categorical data (with discussion). *Journal of the Royal Statistical Society, Series B, 54*, 3–40.

McCullagh, P., & Nelder, J.A. (1989). *Generalized Linear Models*, (2nd ed.). London: Chapman & Hall.

Nelder, J.A., & Wedderburn, R.W.M. (1972). Generalized linear models. *Journal of the Royal Statistical Society, Series A, 135*, 370–384.

Neyman, J. [1923] (1990). On the application of probability theory to agricultural experiments, essays on principles, section 9 (with discussion). *Statistical Science, 4*, 465–480.

Raudenbush, S.W., & Bryk, A.S. (2002). *Hierarchical Linear Models: Applications and Data Analysis Methods*. Thousand Oaks, CA: Sage.

Robins, J.M., Rotnitzky, A., & Zhao, L.P. (1994). Estimation of regression coefficients when some regressors are not always observed. *Journal of the American Statistical Association, 89*, 846–866.

Robins, J.M., Rotnitzky, A., & Zhao, L.P. (1995). Analysis of semiparametric regression models for repeated outcomes in the presence of missing data. *Journal of the American Statistical Association, 90*, 106–121.

Rubin, D.B. (1976). Inference and missing data. *Biometrika, 63*, 581–592.

Rubin, D.B. (1978). Bayesian inference for causal effects: The role of randomization. *Annals of Statistics, 6*, 34–58.

Scharfstein, D.O., Robins, J.M., & Rotnitzky, A. (1999). Rejoinder to adjusting for non-ignorable dropout using semiparametric response models. *Journal of the American Statistical Association, 94*, 1135–1146.

Singer, J.D., & Willett, J.T. (2003). *Applied Longitudinal Data Analysis: Modeling Change and Event Occurrence*. New York: Oxford University Press.

Thall, P.F., & Vail, S.C. (1990). Some covariance models for longitudinal count data with overdispersion. *Biometrics, 46*, 657–671.

Verbeke, G., & Molenberghs, G. (2000). *Linear Mixed Models for Longitudinal Data*. New York: Springer.

Verma, S., Sharma, D., & Larson, R.W. (2002). School stress in India: Effects on time and daily emotions. *International Journal of Behavioral Development, 26*, 500–508.

Wedderburn, R.W.M. (1974). Quasi-likelihood functions, generalized linear models, and the Gauss-Newton method. *Biometrika, 61*, 439–447.

White, H. (1980). A heteroscedasticity-consistent covariance matrix estimator and a direct test for heteroscedasticity. *Econometrica, 48*, 817–838.

Winship, C., & Sobel, M.E. (2004). Causal inference in sociological studies. In M. Hardy (Ed.), *The Handbook of Data Analysis*, Thousand Oaks, CA: Sage.

3

A Local Linear Estimation Procedure for Functional Multilevel Modeling

Runze Li, Tammy L. Root, and Saul Shiffman

Linear mixed models, also referred to as hierarchical linear models (HLM), have been particularly useful for researchers analyzing longitudinal data but they are not appropriate for all types of longitudinal data. For instance, these methods are not able to estimate changes in slope between an outcome variable and possibly time-varying covariates over time. The functional multilevel modeling technique proposed in this chapter addresses this issue by expanding the linear mixed model to allow coefficients, both fixed and random, to vary nonparametrically over time. Estimation of time-varying coefficients is accomplished by adding a local linear regression estimation procedure (Fan & Gijbels, 1996) to the traditional linear mixed model.

The primary motivation for the current work was the methodological challenges often faced by drug use researchers on how to model intensive longitudinal data (i.e., massively multivariate). Ecological Momentary Assessment data (EMA; Shiffman, 1999) (a type of intensive longitudinal data) provides one example where functional multilevel modeling would prove beneficial. Functional multilevel models are particularly well suited for certain types of intensive longitudinal analyses because (1) the technique does not attempt to adopt a static process, (2) it allows for the analysis of intensive multivariate data without complication, and, perhaps most importantly, (3) the powerful exploratory capabilities of functional multilevel models allow researchers to detect relationships that change over time as a result of an intervention (e.g., smoking cessation program) or a natural growth process (e.g., cognitive decline).

There are several methods for collecting intensive longitudinal data (Walls et al., chapter 1, this volume). Using handheld computers for EMA data collection was pioneered in tobacco use research (Shiffman et al., 1996a). Through the collection of EMA data, substance use researchers sought to address the following questions:

1. How does the subjective sensation of nicotine withdrawal vary over a day or a week, and how does it vary according to environmental cues?
2. What is the relationship between mood and tobacco use?
3. How does this relationship change over the process of quitting smoking?
4. What environmental cues trigger smoking?

Because the effects of smoking withdrawal may fluctuate repeatedly over the course of a day or a week across individuals (i.e, a time-varying effect), linear mixed models are not well suited for answering these kinds of questions. It is *very* important to note that time-varying effect is different from a time-varying covariate. Although a time-varying covariate, which changes across time, is allowed in the linear mixed model, its effect is constant. Time-varying effects on the other hand, such as changes in the direction and strength of the relationship between mood and smoking, cannot be directly or efficiently estimated using linear mixed models. Therefore, our goal was to develop a method that could address the aforementioned questions.

In this chapter we propose an estimation procedure for functional multilevel models using a local linear regression technique (Fan & Gijbels, 1996). Although nonparametric regression models, including time-varying coefficient models, have been used extensively in longitudinal research (Chiang et al., 2001; Fan & Zhang, 2001; Hoover et al., 1998; Huang et al., 2002; Martinussen & Scheike, 2001; Wu et al., 1998), the limitation with time-varying coefficient models is that they do not allow for change over time across individuals, and therefore are not appropriate for intensive longitudinal data such as the EMA data. Additionally, although nonparametric mixed-effects models as proposed by Wu and Zhang (2002) have been used in the analysis of longitudinal data, the authors have not considered the setting of time-varying coefficient models. Random varying-coefficient models (Wu & Liang, 2004), on the other hand, do allow effects to change across individuals and time; however, an assumption of smoothing covariates is imposed with this model.

As the reader will see later, the requirement of smoothing covariates is not realistic with our EMA example, as is the case with most intensive longitudinal data. Therefore, a need exists to expand on the current methodologies in order to allow coefficients to vary nonparametrically over time. The time-varying mixed-effects model (i.e., functional multilevel model) illustrated in this chapter is based on nonparametric functional estimation methods, which allow for the hierarchical structure of longitudinal data (i.e., observations at multiple occasions are nested within individuals) as well as the changing relations between dynamic variables. In other words, this method models parameters as functions of time and uses nonparametric methods to estimate these parameters. Prior to introducing the functional multilevel model, we shall illustrate the local linear regression technique (Fan & Gijbels, 1996).

The remainder of this chapter will be organized as follows. In section 3.1, we introduce local linear regression techniques as well as the functional multilevel model, and its estimation procedure. Practical considerations are presented in section 3.2. In section 3.3, results of the empirical analysis are presented. Section 3.4 provides extensions and limitations of the methodology.

3.1 The Model

In this section, we first introduce the fundamental ideas of local linear regression. In section 3.1.2, we explain why the time-varying coefficient model is more appropriate than either the linear regression model or the linear mixed model for intensive longitudinal data in the presence of time-varying effects. The estimation procedure for the functional multilevel model is discussed in section 3.1.3.

3.1.1 Local Linear Regression

Suppose that (x_i, y_i) is a random sample from a regression model

$$y_i = m(x_i) + \varepsilon_i, \tag{3.1}$$

where ε_i is a random error with mean zero and variance $\sigma^2(x_i)$, implying that the model allows a heteroscedastic error, and $m(x)$ is a regression function. Let us assume that $m(x)$ is a smooth function rather than possessing a parametric form such as a polynomial function of x. Under these assumptions, model (3.1) is considered a *nonparametric* regression model.

Estimation of $m(x)$ has been studied extensively in the statistical literature. It is known that various smoothing methods, including smoothing splines (Wahba, 1990), local regression (also referred to as LOESS) (Cleveland et al., 1993), and local polynomial regression (Fan & Gijbels, 1996), can be used to estimate $m(x)$. In this chapter, we shall focus on local linear regression. Let us describe the idea behind this procedure. For any given x_0 in an interval over which data are collected, we want to estimate $m(x_0)$. In a neighborhood of x_0, we locally approximate $m(x)$ by a linear function:

$$m(x) \approx m(x_0) + m'(x_0)(x - x_0) \stackrel{\text{def}}{=} \beta_0 + \beta_1(x - x_0).$$

Typically, a datum point closer to x carries more information about the value of $m(x)$, suggesting the use of a locally weighted linear regression (i.e., local linear regression) which minimizes the following locally weighted least squares function:

$$\sum_{i=1}^{n} w_i\{y_i - \beta_0 - \beta_1(x_i - x_0)\}^2, \tag{3.2}$$

where $w_i = h^{-1}K\{(x_i - x_0)/h\}$, $i = 1, \ldots, n$, are weights. Here $K(x)$ is a probability density function and is called a *kernel* function, and h is a positive number called a *bandwidth* or a *smoothing parameter*.

The minimizer of (3.2) is denoted by $\hat{\beta}_j$ ($j = 0, 1$). The above exposition implies that an estimator for $m(x)$ and for the first-order derivative $m'(x)$ at $x = x_0$ is

$$\hat{m}(x_0) = \hat{\beta}_0 \quad \text{and} \quad \hat{m}'(x_0) = \hat{\beta}_1,$$

respectively. The minimizer of (3.2) is the least squares estimate of the following simple linear regression model:

$$\sqrt{w_i} y_i = \beta_0 \sqrt{w_i} + \beta_1 \sqrt{w_i}(x_i - x_0) + \varepsilon_i^*,$$

where ε_i^* is random error with mean zero. Thus, minimization of (3.2) can be accomplished by regressing $\sqrt{w_i} y_i$ (dependent variable) over two independent variables $\sqrt{w_i}$ and $\sqrt{w_i}(x_i - x_0)$. This procedure can easily be carried out by existing software packages such as SAS and S-Plus.

It is known that the choice of K is not very sensitive, scaled in a canonical form, to the estimate $m(x)$ (Marron & Nolan, 1988). Therefore, it is assumed in this chapter that the kernel function is a symmetric probability density function. The most commonly used kernel function is the Gaussian density function given by

$$K(x) = \frac{1}{\sqrt{2\pi}} \exp(-x^2/2). \tag{3.3}$$

Other popular kernel functions include the symmetric beta family:

$$K(x) = \frac{1}{Beta(1/2, \gamma + 1)} (1 - x^2)_+^\gamma, \quad \gamma = 0, 1, \ldots, \tag{3.4}$$

where $+$ denotes the positive component, which is assumed to be taken before exponentiation. The support of K is $[-1, 1]$, and $Beta(\cdot, \cdot)$ is a beta function. The corresponding kernel functions when $\gamma = 0$, 1, 2, and 3 are the uniform, Epanechnikov, biweight, and triweight kernel functions, respectively. See figure 3.1 for a pictorial representation of these kernel functions.

Using the uniform kernel, the regression curve is estimated by the intercept in the linear regression based on data whose x_i's lie between $x - h$ and $x + h$. Thus, the local linear fit with the uniform kernel coincides with the *running line*, a special case of LOESS (Cleveland et al., 1993).

The parameter h controls the degree of smoothness of the regression function. The choice of bandwidth is an important consideration. For example, if h is too large, then the resulting estimate may not detect fine features of the data. Conversely, if h is too small, then spurious sharp structures may become visible. For additional information on bandwidth selection for model (3.1) the reader is referred to Fan and Gijbels (1996) and Jones et al. (1996a,b). In practice, data-driven methods can be used to select an optimal bandwidth or selection of the bandwidth can be obtained by visual inspection of the estimated regression function.

Let us provide an illustration adapted from Fan and Gijbels (2000) on how to fit a local linear regression. The relationship between the concentration of nitric

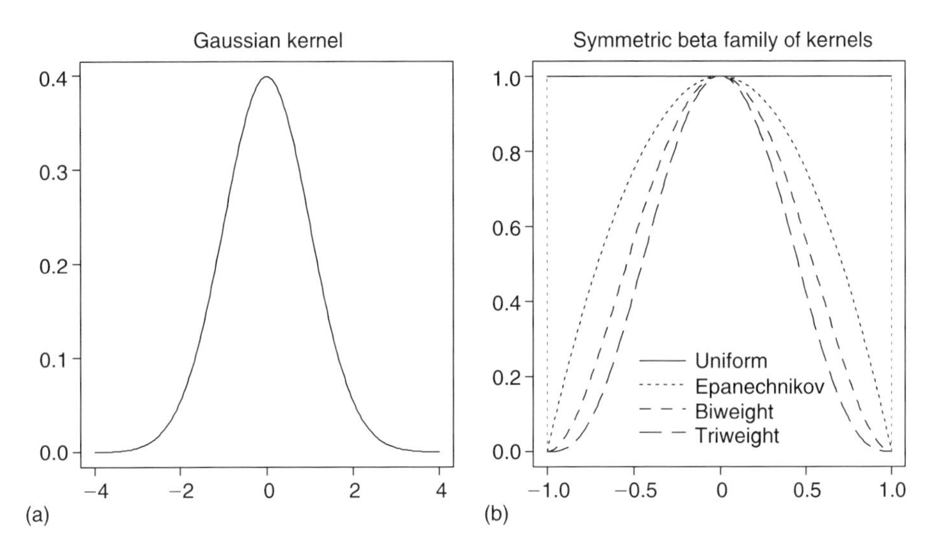

Figure 3.1 Commonly used kernels. (a) Gaussian kernel; (b) symmetric beta family of kernels that are renormalized to have maximum height 1.

oxides in engine exhaust (dependent variable) and the equivalence ratio (independent variable), a measure of the richness of the air/ethanol mix, was examined for the burning of ethanol in a single-cylinder automobile test engine. The scatterplot is presented in figure 3.2, from which we can see that the relationship is clearly nonlinear.

Let us take the bandwidth h to be 0.051 and the kernel to be an Epanechnikov kernel. To estimate the regression function at the point $x_0 = 0.8$, we use the local data in the strip $x_0 \pm h$ to fit a regression line (cf. figure 3.2). The local

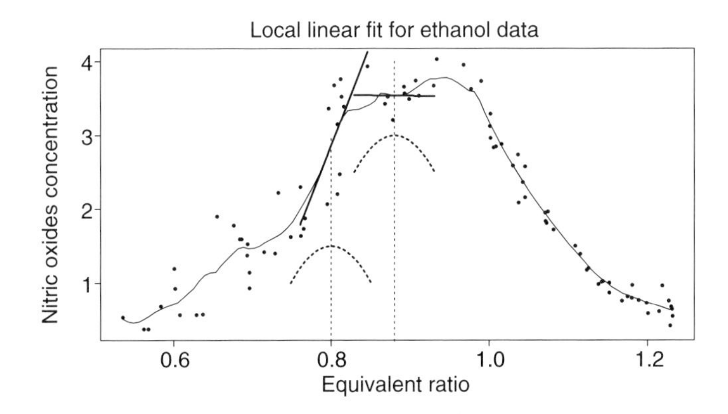

Figure 3.2 Illustration of the local linear fit. For each given x_0, a linear regression is fitted through the data contained in the strip $x_0 \pm h$, using the weight function indicated at the bottom of the strip. The interactions of the fitted lines and the short dashed lines are the local linear fits. (Adapted from Fan and Gijbels, 2000.)

linear estimate at x_0 is simply the intersection of the fitted line and the line $x = x_0$. If we wanted to estimate the regression function at a different point, say, $x_0 = 0.88$, then another line would be fitted using the data in the window 0.88 ± 0.051. The entire curve is obtained by estimating the regression function in a grid of points. The curve in figure 3.2 was obtained by 100 local linear regressions, taking the 100 grid points from 0.0535 to 1.232.

3.1.2 Functional Multilevel Model

Before we discuss the functional multilevel model, let us illustrate the presence of time-varying effects using the EMA data previously discussed (see section 3.3 for a thorough description of the data). Let $y_i(t_{ij})$ be the score of *urge to smoke* of the ith subject at time t_{ij}, and $x_{i1}(t_{ij})$, $x_{i2}(t_{ij})$, and $x_{i3}(t_{ij})$ be the centered score of *negative affect*, *arousal*, and *attention* of the ith subject at time t_{ij}, respectively. For the purpose of illustration of time-varying effects, we consider a linear regression model

$$y_i(t_{ij}) = \beta_0 + \beta_1 x_{i1}(t_{ij}) + \beta_2 x_{i2}(t_{ij}) + \beta_3 x_{i3}(t_{ij}) + \beta_4 x_{i1}(t_{ij}) x_{i2}(t_{ij}) + \varepsilon_i(t_{ij}),$$

where $\varepsilon_i(t_{ij})$ is a random error with mean zero. To examine possible time-varying effects, we first fit the model to the data day by day (before and after quitting smoking) using PROC REG in SAS. That is, for day k, we estimate the regression coefficients using data collected at day k rather than the entire data set. Figure 3.3 displays a plot of the resulting estimates against day.

It is apparent from figure 3.3 that the plots of the resulting estimates have not been smoothed. Additionally, we see that the intercept dramatically changes over time, and the effect of negative affect seems to be constant before quitting smoking while varying over time after quitting smoking. Thus, it is not appropriate to use a linear regression model to fit these data. Note that the effect in a linear mixed model is also held constant, and therefore is not appropriate for this situation. As an extension of linear regression models, time-varying coefficient models allow the effects in a linear regression model to vary across time. The functional multilevel model, which is described next, allows the regression coefficients in the linear mixed model to change across time, and therefore represents the appropriate methodology based on these preliminary results.

Suppose that intensive longitudinal data $\{\mathbf{x}_i(t), y_i(t), \mathbf{z}_i(t)\}$ were collected at $t = t_{ij}, j = 1, \ldots, n_i$, for the ith subject, where $y_i(t)$ is the response variable, $\mathbf{x}_i(t)$ is a p-dimensional covariate vector, and $\mathbf{z}_i(t)$ is a q-dimensional covariate vector. A functional multilevel model allows its coefficients to depend on time,

$$y_i(t_{ij}) = \boldsymbol{\beta}^T(t_{ij})\mathbf{x}_i(t_{ij}) + \boldsymbol{\gamma}_i^T(t_{ij})\mathbf{z}_i(t_{ij}) + \varepsilon_i(t_{ij}), \tag{3.5}$$

where $\varepsilon_i(t_{ij})$ is an error process, $\boldsymbol{\beta}(t)$ consists of unknown coefficient functions, and $\boldsymbol{\gamma}_i(t)$ consists of random effect functions. It is assumed in model (3.5) that

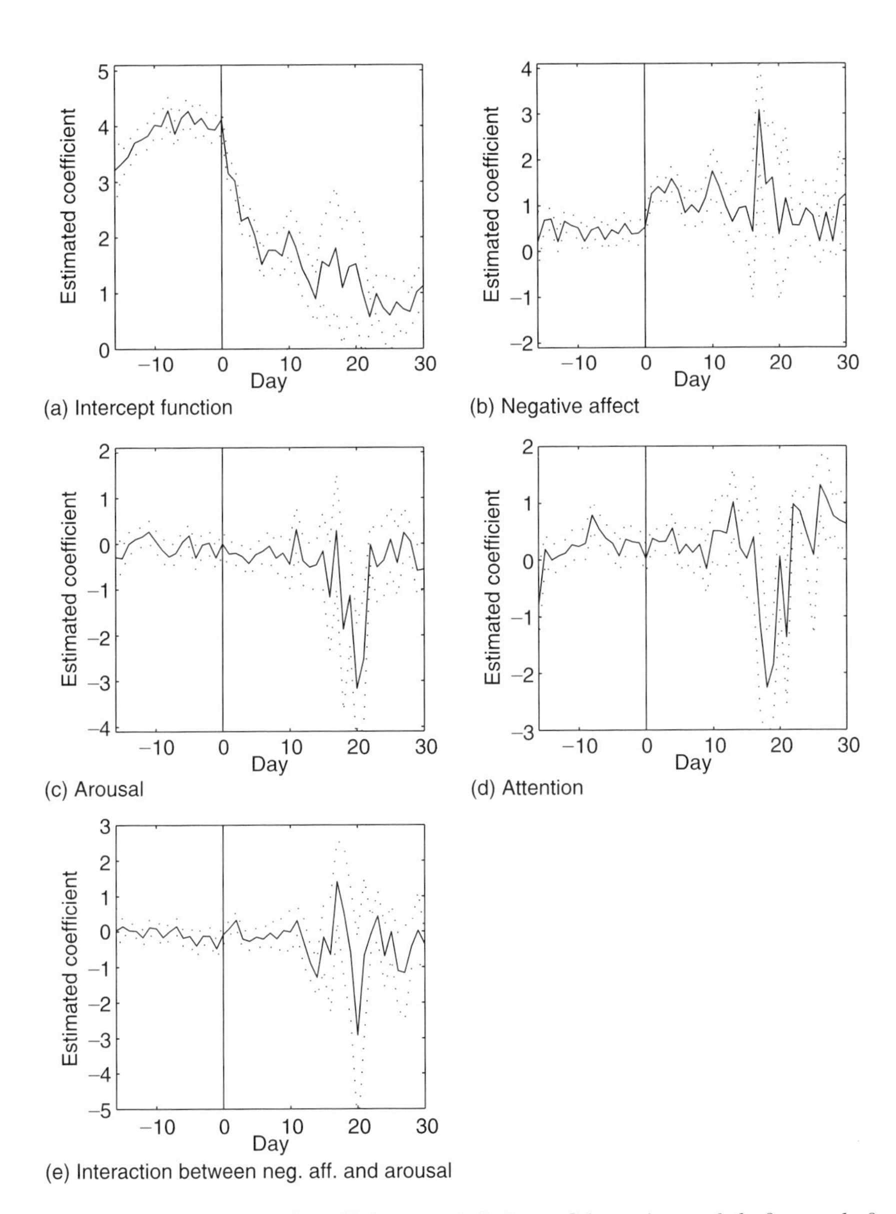

Figure 3.3 Plots of estimated coefficients and their confidence intervals before and after quitting smoking. The solid line stands for the estimated coefficient and the dotted lines for 95% pointwise confidence interval. We have aligned the data so that all subjects have quit day at day 0, highlighted by the vertical solid line.

$\boldsymbol{\beta}(t)$ and $\boldsymbol{\gamma}_i(t)$ are smoothed functions of t. To include an intercept in the model, we can simply set $x_{i1}(t_{ij}) = 1$. Similarly, to include a random intercept into the model, we set $z_{i1}(t_{ij}) = 1$. Note that the nonparametric mixed-effects model proposed by Wu and Zhang (2002) is a special case of model (3.5). Similarly, the random varying-coefficient model proposed by Wu and Liang (2004) is akin to model (3.5) with smoothing covariates. It should be noted that the requirement of smoothing covariates with the random varying-coefficient model is too strong to implement with our EMA data.

It is important to note that the word "functional" in functional multilevel model is used because the regression coefficients are functions of time t. Conditioning on t_{ij}, model (3.5) is a multilevel model, and therefore the coefficients $\boldsymbol{\beta}(t_{ij})$ and $\boldsymbol{\gamma}_i(t_{ij})$ have the same interpretation as those in a linear mixed model. Since $\boldsymbol{\beta}(t)$ and $\boldsymbol{\gamma}_i(t)$ are nonparametric smoothing functions, model (3.5) is considered a nonparametric model. Next, we describe the estimation procedure for model (3.5) using the local linear regression procedure described in section 3.1.1.

3.1.3 Estimation Procedure

Similar to local linear regression, we locally and linearly approximate $\beta_u(t)$, $u = 1, \ldots, p$, and $\gamma_{iv}(t)$, $v = 1, \ldots, q$, in a neighborhood of t_0:

$$\beta_u(t) \approx \beta_{u0} + \beta_{u1}(t - t_0) \tag{3.6}$$

and

$$\gamma_{iv}(t) \approx \gamma_{iv0} + \gamma_{iv1}(t - t_0). \tag{3.7}$$

We propose an estimation procedure for $\boldsymbol{\beta}(t)$ and $\boldsymbol{\gamma}(t)$ by directly applying the estimation procedure used for mixed-effects models. Let w_{ij} be $h^{-1}K\{(t_{ij} - t_0)/h\}$, and define

$$\begin{aligned}
&y_i^*(t_{ij}) = \sqrt{w_{ij}}y_i(t_{ij}), \quad \varepsilon_i^*(t_{ij}) = \sqrt{w_{ij}}\varepsilon_i(t_{ij}),\\
&\mathbf{x}_i^*(t_{ij}) = \sqrt{w_{ij}}[x_{i1}(t_{ij}), (t_{ij} - t_0)x_{i1}(t_{ij}), \ldots, x_{ip}(t_{ij}), (t_{ij} - t_0)x_{ip}(t_{ij})]^T,\\
&\boldsymbol{\beta}^* = (\beta_{10}, \beta_{11}, \ldots, \beta_{p0}, \beta_{p1})^T,\\
&\mathbf{z}_i^*(t_{ij}) = \sqrt{w_{ij}}[z_{i1}(t_{ij}), (t_{ij} - t_0)z_{i1}(t_{ij}), \ldots, z_{iq}(t_{ij}), (t_{ij} - t_0)x_{iq}(t_{ij})]^T,\\
&\boldsymbol{\gamma}_i^* = (\gamma_{10}, \gamma_{11}, \ldots, \gamma_{q0}, \gamma_{q1})^T.
\end{aligned}$$

We estimate $\boldsymbol{\beta}^*$ and $\boldsymbol{\gamma}_i^*$ by regarding them as regression coefficients in the following model:

$$y_i^*(t_{ij}) = \boldsymbol{\beta}^{*T}\mathbf{x}_i^*(t_{ij}) + \boldsymbol{\gamma}_i^{*T}\mathbf{z}_i^*(t_{ij}) + \epsilon_i^*(t_{ij}), \tag{3.8}$$

which can be viewed as a mixed model. Using existing software packages, such as PROC MIXED in SAS and LME in S-plus, we are able to estimate $\boldsymbol{\beta}^*$ and $\boldsymbol{\gamma}_i^*$. We can further obtain the standard error of the resulting estimates.

Let $\hat{\beta}_{ul}$ and $\hat{\gamma}_{vl}$ be estimates of β_{ul} and γ_{ivl}, respectively. Then, from (3.6) and (3.7),

$$\hat{\beta}_u(t_0) = \hat{\beta}_{u0} \quad \text{and} \quad \hat{\gamma}_{iv}(t_0) = \hat{\gamma}_{iv0}. \tag{3.9}$$

The procedure for implementing PROC MIXED in SAS will be discussed later.

3.2 Practical Considerations

Equation (3.9) gives us an estimate for $\beta_u(t)$ and $\gamma_{iv}(t)$ only at $t = t_0$. In practice, we may wish to estimate $\beta_u(t)$ and $\gamma_{iv}(t)$ over an interval of t. It is impossible to evaluate $\hat{\beta}_u(t)$ and $\hat{\gamma}_{iv}(t)$ for all t across an interval. Thus, we only evaluate $\hat{\beta}_u(t)$ and $\hat{\gamma}_{iv}(t)$ over a certain number of grid points for t. For instance, if we wanted to obtain $\hat{\beta}_u(t)$ over $[a, b]$, we would calculate $\beta_u(t)$ at $t_k = a + (b - a) * k/N$, for $k = 1, \ldots, N$. Here, N is the number of grid points. If N is too small, then the plot of resulting estimates may not appear smooth. On the other hand, if N is too big, the computation requirements could become quite burdensome. In practice, N is typically set to 100, 200, or 400. We then plot $\hat{\beta}_u(t)$ against t over the specified grid points.

A summary of the local linear estimation procedure for functional multilevel models is provided below.

Step 1. Set bandwidth h and take grid points $\mathcal{G} = \{t_1, \cdots, t_N\}$. Usually, $t_k = a + (b - a) * k/N$, where $[a, b]$ is the interval over which $\hat{\boldsymbol{\beta}}(t)$ and $\hat{\boldsymbol{\gamma}}(t)$ are evaluated.

Step 2. Set $k = 1$.

Step 3. Compute $w_{ij} = h^{-1}K\{(t_{ij} - t_k)/h\}$ for $i = 1, \ldots, n$ and $j = 1, \ldots, n_i$.

Step 4. Define a response $y_i^*(t_{ij}) = \sqrt{w_{ij}}y_i(t_{ij})$, and independent variables $x_{iju}^* = \sqrt{w_{ij}}x_{iu}(t_{ij})$, $tx_{iju}^* = (t_{ij} - t_k) * x_{iju}^*$ for $u = 1, \ldots, p$, $z_{ijv}^* = \sqrt{w_{ij}}z_{iv}(t_{ij})$, $tz_{ijv}^* = (t_{ij} - t_k) * z_{ijv}^*$, for $u = 1, \ldots, q$,

Step 5. Run a mixed model (using PROC MIXED with output covtest) with the following model specification:

$$\text{model } y^* = x_1^* - x_p^* \quad tx_1^* - tx_p^*/\text{noint solution;}$$

$$\text{random } z_1^* - z_q^* \quad tz_1^* - tz_q^*/\text{subj} = \text{subjectid solution;}$$

where the subscript ij for y, w, t, x, tx, z, and tz is suppressed.

Step 6. Go back to Step 2 and set $k = 2, \ldots, N$.

After obtaining estimates of both time-varying fixed effects and time-varying random effects, we may plot the estimate against time using any kind of graphical software, such as Excel in Microsoft Office or MATLAB. All figures in this chapter are produced using MATLAB code.

3.3 Application: Smoking Cessation Study

Having introduced the procedure for functional multilevel modeling and the justification for using this technique with EMA data presented in section 3.1.2, we shall illustrate its application with these data. Prior to discussing the analyses, a description of the data collection procedure, research questions to be addressed, and characteristics of the variables are presented.

3.3.1 Data Collection

As previously mentioned, EMA can be thought of as a comprehensive sampling of participant behavior. The procedure used by Shiffman et al. (1996b) involved a handheld palmtop computer that beeped at random times (approximately 250 occasions over the course of the study). Following these prompts, participants, who were smokers enrolled in a smoking cessation study, were asked to record directly onto the electronic diary answers to a series of 50 questions pertaining to their current setting and activities, as well as current mood and urge to smoke. Monitoring first occurred for a two-week interval when participants were engaged in their regular smoking behavior. Participants were asked to record all smoking occasions during this period. A targeted quit date was then introduced and participants attempted to abstain from smoking cigarettes during a subsequent four-week period. After the quit date, participants were asked to record any temptations or lapses they experienced as well as respond to the random assessment prompts. Compliance rates regarding response to the random prompts were high, ranging from 88% to 90% (Shiffman et al., 1996b, 2002).

3.3.2 Research Questions

Most theories of smoking behavior and relapse suggest that affect and arousal (i.e., mood or emotion) are important triggers related to smoking (Shiffman et al., 2002). For example, it has been shown that negative affect is associated with smoking relapse (Shiffman, 1982; Shiffman et al., 1996a). However, research has also shown a relation between positive affect and smoking (Baker et al., 1987; Robinson & Berridge, 1993; Stewart et al., 1984). Most theories posit that the influence of emotional states on smoking is impacted through craving or the urge to smoke. This relationship can change over time and depends on smoking status. For example, Baker et al. (1987) report that smoking behavior and urge to smoke are related to positive affect under conditions of smoking and to negative affect under conditions of deprivation.

Motivated by this discrepancy in the extant literature, Shiffman et al. (2002) evaluated the effect of positive and negative affect as an antecedent to smoking behavior using EMA data. It was found that affect (positive or negative) was not related to smoking behavior during unrestricted smoking time, but urge to smoke was significantly associated with smoking behavior. Therefore, the question arises as to whether positive or negative affect has an impact on the urge to smoke. To address this question, the relationship between urge to smoke and mood was examined. Consistent with Baker's model, it is hypothesized that (1) urge to smoke will be negatively correlated with negative affect during smoking (baseline), and (2) urge to smoke will be positively correlated with negative affect during abstinence (after quitting but before lapsing). Functional multilevel modeling was used to address the aforementioned research questions.

3.3.3 Data

The dataset consists of 304 smokers enrolled in a smoking cessation program. Day 17 of the study was the designated quit day, when subjects were instructed to stop smoking, thus entering the *trying to quit* phase of the study. Subjects were considered to have achieved an *abstinent* state when the electronic diary (ED) records showed they had abstained for 24 hours. Once subjects were abstinent, they were asked to record any episodes of smoking ("lapses") as well as episodes of a strong temptation to smoke ("temptations"). Subjects continued to be sampled at random for assessment of affect and urge to smoke. 149 subjects lapsed during the observation period, at which point the abstinent state was terminated. Our analysis focuses specifically on the randomly scheduled assessment data. Data on the intensity of subjects' urge to smoke was scored on a scale ranging from 0 to 10.

3.3.4 Definition of States of Smoking

The various "states" for each subject are defined by the following variables:

- *WAITDAY*: the designated day for subjects to quit smoking.
- *QUITDAY*: the day on which a subject actually quits smoking (defined as 24 hours without smoking).
- *LAPSDAY*: the day on which a subject lapsed (defined as any smoking at all, even a single puff; typically, these are very limited exposures).
- *RELAPSDY*: the day on which a subject is defined as having resumed smoking (defined as having smoked at least 5 cigarettes per day for 3 consecutive days).
- *LASTDAY*: the last day of observation in the study.

The five states were defined as follows:

1. *Smoking ad lib*: from day 1 to WAITDAY or LASTDAY.
2. *Try to quit*: from WAITDAY to QUITDAY or LASTDAY.
3. *Abstinent*: from QUITDAY to LAPSDAY or LASTDAY.
4. *Lapsed*: from LAPSDAY to RELAPSDY or LASTDAY.
5. *Relapsed*: from RELAPSDY to LASTDAY.

It is worth noting that the five states are sequentially ordered, suggesting that, if they occur, they must occur in the sequence described above. For our analysis, we are interested in comparing effects for *negative affect* on *urge to smoke* between the smoking ad lib and the abstinent states.

3.3.5 Preliminary Analysis

Initially, means, standard deviations, minimum, and maximum of urge to smoke, negative affect, arousal, and attention for the smoking ad lib, try to quit, and abstinent states were calculated. As evidenced from table 3.1, affect, arousal, and attention yielded similar results. However, the mean for urge to smoke changed from the smoking ad lib state to the abstinent state. Therefore, it is of interest to examine the correlations between urge to smoke and affect from these two states. Correlations between urge to smoke, affect, arousal, and attention across the three states are presented in table 3.2. It can be seen that the positive correlation between urge to smoke and negative affect becomes stronger from the smoking ad lib state to the abstinent state. This finding is somewhat inconsistent with the conjecture of Baker's theory (Baker et al., 1987), which predicts that the correlation should be negative during ad lib smoking. Overall, the correlations between urge to smoke, arousal, and attention were small across the three states.

Table 3.1 Summary of the three states

Variable	Mean	SD	Minimum	Maximum
State of smoking ad lib: Total number of observations = 9931				
Urge to smoke	3.87876	3.13124	0	10
Negative affect	−0.03301	0.99027	−2.65178	4.43822
Arousal	0.01538	1.00065	−2.99206	2.75794
Attention	−0.03809	0.92514	−2.23320	4.65680
State of trying to quit: Total number of observations = 850				
Urge to smoke	4.26471	1.90271	0	10
Negative affect	0.14522	1.02896	−2.33178	4.47822
Arousal	−0.13891	1.04380	−2.90206	2.40794
Attention	−0.0003970	1.09960	−2.08320	4.60680
State of abstinent: Total number of observations = 3395				
Urge to smoke	2.14433	1.88981	0	10
Negative affect	0.06021	1.00590	−1.99178	4.34822
Arousal	−0.01022	0.93918	−2.55206	2.64794
Attention	0.11152	1.00625	−2.15320	5.09680

Table 3.2 Correlation coefficients between urge to smoke and negative affect, arousal and attention

State	Negative affect	Arousal	Attention
Smoking ad lib	0.14466	−0.00979	0.06493
Try to quit	0.18124	0.06180	0.04067
Abstinent	0.46761	−0.11572	0.14600

3.3.6 Data Alignment, Model Specification, and Outputs

Because different subjects may have a different designated day for subjects to quit smoking (WAITDAY), data alignment was necessary. We set the WAITDAY as the time origin 0 in our analysis, given that most subjects did not smoke in the *try to quit* state. In what follows, we report the results from the final functional multilevel model,

$$\begin{aligned} y_i(t_{ij}) &= \beta_0(t_{ij}) + \beta_1(t)x_{i1}(t_{ij}) + \beta_2(t)x_{i2}(t_{ij}) + \beta_3(t)x_{i3}(t_{ij}) + \beta_4(t_{ij})x_{i1}(t_{ij})x_{i2}(t_{ij}) \\ &\quad + \gamma_{1i}(t_{ij})x_{i1}(t_{ij}) + \gamma_{2i}(t_{ij})x_{i2}(t_{ij}) + \gamma_{3i}(t_{ij})x_{i3}(t_{ij}) + \epsilon_i(t_{ij}), \end{aligned} \tag{3.10}$$

where

$y_i(t_{ij})$ = the score of *urge to smoke* of the ith subject at time t_{ij};
$x_{i1}(t_{ij})$ = the centered score of *negative affect* of the ith subject at time t_{ij};
$x_{i2}(t_{ij})$ = the centered score of *arousal* of the ith subject at time t_{ij};
$x_{i3}(t_{ij})$ = the centered score of *attention* of the ith subject at time t_{ij}.

In model (3.10), $\beta_j(t), j = 0, 1, 2, 3$, are fixed effects, and $\gamma_{ij}(t), j = 1, 2, 3$, are random effects with $\gamma_{ij}(t) \sim N(0, \sigma_j^2(t))$. Subjects with missing values on WAITDAY and QUITDAY were excluded from the analyses. The SAS code is presented in the companion website to this book.

Remark. *In our initial analysis, we considered a functional multilevel model that included fixed and random effects of the linear terms, interaction terms, and squared terms. We found that introducing too many random effects usually led the optimization algorithm for the mixed model to not converge. As a result, we only included the linear terms as covariates for random effects. We plotted the resulting estimates of the full quadratic model and found that most interaction terms and squared terms were nonsignificant and, therefore, excluded them from our final model.*

The estimated fixed effect coefficient functions are depicted in figure 3.4. It is clear that the fixed effects are close to being time-independent during the pre-quit period. The effects of arousal and attention appear to be constant and, therefore, may not be significant. It is also clear that the intercept function is

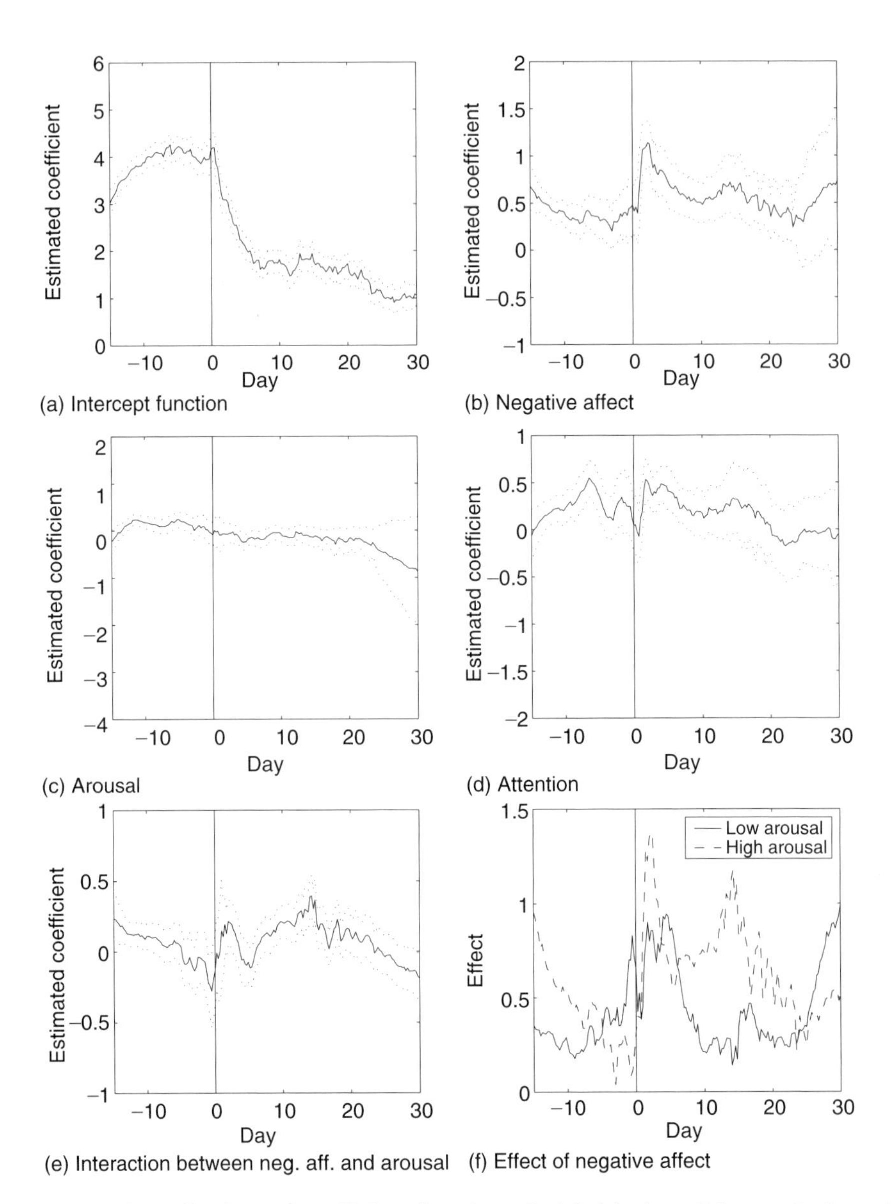

Figure 3.4 Plots of estimated coefficient functions. In (a)–(e), the solid curve is the estimated coefficient function and the dotted curves are 95% pointwise confidence intervals. In (f), the solid curve is the effect of negative affect conditioning on a low arousal score, and the dashed curve is the effect of negative affect conditioning on a high arousal score.

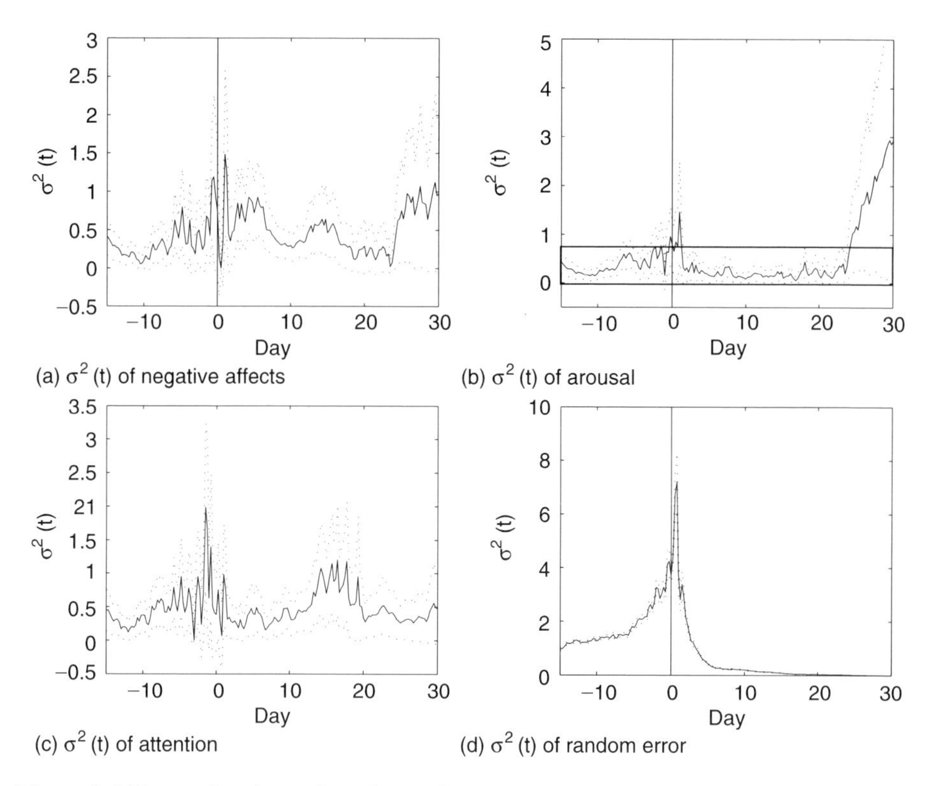

Figure 3.5 Plots of estimated variance functions. The solid curve is the estimated coefficient function and the dotted curves are 95% pointwise confidence intervals.

decreasing over time during the post-quit period. The effect of negative affect also appears to be constant during the post-quit period, with the exception of a significant jump for the pre-quit period. The interaction between negative affect and arousal is close to being time-independent as well. Figure 3.4(f) depicts the effect of negative affect conditional on low arousal (negative twice standard deviation) and high arousal (twice the standard deviation) as the score of arousal was centralized. The overall trends of the resulting estimates in figure 3.4 are consistent with those depicted in figure 3.3. But the confidence intervals in figure 3.4 are certainly narrower than those in figure 3.3. This implies that the proposed estimation procedure yielded more efficient estimates than that in section 3.1.2.

The variance functions of the three random effects are depicted in figure 3.5(a)–(c), from which it can be seen that they are significantly different from zero. It is possible that these effects are time-independent. The variance function of random error $\varepsilon(t)$ is depicted in figure 3.5(d), which clearly indicates that the variance function significantly changes over time. Additionally, because

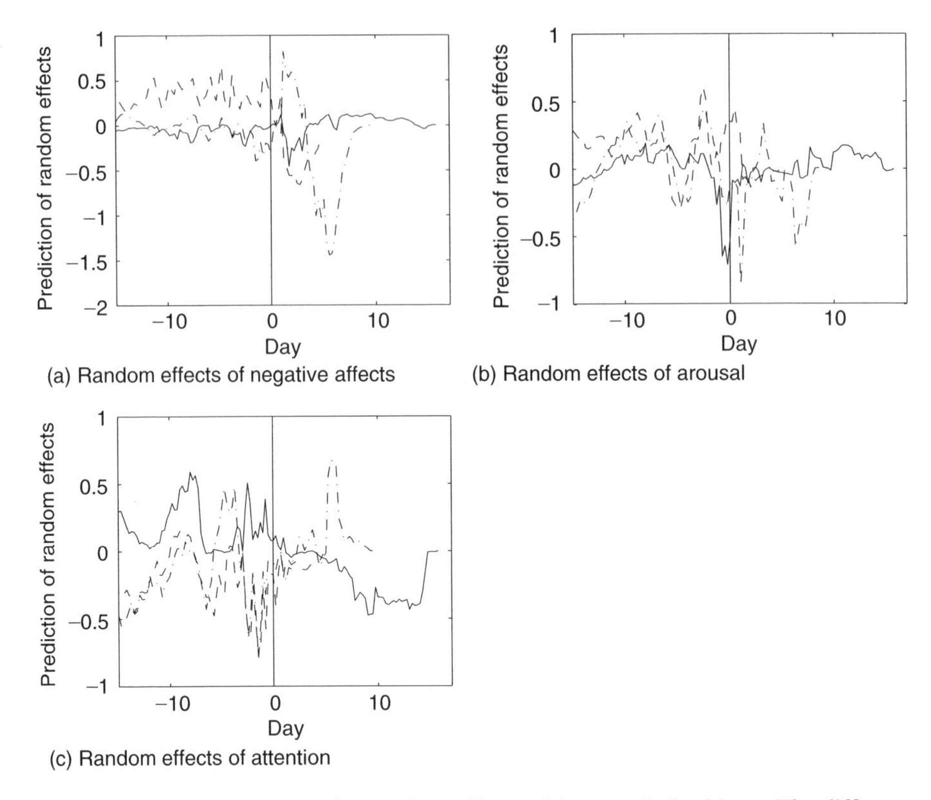

Figure 3.6 Plots of prediction for random effects of three typical subjects. The difference in curves represent the random effects of the individual subjects.

only a few subjects maintained an abstinent state until the end of the study, the variance function is close to zero.

Finally, figure 3.6 depicts random effects of three typical subjects, one of which could not maintain a state of abstinence three days after the quit date. From figure 3.6, we can see that the three random effects are changing over time around 0 since their means are zero.

3.3.7 Results

The results of the EMA data presented herein provide useful information on understanding the effects of mood and emotion on smoking behavior. For example, contrary to Baker et al. (1987), we did not find that urge to smoke during ad lib smoking was associated with positive affect. In fact, urge to smoke was significantly associated with negative affect, contradicting Baker et al.'s hypothesized model. However, consistent with Baker's theory, the association between urge to smoke and negative affect did increase after quitting smoking. The increased association of urge to smoke and negative affect during abstinence was particularly evident (figure 3.4(b)) during the first week of abstinence, which intuitively makes sense because this is the period during which nicotine withdrawal—which is itself largely characterized by negative affect—is at its peak (Shiffman & Jarvik, 1976). This finding suggests that withdrawal may drive both urge to smoke and negative affect during early periods of abstinence, accounting for the higher intercorrelation. This early abstinence period is also the time when most of our subjects lapsed. It is possible that the drop in the association evident in figure 3.4(b) between negative affect and urge to smoke reflects the loss from the abstinent sample of smokers for whom urges and negative affect were most linked, suggesting that this pattern might be a risk factor for lapsing. Further analysis introducing a variable to designate those who lapsed or continuing the analysis into the lapsed phase may help address this concern.

The analysis also showed interactions between negative affect and arousal in producing urge to smoke intensity. It appears that urge to smoke is associated with states of high negative affect and high arousal (e.g., states such as anger and anxiety), which have typically been associated with nicotine withdrawal (American Psychiatric Association, 1994).

Emotional states associated with negative affect and low arousal (i.e., depressive emotions) were less associated with urge to smoke, which seems at odds with the extant literature linking smoking and smoking relapse to depression (Kassel et al., 2003).

Results (figure 3.4(f)) also suggested that the effect of negative affect (conditional on arousal) varied in unexpected ways throughout the abstinent period, which is consistent with the findings presented by Piasecki et al. (2003), who reported heterogeneity in the course of affective disturbance following smoking cessation, including some subgroups whose symptoms grew worse over time, even as average symptoms tended to decline. The surprising variation in associations among negative affect, arousal, and urge to smoke warrants further attention.

The analysis also indicated that urge to smoke was associated with difficulty concentrating during both the ad lib smoking and the abstinent states, suggesting that urge to smoke may be linked to nicotine withdrawal, which then may lead to increased difficulty concentrating (American Psychiatric Association, 1994). This result also suggests that smokers may be motivated to smoke (as indicated by high urge levels) in order to restore normal attention. This association was relatively similar during ad lib smoking and abstinence, but showed some unexplained variation within the baseline ad lib smoking period.

In summary, the analysis only partially confirmed Baker et al.'s (1987) hypothesis that the association between affect and urge to smoke changes during ad lib smoking to abstinence. Day-by-day tracking of the association suggested that the relation between affect and urge to smoke peaked during the first 5–7 days of abstinence, which is when nicotine withdrawal is also at its peak.

3.4 Discussion

In this chapter, we introduced functional multilevel models for intensive longitudinal data. We proposed a local linear estimation procedure for the models and illustrated the estimation procedure by conducting an empirical analysis of the EMA data. The proposed models are highly useful in the presence of time-varying effects. Below we briefly summarize the strengths, limitations, and possible extensions of the developed procedures.

3.4.1 Strengths

In this chapter, we illustrated the use of functional multilevel modeling for analyzing intensive longitudinal data. By allowing regression coefficients to be time-dependent, the functional multilevel model can be viewed as a natural extension of the linear mixed model. The functional multilevel model inherits the advantages of the linear mixed model, including the interpretation of the regression coefficients, but also, and perhaps most importantly, provides us with a graphical tool that allows researchers to examine whether regression coefficients are truly time-dependent or not. Thus, it can provide information as to whether a linear mixed model is an appropriate method.

3.4.2 Limitations

Although functional multilevel models are more flexible than linear mixed models, one limitation of this technique is that its estimation procedure requires us to select the smoothing parameter h. Different smoothing parameters may yield slightly different estimates. As a result, interpretation of the resulting estimates usually requires more advanced statistical knowledge in order to draw concrete conclusions.

3.4.3 Extensions

From our EMA analysis, is was found that some effects were truly time-dependent, while others were not. Functional multilevel models allowed for the inclusion of time-dependent effects. Three alternative models to the functional multilevel models are (1) the semi-varying coefficient model, (2) the generalized semi-varying coefficient model, and (3) the dynamical systems model. Each has its own applicability depending on the research questions to be addressed. The first alternative model is the following semi-varying coefficient model:

$$y_i(t_{ij}) = \boldsymbol{\beta}_1^T(t)\mathbf{x}_{i1}(t_{ij}) + \boldsymbol{\beta}_2^T\mathbf{x}_{i2}(t_{ij}) + \boldsymbol{\gamma}_{i1}^T(t_{ij})\mathbf{z}_{i1}(t_{ij}) + \boldsymbol{\gamma}_{i2}^T\mathbf{z}_{i2}(t_{ij}) + \varepsilon_i(t_{ij}), \qquad (3.11)$$

where $\boldsymbol{\beta}_1(t)$ consists of time-dependent fixed effects, $\boldsymbol{\beta}_2$ is time-independent fixed effects, $\boldsymbol{\gamma}_{i1}(t)$ is time-dependent random effects, and $\boldsymbol{\gamma}_{i2}$ is time-independent random effects. This model also allows the variance function of $\varepsilon(t)$ to be time-dependent and is denoted by $\sigma^2(t)$. The semi-varying coefficient regression model (3.11) is a combination of a linear mixed model and a functional multilevel model. The limitation is that it is challenging to obtain an accurate (most efficient) estimate of the time-independent effect in model (3.11).

In situations where the response variable $y(t)$ is discrete (e.g., binary, categorical, count), model (3.11) may not be appropriate. In this case, one should consider the second alternative model, the generalized semi-varying coefficient model, given by

$$g[E\{y_i(t_{ij})|\mathbf{x}_{i1}(t_{ij}), \mathbf{x}_{i2}(t_{ij}), \mathbf{z}_{i1}(t_{ij}), , \mathbf{z}_{i2}(t_{ij})\}]$$
$$= \boldsymbol{\beta}_1^T(t)\mathbf{x}_{i1}(t_{ij}) + \boldsymbol{\beta}_2^T\mathbf{x}_{i2}(t_{ij}) + \boldsymbol{\gamma}_{i1}^T(t)\mathbf{z}_{i1}(t_{ij}) + \boldsymbol{\gamma}_{i2}\mathbf{z}_{i2}(t_{ij}), \qquad (3.12)$$

where $g(\cdot)$ is a known link function (e.g., $g(\cdot)$ is the logit link function for binary response data). Finally, the dynamical systems approach is the third alternative for analyzing the intensive longitudinal data. A nice explanation and example of dynamical systems, Boker and Laurenceau, can be found in this volume (chapter 9).

3.4.4 Further Research

It is known that some participants succeeded in quitting smoking, while others did not. Of course, different people may have different smoking patterns. Hence, the data may consist of several clusters. Although intuitively it makes sense to apply a cluster analytic technique to these data, a challenge arises due to the irregularity of observational time points in these data. Future research on implementing cluster analysis with intensive longitudinal data is therefore needed.

In this chapter, we focused on local linear estimation techniques. Other smoothing techniques (Hastie et al., 2001) can also be used to estimate the time-varying coefficients. For example, spline smoothing has been proposed for the analysis of intensive longitudinal data (Fok & Ramsay, this volume, chapter 5).

ACKNOWLEDGMENTS

Runze Li's research was supported by National Institute on Drug Abuse (NIDA) Grant P50-DA10075 and DMS-0348869. Tammy Root's research was supported by National Institute on Drug Abuse (NIDA) Grant P50-DA10075.

References

American Psychiatric Association (1994). *Diagnostic and Statistical Manual of Mental Disorders* (4th ed.). Washington, DC: APA.

Baker, T.B., Morse, E., & Sherman, J.E. (1987). The motivation to use drugs: A psychobiological analysis of urges. In P.C. Rivers (Ed.), *The Nebraska Symposium on Motivation: Alcohol Use and Abuse* (pp. 257–323). Lincoln: University of Nebraska Press.

Chiang, C.-T., Rice, J.A., & Wu, C.O. (2001). Smoothing spline estimation for varying coefficient models with repeatedly measured dependent variables. *Journal of American Statistical Association, 96*, 605–619.

Cleveland, W.S., Grosse, E., & Shyu, W.M. (1993). Local regression models. In J.M. Chambers & T.J. Hastie (Eds.), *Statistical Models in S*. New York: Chapman & Hall.

Fan, J., & Gijbels, I. (1996). *Local Polynomial Modelling and Its Applications*. London: Chapman & Hall.

Fan, J., & Gijbels, I. (2000). Local polynomial fitting. In M.G. Schimek (Ed.), *Smoothing and Regression: Approaches, Computation and Application* (pp. 228–275). New York: John Wiley.

Fan, J., & Zhang, J. (2000). Two-step estimation of functional linear models with applications to longitudinal data, *Journal of the Royal Statistical Society, Series B, 62*, 303–322.

Hastie, T.J., Tibshirani, R., & Friedman, J. (2001). *The Elements of Statistical Learning*. New York: Springer.

Hoover, D.R., Rice, J.A., Wu, C.O., & Yang, L.P. (1998). Nonparametric smoothing estimates of time-varying coefficient models with longitudinal data. *Biometrika, 85*, 809–822.

Huang, J.Z., Wu, C.O., & Zhou, L. (2002). Varying-coefficient models and basis function approximations for the analysis of repeated measurements. *Biometrika, 89*, 111–128.

Jones, M.C., Marron, J.S., & Sheather, S.J. (1996a). A brief survey of bandwidth selection for density estimation, *Journal of the American Statistical Association, 91*, 401–407.

Jones, M.C., Marron, J.S. & Sheather, S.J. (1996b). Progress in data-based bandwidth selection for kernel density estimation, *Computational Statistics, 11*, 337–381.

Kassel, J.D., Stroud, L.R., & Paronis, C.A. (2003). Smoking, stress, and negative affect: Correlation, causation, and context across stages of smoking. *Psychological Bulletin, 129*, 270–304.

Marron, J.S., & Nolan, D. (1988). Canonical kernels for density estimation. *Statistical Probability Letters, 7*, 195–199.

Martinussen, T., & Scheike, T.H. (2001). Sampling adjusted analysis of dynamic additive regression models for longitudinal data. *Scandinavian Journal of Statistics, 28*, 303–323.

Piasecki, T.M., Jorenby, D.E., Smith, S.S., Fiore, M.C., & Baker, T.B. (2003). Smoking withdrawal dynamics II. *Journal of Abnormal Psychology, 112*, 14–37.

Robinson, T.E., & Berridge, K.C. (1993). The neural basis of drug craving: An incentive-sensitization theory of addiction. *Brain Research Reviews, 18*, 247–291.

Shiffman, S. (1982). Relapse following smoking cessation: A situational analysis. *Journal of Consulting and Clinical Psychology, 50*, 71–86.

Shiffman, S. (1999). Real-time self-report of momentary states in the natural environment: Computerized Ecological Momentary Assessment. In A.A. Stone, J. Turkkan, J. Jobe, C. Bachrach, H. Kurtzman, & V. Cain (Eds.), *The Science of Self-Report: Implications for Research and Practice* (pp. 277–296). Mahwah, NJ: Lawrence Erlbaum.

Shiffman, S., Gwaltney, C.J., Balabanis, M.H., Liu, K.S., Paty, J.A., Kassel, J.D., Hickcox, M., & Gnys, M. (2002). Immediate antecedents of cigarettes smoking: An analysis from ecological momentary assessment. *Journal of Abnormal Psychology, 111*, 531–545.

Shiffman, S., Hickcox, M., Paty, J.A., Gnys, M., Kassel, J.D., & Richards, T.J. (1996a). Progression from a smoking lapse to relapse: Prediction from abstinence violation effects, nicotine dependence, and lapse characteristics. *Journal of Consulting and Clinical Psychology, 64*, 366–379.

Shiffman, S., & Jarvik, M.E. (1976). Trends in withdrawal symptoms in abstinence from cigarette smoking. *Psychopharmacologia, 50*, 35–39.

Shiffman, S., Paty, J., Kassel, J., & Hickcox, M. (1996b). First lapses to smoking: Within-subjects analysis of real-time reports. *Journal of Consulting and Clinical Psychology, 62*, 366–379.

Stewart, J., de Wit, H., & Eikelboom, R. (1984). Role of unconditioned and conditioned drug effects in the self-administration of opiates and stimulants. *Psychological Review, 91*, 251–268.

Wahba, G. (1990). *Spline Model for Observational Data*. Philadelphia: Society for Industrial and Applied Mathematics.

Wu, C.O., Chiang, C.-T., & Hoover, D.R. (1998). Asymptotic confidence regions for kernel smoothing of a varying-coefficient model with longitudinal data. *Journal of the American Statistical Association, 93*, 1388–1402.

Wu, F., & Zhang, J. (2002). Local polynomial mixed-effects models for longitudinal data. *Journal of the American Statistical Association, 97*, 883–897.

Wu, H., & Liang, H. (2004). Random varying-coefficient models with smoothing covariates, applications to an AIDS clinical study. *Scandinavian Journal of Statistics*, 31, 3–19.

4

Application of Item Response Theory Models for Intensive Longitudinal Data

Donald Hedeker, Robin J. Mermelstein, and Brian R. Flay

Item response theory (IRT) or latent trait models provide a statistically rich class of tools for analysis of educational test and psychological scale data. In the simplest case, these data are composed of a sample of subjects responding dichotomously to a set of test or scale items. Interest is in estimation of characteristics of the items and subjects. These methods were largely developed in the 1960s through 1980s, though, as Bock (1997) notes in his brief historical review of IRT, the seeds for these models began with Thurstone in the 1920s (Thurstone, 1925, 1926, 1927). A seminal reference on IRT is the book by Lord and Novick (1968), and in particular the chapters written by Birnbaum in that book. More recent texts and collections include Embretson and Reise (2000), Hambleton et al. (1991), Heinen (1996), Hulin et al. (1983), Lord (1980), and van der Linden and Hambleton (1997).

Prior to the development of IRT, classical test theory (Novick, 1966; Spearman, 1904) was used to estimate an individual's score on a test. IRT models overcome several limitations of classical test theory for analysis of such data. In classical test theory, the test was considered the unit of analysis, whereas IRT analysis focuses on the test item. A major challenge for classical test theory is how to score individuals who complete different versions of a test. By focusing on the item as the unit of analysis, IRT effectively solved this problem.

Whereas IRT was originally developed for dichotomous items, extensions for polytomous items, ordinal (Samejima, 1969) and nominal (Bock, 1972) for instance, soon emerged. Thissen and Steinberg (1986) present a taxonomy of many ordinal and nominal IRT models, succinctly describing the various ways in which these models relate to each other. A similar synthesis of IRT models for polytomous items can be found in Mellenbergh (1995). Also, the collection by van der Linden and Hambleton (1997) contains several articles describing IRT models for polytomous items.

In its original form, an IRT model assumes that an individual's score on the test is a unidimensional latent "ability" variable or trait, often denoted as θ. Of course, for psychological scales this latent variable might be better labeled as "mood" or "severity," depending on what the scale is intended to measure. Some examples of using IRT with psychological scales can be found in Thissen and Steinberg (1988) and Schaeffer (1988). This latent variable is akin to a factor in a factor analysis model for continuous variables, and so IRT and factor analysis models are also very much related; a useful source for this is Bartholomew and Knott (1999). Because a test or scale can measure more than one latent factor, multidimensional IRT models have also been developed (Bock et al., 1988; Gibbons & Hedeker, 1992); the collection by van der Linden and Hambleton (1997) presents some of these developments.

Typically, a sample of N subjects respond to n_i items at one occasion, though the amount of actual time that one occasion represents can vary. In some situations, subjects are assessed at multiple occasions, perhaps with the same set or a related set of items. Several articles have described applications and developments of IRT modeling to such longitudinal situations in psychology (Adams et al., 1997a; Anderson, 1985; Embretson, 1991, 2000; Fischer & Pononcy, 1994).

In this chapter the basic IRT model for dichotomous items will be described. It will be shown how this model can be viewed as a mixed model for dichotomous data, which can be fitted using standard software. Illustration of IRT modeling will be done using Ecological Momentary Assessment (EMA) data from a study of adolescent smoking. Our example will be a somewhat nontraditional IRT illustration, in the sense that we shall not examine responses to test or questionnaire items per se. Instead, as described more fully later, we shall examine whether a subject records a smoking report (that is, smokes a cigarette) in specific time periods defined by day of week and hour of day.

The basic premise that we are examining with these data is that patterns of smoking behavior, over different times of the day and days of the week, may be early markers of the development of nicotine dependence. In all, we shall examine 35 time periods based on data collected over one week (seven days crossed with five time-of-day intervals). Thus, our "items" are these 35 time periods, and our dichotomous response is whether or not a subject smokes, which is ascertained for each of these periods. A subject's latent "ability" can then be construed as their underlying level of smoking during this one-week reporting period, and our interest is to see how this behavior relates to these day-of-week and time-of-day periods. Our aim is to show how IRT models can be applied to analysis of intensive longitudinal data to address key research questions. Specifically, we hypothesized that both early-morning and midweek smoking would be key determinants of the development of dependence or smoking level.

4.1 IRT Model

To set notation, let Y_{ij} be the dichotomous response of subject i ($i = 1, 2, \ldots, N$ subjects) to item j ($j = 1, 2, \ldots, n_i$ items). Note that subject i is measured on n_i

items, so we do not necessarily assume that all subjects are measured on all items. Also, although the notation might imply that items are nested within subjects, in a typical testing situation it is more common for subjects and items to be crossed, because all N subjects are given the same n items. An exception to this is in computerized adaptive testing where the items that a subject receives are selectively chosen from a large pool of potential items based on their sequential item responses. Here, we shall simply assume that there is a set of n items in total, but that not all subjects necessarily respond to all of these items.

Denote a "correct" or positive response as $Y_{ij} = 1$ and an "incorrect" or negative response as $Y_{ij} = 0$. A popular IRT model is the one-parameter logistic model, which is commonly referred to as the Rasch model (Rasch, 1960; Thissen, 1982; Wright, 1977). This model specifies the probability of a correct response to item j ($Y_j = 1$) conditional on the "ability" of subject i (θ_i) as

$$P(Y_{ij} = 1 \mid \theta_i) = \frac{1}{1 + \exp[-a(\theta_i - b_j)]}. \tag{4.1}$$

The parameter θ_i denotes the level of the latent trait for subject i; higher values reflect a higher level on the trait being measured by the items. Trait values are usually assumed to be normally distributed in the population of subjects with a mean of 0 and a variance of 1.

The parameter b_j is the item difficulty, which determines the position of the logistic curve along the ability scale. The further the curve is to the right, the more difficult the item. The parameter a is the slope or discriminating parameter; it represents the degree to which the item response varies with ability θ. In the Rasch model all items are assumed to have the same slope, and so a does not carry the j subscript in this model. In some representations of the Rasch model the common slope parameter a does not explicitly appear in the model. Notice that the function $\{1 + \exp[-(z)]\}^{-1}$ is simply the cumulative distribution function (cdf) for the logistic distribution. Thus, if a subject's trait level θ_i exceeds the difficulty of the item b_j, then the probability of a correct response is greater than 0.5 (similarly, if $\theta_i < b_j$, the probability of a correct response is less than 0.5).

Figure 4.1 provides an illustration of the model with three items. In this figure the values of b_j have been set equal to -1, 0, and 1 to represent a relatively easy, moderate, and difficult item, respectively. Also, the value of a equals unity for these curves. As can be seen, the item difficulty b is the trait level where the probability of a correct response is 0.5. So, for example, the first item is one that does not require a high ability level (i.e., -1) to yield a 50:50 chance of getting the item correct. For a test to be an effective measurement tool it is useful to have items with a range of difficulty levels.

As noted, the Rasch model assumes that all items are equally discriminating. To relax this assumption, several authors described various two-parameter models that included both item difficulty and discrimination parameters (Birnbaum, 1968; Bock & Aitkin, 1981; Lawley, 1943; Richardson, 1936). Here, consider the two-parameter logistic model that specifies the probability of a correct response

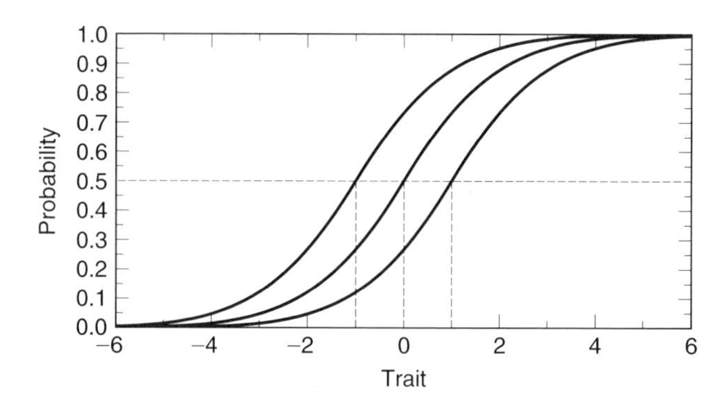

Figure 4.1 Rasch model with three items.

to item j ($Y_j = 1$) conditional on the ability of subject i (θ_i) as

$$P(Y_{ij} = 1 \mid \theta_i) = \frac{1}{1 + \exp[-a_j(\theta_i - b_j)]}, \tag{4.2}$$

where a_j is the slope parameter for item j, and b_j is the difficulty parameter for item j.

Figure 4.2 provides an illustration of the model with three items. In this figure the values of a_j have been set equal to 0.75, 1, and 1.25, respectively. As can be seen, the lower the value of a, the lower the slope of the logistic curve and the less the item discriminates levels of the trait. In an extreme case, if $a = 0$, the slope is horizontal and the item is not able to discriminate any levels of the trait. These item discrimination parameters are akin to factor loadings in a factor analysis model. They indicate the degree to which an item "loads" on the latent trait θ.

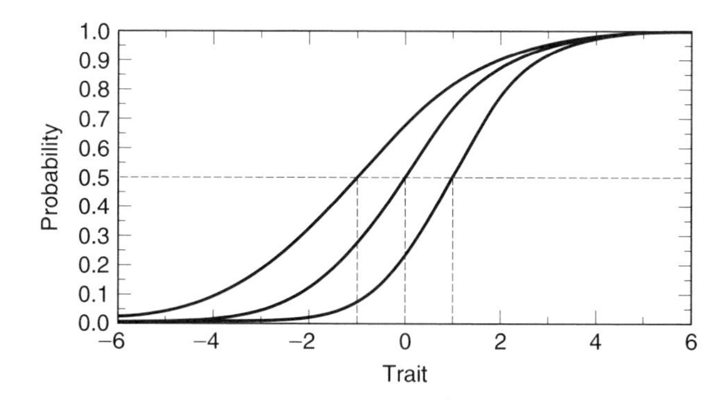

Figure 4.2 Two-parameter model with three items.

As noted by Bock & Aitkin (1981), it is convenient to represent the two-parameter model as

$$P(Y_{ij} = 1 \mid \theta_i) = \frac{1}{1 + \exp[-(c_j + a_j\theta_i)]}, \tag{4.3}$$

where $c_j = -a_j b_j$ represents the item-intercept parameter. Similarly, the Rasch model can be written as

$$P(Y_{ij} = 1 \mid \theta_i) = \frac{1}{1 + \exp[-(c_j + a\,\theta_i)]}, \tag{4.4}$$

with $c_j = -a\,b_j$. In this form, it is easy to see that these models are simply variants of mixed-effects logistic regression models. For example, the Rasch model written in terms of the log odds, or logit, of response is

$$\log\left[\frac{P(Y_{ij} = 1 \mid \theta_i)}{1 - P(y_{ij} = 1 \mid \theta_i)}\right] = c_j + a\,\theta_i. \tag{4.5}$$

4.1.1 IRT in Mixed Model Form

These IRT models can also be represented using notation that is more common in mixed-effects regression models. For this, let $\boldsymbol{\lambda}_i$ represent the $n_i \times 1$ vector of logits for subject i. The Rasch model can then be written as

$$\boldsymbol{\lambda}_i = \boldsymbol{X}_i\boldsymbol{\beta} + \mathbf{1}_i\sigma_\upsilon\theta_i, \tag{4.6}$$

where $\boldsymbol{X}_i$ is an $n_i \times n$ item indicator matrix obtained from $\boldsymbol{I}_n$ (i.e., the $n \times n$ identity matrix), $\boldsymbol{\beta}$ is the $n \times 1$ vector of item difficulty parameters (i.e., the b_j parameters in the IRT notation), $\mathbf{1}_i$ is a $n_i \times 1$ vector of ones, and θ_i is the latent trait (i.e., random effect) value of subject i, which is distributed $\mathcal{N}(0, 1)$. The parameter σ_υ is the common slope parameter (i.e., the a parameter in the IRT notation) which indicates the heterogeneity of the random subject effects. Note that it is common to write the random-intercept/mixed logistic model as

$$\boldsymbol{\lambda}_i = \boldsymbol{X}_i\boldsymbol{\beta} + \mathbf{1}_i\upsilon_i, \tag{4.7}$$

where the random subject effects υ_i are distributed in the population of subjects as $N(0, \sigma_\upsilon^2)$. This is equivalent to the above since $\theta_i = \upsilon_i/\sigma_\upsilon$.

Mixed models are often described as multilevel (Goldstein, 1995) or hierarchical linear models (HLM; Raudenbush & Bryk, 2002). IRT models can also be thought of in this way by considering the level 1 item responses as nested within the level 2 subjects. The usual IRT notation is a little different from the usual multilevel or HLM notation, but the Rasch model is simply a random-intercept logistic regression model with item indicator variables as level 1 regressors. As such it can be estimated by several software programs for multilevel or mixed modeling, for example, SAS PROC NLMIXED, Stata (StataCorp, 1999), HLM (Raudenbush et al., 2004), MLwiN (Rasbash et al., 2004), EGRET (Corcoran et al., 1999), LIMDEP (Greene, 1998), GLLAMM (Rabe-Hesketh et al., 2001),

Mplus (Muthén and Muthén, 2001), and MIXOR (Hedeker & Gibbons, 1996). This is in addition to the many software programs that are specifically implemented for IRT analysis, for example, BILOG (Mislevy and Bock, 1998), MULTILOG (Thissen, 1991), NOHARM (Fraser, 1988), and ConQuest (Wu et al., 1998).

Just to be entirely explicit, because we are not assuming that all subjects provide responses to all n items, the n_i rows of the item indicator matrix $\boldsymbol{X}_i$ are obtained from $\boldsymbol{I}_n$ depending on which items subject i responded to. For example, if $n = 3$ and subject i answered items 1 and 3, then $\boldsymbol{X}_i$ equals

$$\boldsymbol{X}_i = \begin{bmatrix} 1 & 0 & 0 \\ 0 & 0 & 1 \end{bmatrix}. \tag{4.8}$$

Thus, the number of rows in $\boldsymbol{X}_i$ is simply the number of items for that subject. Likewise for the vectors $\boldsymbol{\lambda}_i$ and $\mathbf{1}_i$. The dimension of the parameter vector $\boldsymbol{\beta}$ is equal to the total number of items.

The two-parameter model can also be represented in matrix form as

$$\boldsymbol{\lambda}_i = \boldsymbol{X}_i{}'\boldsymbol{\beta} + \boldsymbol{X}_i{}'\boldsymbol{T}\theta_i, \tag{4.9}$$

where $\boldsymbol{T}$ is a vector of standard deviations,

$$\boldsymbol{T}' = \begin{bmatrix} \sigma_{\upsilon_1} & \sigma_{\upsilon_2} & \dots & \sigma_{\upsilon_n} \end{bmatrix}. \tag{4.10}$$

These standard deviations correspond to the discrimination parameters of the IRT model (the Rasch model is simply a special case of this model, where these standard deviations are all the same). As can be seen, the two-parameter model is a mixed-effects logistic regression model that allows the random effect variance terms to vary across the items at the first level. Goldstein (1995) refers to such multilevel models as having complex variance structure. Several of the aforementioned mixed or multilevel software programs can fit such models, and so can fit two-parameter IRT models.

An aspect that can be confusing concerns the different parameterizations used in IRT and mixed models. In the two-parameter IRT model, the following parameterizations are typically used (see Bock & Aitkin, 1981):

$$\sum_{j=1}^{n} b_j = 0 \quad \text{and} \quad \prod_{j=1}^{n} a_j = 1. \tag{4.11}$$

These parameterizations center the item difficulty parameters around 0, and multiplicatively center the item discrimination parameters around 1. As a result, mixed model estimates need to be rescaled and/or standardized to be consistent with IRT results. To illustrate this, we consider the oft-analyzed LSAT section 6 data published in Thissen (1982). This dataset consists of responses from 1,000 subjects to five dichotomous items from section 6 of the LSAT exam. In his article, Thissen published results for the Rasch model applied to these data. These results are presented in table 4.1 (labeled as IRT). We analyzed these data using a random-intercept logistic model with item indicators as fixed regressors using

Table 4.1 Rasch model estimates for the LSAT-6 data

Item	IRT ($\hat{b}_j$)	NLMIXED Raw ($\hat{\beta}_j$)	NLMIXED Transformed ($\hat{b}_j$)
1	−1.255	2.730	−1.255
2	0.476	0.999	0.476
3	1.234	0.240	1.235
4	0.168	1.306	0.168
5	−0.624	2.099	0.625
	$\hat{a} = 0.755$	$\hat{\sigma}_v = \hat{a} = 0.755$	

MIXOR and SAS PROC NLMIXED (see the accompanying website to this book for the SAS PROC NLMIXED code). Both programs gave near-identical results; the NLMIXED estimates are listed in table 4.1 in both the raw and translated forms.

To get to the IRT formulation, first the sign of the mixed model difficulty parameter estimates is reversed, and then the mean of the estimates is subtracted from each, namely,

$$b_j = -\beta_j - \frac{1}{n} \sum_{j'=1}^{n} -\beta_{j'}.$$

This ensures that the mean of these transformed estimates equals zero. As can be seen, this yields item difficulty estimates nearly identical to those reported in Thissen (1982). In either representation it is clear that relatively many subjects got item 1 correct and relatively fewer got item 3 correct. This is because in the mixed model (using the formulation described herein) a positive regression coefficient indicates increased probability of response (i.e., a correct response) with increasing values of the regressor, whereas in the IRT a negative difficulty indicates an easier item (i.e., more subjects got the item correct).

Transitioning between the mixed and two-parameter model estimates is slightly more complicated. Bock & Aitkin (1981) list results for a two-parameter probit analysis of these same data; these are given in table 4.2 (labeled as IRT). This same model was estimated using MIXOR and SAS PROC NLMIXED with similar results. Table 4.2 lists those from MIXOR; the PROC NLMIXED code for this run can be found in the accompanying website to this book.

Here are the steps necessary to go from the mixed to the IRT model estimates:

1. Transform σ_j estimates so that their product equals 1. This yields the a_j estimates, and can be done by

$$a_j = \exp\left[\log \sigma_j - \frac{1}{n} \sum_{j'=1}^{n} \log \sigma_{j'}\right].$$

Table 4.2 Two-parameter probit model estimates for the LSAT-6 data

	IRT		MIXOR raw		MIXOR transformed	
	Difficulty	Discrimination	Difficulty	Discrimination	Difficulty	Discrimination
Item	$\hat{b}_j$	$\hat{a}_j$	$\hat{\beta}_j$	$\hat{\sigma}_j$	$\hat{b}_j$	$\hat{a}_j$
1	−0.6787	0.9788	1.5520	0.4169	−0.6804	0.9779
2	0.3161	1.0149	0.5999	0.4333	0.3165	1.0164
3	0.7878	1.2652	0.1512	0.5373	0.7867	1.2603
4	0.0923	0.9476	0.7723	0.4044	0.0926	0.9487
5	−0.5174	0.8397	1.1966	0.3587	−0.5154	0.8415

2. Reverse the sign of the β_j estimates, and transform so that $\sum -\beta_j/a_j = 0$, using the standardized a_j estimates from the previous step:

$$b_j = -(\beta_j/a_j) - \frac{1}{n}\sum_{j'=1}^{n}(-\beta_{j'}/a_{j'}).$$

As table 4.2 shows, the IRT and translated mixed model estimates agree closely. Also, as can be seen from the discrimination parameter estimates, items 3 and 5 are the most and least discriminating items, respectively. Item 3 is therefore both a difficult and a discriminating item.

Unlike traditional IRT models, the mixed or multilevel model formulation easily allows multiple covariates at either level (i.e., items or subjects). This and other advantages of casting IRT models as multilevel models are described by Adams et al. (1997b) and Reise (2000). In particular, multilevel models are well suited for examining whether item parameters vary by subject characteristics, and also for estimating ability in the presence of such item by subject interactions. Interactions between item parameters and subject characteristics, often termed item bias (Camilli and Shepard, 1994), is an area of active psychometric research.

4.1.2 Generalized Linear Mixed Models

IRT models have connections with the general class of models known as generalized linear mixed models (GLMMs; McCulloch & Searle, 2001). GLMMs extend generalized linear models (GLMs) by inclusion of random effects, and are commonly used for analysis of correlated nonnormal data. An excellent and comprehensive source on GLMM applications is the text by Skrondal and Rabe-Hesketh (2004); a more condensed review can be found in Hedeker (2005).

There are three specifications in a GLMM. First, the linear predictor, denoted as η_{ij} (as before, i and j represent subjects and items, respectively), of a GLMM is of the form

$$\eta_{ij} = \boldsymbol{x}'_{ij}\boldsymbol{\beta} + \boldsymbol{z}'_{ij}\boldsymbol{v}_i, \tag{4.12}$$

where $\boldsymbol{x}_{ij}$ is the vector of regressors for unit ij with fixed effects $\boldsymbol{\beta}$, and $\boldsymbol{z}_{ij}$ is the vector of variables having random effects which are denoted $\boldsymbol{\upsilon}_i$. The random effects are usually assumed to be multivariate normally distributed, that is, $\boldsymbol{\upsilon}_i \sim (\mathbf{0}, \Sigma_{\upsilon})$. Note that they could be expressed in standardized form (as is typical in IRT models) as $\boldsymbol{\theta}_i \sim (\mathbf{0}, \boldsymbol{I})$, where $\boldsymbol{\upsilon}_i = \boldsymbol{T}\boldsymbol{\theta}_i$ and $\boldsymbol{T}$ is the Cholesky factor of the variance–covariance matrix Σ_{υ}, namely, $\boldsymbol{TT}' = \Sigma_{\upsilon}$. Here, we have also allowed for multiple random effects, whereas the IRT models considered in this chapter have been in terms of a single random effect.

The second specification of a GLMM is the selection of a link function $g(\cdot)$ which converts the expected value μ_{ij} of the outcome variable Y_{ij} to the linear predictor η_{ij}:

$$g(\mu_{ij}) = \eta_{ij}. \tag{4.13}$$

Here, the expected value of the outcome is conditional on the random effects (i.e., $\mu_{ij} = E[Y_{ij} \mid \boldsymbol{\upsilon}_i]$). Finally, a specification for the form of the variance in terms of the mean μ_{ij} is made. These latter two specifications usually depend on the distribution of the outcome Y_{ij}, which is assumed to fall within the exponential family of distributions.

The Rasch model is seen as a GLMM by specifying item indicator variables as the vector of regressors $\boldsymbol{x}_{ij}$, and by setting $\boldsymbol{z}_{ij} = 1$ and $\boldsymbol{\upsilon}_i = \upsilon_i$ for the random effects part. The link function is specified as the logit link, namely,

$$g(\mu_{ij}) = \text{logit}(\mu_{ij}) = \log\left[\frac{\mu_{ij}}{1 - \mu_{ij}}\right] = \eta_{ij}. \tag{4.14}$$

The conditional expectation $\mu_{ij} = E(Y_{ij} \mid \upsilon_i)$ equals $P(Y_{ij} = 1 \mid \upsilon_i)$, namely, the conditional probability of a response given the random effects. Since Y is dichotomous, the variance function is simply $\mu_{ij}(1 - \mu_{ij})$.

Rijmen et al. (2003) present an informative overview and bridge between IRT models, multilevel models, mixed models, and GLMMs. As they point out, the Rasch model, and variants of it, belong to the class of GLMMs. However, the more extended two-parameter model is not within the class of GLMMs because the predictor is no longer linear, but includes a product of parameters.

4.2 Estimation

Parameter estimation for IRT models and GLMMs typically involves maximum likelihood (ML) or variants of ML. Additionally, the solutions are usually iterative ones that can be numerically quite intensive. Here, the solution is merely sketched; further details can be found in McCulloch & Searle (2001) and Fahrmeir and Tutz (2001). Let $\mathbf{Y}_i$ denote the vector of responses from subject i. The probability of any response pattern $\mathbf{Y}_i$ (of size n_i), conditional on $\boldsymbol{\upsilon}$, is equal to the product of the probabilities of the level 1 responses:

$$\ell(\mathbf{Y}_i \mid \boldsymbol{v}_i) = \prod_{i=1}^{n_i} P(Y_{ij} = 1 \mid \boldsymbol{v}_i). \tag{4.15}$$

The assumption that a subject's responses are independent given the random effects (and therefore can be multiplied to yield the conditional probability of the response vector) is known as the *conditional independence* assumption. The marginal density of $\mathbf{Y}_i$ in the population is expressed as the following integral of the conditional likelihood $\ell(\cdot)$:

$$h(\mathbf{Y}_i) = \int_{\boldsymbol{v}} \ell(\mathbf{Y}_i \mid \boldsymbol{v}_i) f(\boldsymbol{v})\, d\boldsymbol{v}, \tag{4.16}$$

where $f(\boldsymbol{v})$ represents the distribution of the random effects, often assumed to be a multivariate normal density. Whereas (4.15) represents the conditional probability, (4.16) indicates the unconditional probability for the response vector of subject i. The marginal log-likelihood from the sample of N subjects is then obtained as $\log L = \sum_i^N \log h(\mathbf{Y}_i)$. Maximizing this log-likelihood yields ML estimates (which are sometimes referred to as maximum marginal likelihood estimates) of the regression coefficients $\boldsymbol{\beta}$ and the variance–covariance matrix of the random effects Σ_υ (or the Cholesky of this matrix, denoted $\boldsymbol{T}$).

4.2.1 Integration over the Random-Effects Distribution

In order to solve the likelihood solution, integration over the random-effects distribution must be performed. As a result, estimation is much more complicated than in models for continuous normally distributed outcomes where the solution can be expressed in closed form. Various approximations for evaluating the integral over the random-effects distribution have been proposed in the literature; many of these are reviewed in Rodríguez and Goldman (1995). Perhaps the most frequently used methods are based on first- or second-order Taylor expansions. Marginal quasi-likelihood (MQL) involves expansion around the fixed part of the model, whereas penalized or predictive quasi-likelihood (PQL) additionally includes the random part in its expansion (Goldstein & Rasbash, 1996). Unfortunately, these procedures yield estimates of the regression coefficients and random-effects variances that are biased toward zero in certain situations, especially for the first-order expansions (Breslow and Lin, 1995). To remedy this, Raudenbush et al. (2000) proposed an approach that uses a combination of a fully multivariate Taylor expansion and a Laplace approximation. This method yields accurate results and is computationally fast. Also, as opposed to the MQL and PQL approximations, the deviance obtained from this approximation can be used for likelihood-ratio tests.

Numerical integration can also be used to perform the integration over the random-effects distribution (Bock & Lieberman, 1970). Specifically, if the assumed distribution is normal, Gauss–Hermite quadrature can approximate the above integral to any practical degree of accuracy. Additionally, like the Laplace approximation, the numerical quadrature approach yields a deviance that can be readily used for likelihood-ratio tests. The integration is approximated by a summation on a specified number of quadrature points for each dimension of the integration. An issue with the quadrature approach is that it can involve summation over a large number of points, especially as the number of random effects is increased. To address this, methods of adaptive quadrature have been developed that use a small number of points per dimension that are adapted to the location and dispersion of the distribution to be integrated (Rabe-Hesketh et al., 2002).

More computer-intensive methods, involving iterative simulations, can also be used to approximate the integration over the random-effects distribution. Such methods fall under the rubric of Markov chain Monte Carlo (MCMC; Gilks et al., 1997) algorithms. Use of MCMC for estimation of a wide variety of models has exploded in the last ten years or so. MCMC solutions are described in Patz and Junker (1999) and Kim (2001) for IRT models, and in Clayton (1996) for GLMMs.

4.2.2 Estimation of Random Effects

In many cases, it is useful to obtain estimates of the random effects. This is particularly true in IRT, where estimation of the random effect or latent trait is of primary importance, since these are the test ability scores for individuals. The random effects υ_i can be estimated using empirical Bayes methods. For the univariate case, this estimator υ_i is given by

$$\hat{\upsilon}_i = E(\upsilon_i \mid \mathbf{Y}_i) = h_i^{-1} \int_{\upsilon} \upsilon_i \, \ell_i \, f(\upsilon) \, d\upsilon, \tag{4.17}$$

where ℓ_i is the conditional probability for subject i under the particular model and h_i is the analogous marginal probability. This is simply the mean of the posterior distribution. Similarly, the variance of the posterior distribution is obtained as

$$V(\hat{\upsilon}_i \mid \mathbf{Y}_i) = h_i^{-1} \int_{\upsilon} (\upsilon_i - \hat{\upsilon}_i)^2 \, \ell_i \, f(\upsilon) \, d\upsilon. \tag{4.18}$$

These quantities may then be used, for example, to evaluate the response probabilities for particular subjects, or for ranking subjects in terms of their abilities. Embretson (2000) thoroughly describes uses and properties of these estimators within the IRT context; additionally, the edited collection of Thissen and Wainer (2001) contains a wealth of material on this topic.

4.3 Application: Adolescent Smoking Study

Data for the analyses reported here come from a longitudinal, natural history study of adolescent smoking (Mermelstein et al., in press). Students included in the longitudinal study were either in 8th or 10th grade at baseline, and either had never smoked, but indicated a probability of future smoking, or had smoked in the past 90 days, but had not smoked more than 100 cigarettes in their lifetime. Written parental consent and student assent were required for participation. A total of 562 students completed the baseline measurement wave. The longitudinal study utilized a multimethod approach to assess adolescents at three time points: baseline, 6 months, and 12 months. The data collection modalities included self-report questionnaires, a week-long time/event sampling method via palmtop computers (Ecological Momentary Assessment), and in-depth interviews.

Data for the current analyses came from the ecological momentary assessments obtained during the baseline time point. Adolescents carried the handheld computers with them at all times during the one-week data collection period and were trained both to respond to random prompts from the computers and to event record (initiate a data collection interview) smoking episodes. Immediately after smoking a cigarette, participants completed a series of questions on the handheld computers. Questions pertained to place, activity, companionship, mood, and specifics about smoking access and topography. The handheld computers dated and time-stamped each entry. For inclusion in the analyses reported here, adolescents must have smoked at least one cigarette during the seven-day baseline data collection period; 152 adolescents met this inclusion criterion. In the analyses reported here, we shall focus on the timing, both in terms of time of day and day of week, of the smoking reports. Thus, we are not analyzing data from the random prompts, but only from the self-initiated smoking reports.

The assessment week was divided into 35 periods, and whether an individual provided a smoking report in each of these 35 periods was determined as follows. First, we recorded the day of the week for a given cigarette report. Then, for each day, the time of day was classified into five bins: 3 A.M. to 8:59 A.M., 9 A.M. to 1:59 P.M., 2 P.M. to 5:59 P.M., 6 P.M. to 9:59 P.M., and 10 P.M. to 2:59 A.M.. Technically, part of the last bin (i.e., the time past midnight) belonged to the next day, but we treated this entire period as a late-night bin. In this way, subjects who, for example, smoked a cigarette at 1 A.M. during a Saturday night party would be classified as having smoked the cigarette on Saturday night and not on Sunday morning. Thus, for each of these 152 subjects, we created a 35×1 dichotomous response vector representing whether or not a cigarette report was made (yes/no) for each of these 35 time periods in the week. Some subjects did report more than one smoking event during a given time period; however this was relatively rare and so we simply treated the outcome as dichotomous (no report versus one or more smoking reports).

In terms of the IRT vernacular our "items" are these 35 time periods, and our dichotomous response is whether or not the subjects recorded smoking in these periods. A subject's latent "ability" can then be construed as their underlying degree of smoking behavior, and our interest is to see how this behavior relates to these day-of-week and time-of-day periods. An item's "difficulty" indicates the relative frequency of smoking reports during the time period, and an item's "discrimination" refers to the degree to which the time period distinguishes levels of the latent smoking behavior variable. The discrimination parameters are akin to factor loadings in a factor analysis or structural equation model. In this light, Rodriguez and Audrain-McGovern (2004) describe a factor analysis of smoking-related sensations among adolescents, while a Rasch analysis of smoking behaviors among more advanced smokers is presented in Bretelera et al. (2004). Admittedly, our example is a little different in that we are examining instances of smoking behavior, rather than smoking-related sensations and/or behaviors; however, we feel that the IRT approach we describe addresses some very interesting smoking-related questions. Specifically, we shall examine the degree to which time periods relate to the occurrence of smoking behavior, and the degree to which these periods differentially distinguish subjects of varying underlying smoking behavior levels. Our hypothesis is that time- and day-related characteristics of smoking among these beginning smokers may serve as early behavioral markers of dependence development.

Table 4.3 lists the proportion and number of smoking reports for each of these 35 time periods. Note that each of these proportions is based on the 152 subjects. In other words, each represents the proportion of the 152 subjects who smoked during the particular time period. As can be seen, the later week and weekend days represent relatively popular smoking days. In terms of time periods, the late afternoon and evening hours are also more common times of smoking reports. Of the 152 subjects, approximately 41%, 24%, 14%, 15%, and 6% provided 1, 2, 3 or 4, 5–9, and 10 or more smoking reports, respectively.

Because the data are rather sparse, it did not seem reasonable to estimate separate item parameters for each of the 35 cells produced by the crossing of day of week and time of day. Instead, we chose a "main effects" type of analysis in which the item parameters varied across these two factors but not their interaction. In terms of item difficulty (the location parameter), all models included effects for time of day and day of week. The referent cell was selected to be Monday from 10 P.M. to 3 A.M., since this was thought to be a time of low smoking reports. In the context of the present example, item difficulty primarily refers to the degree to which smoking reports were made during the particular time periods. In other

Table 4.3 Proportion (and n) of smoking reports by day of week and time of day

Day of week	3 A.M. to 8:59 A.M.	9 A.M. to 1:59 P.M.	2 P.M. to 5:59 P.M.	6 P.M. to 9:59 P.M.	10 P.M. to 2:59 A.M.
Monday	1.3 (2)	4.6 (7)	4.0 (6)	9.2 (14)	2.6 (4)
Tuesday	1.3 (2)	6.6 (10)	15.8 (24)	14.5 (22)	3.3 (5)
Wednesday	5.9 (9)	9.2 (14)	21.7 (33)	11.8 (18)	3.3 (5)
Thursday	5.3 (8)	9.2 (14)	19.7 (30)	17.1 (26)	1.3 (2)
Friday	9.2 (14)	9.9 (15)	21.7 (33)	19.1 (29)	6.6 (10)
Saturday	0.0 (0)	10.5 (16)	16.5 (25)	11.8 (18)	13.2 (20)
Sunday	0.7 (1)	4.0 (6)	4.0 (6)	9.9 (15)	2.6 (4)

Note: Proportion calculated as $n/152$, where 152 is the sample size.

words, the difficulty estimates reflect the proportion of reports made during the time intervals. For item discrimination (or item slopes), models were sequentially fitted assuming these slopes were equal (i.e., a Rasch model), or varied by time of day, day of week, or both time of day and day of week. Again, in the context of the present example, item discrimination refers to the degree to which a given time period distinguishes subjects with varying levels of underlying smoking behavior.

The results of these analysis are presented in tables 4.4 and 4.5. Table 4.4 lists the estimated difficulty parameters under these four models, and table 4.5 lists the corresponding item discrimination parameters. These estimates correspond to the mixed model representation of the parameters. Note that, in table 4.5, for the Rasch model (first column) a single common discrimination parameter is estimated, while in the next two models separate discrimination parameters were estimated for time of the day (second column) or each day of the week (third column). In the last of these models (fourth column), separate discrimination parameters were estimated for each day of week and time of day period under the constraint that the Saturday and 10 P.M. to 3 A.M. estimates were

Table 4.4 IRT difficulty estimates (standard errors)

Variable	Common slopes	Time-varying slopes	Day-varying slopes	Day- and time-varying slopes
Intercept	−4.16	−4.02	−4.35	−4.15
	(0.247)	(0.250)	(0.324)	(0.329)
Tuesday	0.748	0.747	1.01	0.981
	(0.227)	(0.232)	(0.322)	(0.321)
Wednesday	1.03	1.03	0.988	0.945
	(0.206)	(0.211)	(0.334)	(0.332)
Thursday	1.04	1.04	1.07	1.02
	(0.217)	(0.221)	(0.321)	(0.324)
Friday	1.34	1.34	1.58	1.55
	(0.190)	(0.194)	(0.309)	(0.307)
Saturday	1.03	1.03	1.31	1.31
	(0.208)	(0.215)	(0.319)	(0.321)
Sunday	−0.034	−0.034	0.359	0.318
	(0.248)	(0.255)	(0.374)	(0.375)
3 A.M. to 9 A.M.	−0.360	−1.09	−0.363	−1.16
	(0.221)	(0.388)	(0.227)	(0.385)
9 A.M. to 2 P.M.	0.564	0.455	0.569	0.400
	(0.191)	(0.233)	(0.199)	(0.230)
2 P.M. to 6 P.M.	1.37	1.25	1.38	1.21
	(0.181)	(0.210)	(0.186)	(0.212)
6 P.M. to 10 P.M.	1.24	1.11	1.25	1.10
	(0.175)	(0.220)	(0.184)	(0.221)

Table 4.5 IRT discrimination estimates (standard errors)

Variable	Common slopes	Time-varying slopes	Day-varying slopes	Day- and time-varying slopes
Intercept	0.828 (0.105)			
Monday			1.05 (0.264)	0.740 (0.343)
Tuesday			0.694 (0.215)	0.408 (0.303)
Wednesday			1.17 (0.243)	0.910 (0.316)
Thursday			1.08 (0.225)	0.834 (0.304)
Friday			0.711 (0.180)	0.427 (0.278)
Saturday			0.625 (0.141)	0.286 (0.194)
Sunday			0.472 (0.188)	0.211 (0.273)
3 A.M. to 9 A.M.		1.42 (0.239)		1.190 (0.290)
9 A.M. to 2 P.M.		0.785 (0.180)		0.564 (0.226)
2 P.M. to 6 P.M.		0.802 (0.156)		0.573 (0.199)
6 P.M. to 10 P.M.		0.805 (0.128)		0.518 (0.161)
10 P.M. to 3 A.M.		0.621 (0.199)		0.286 (0.194)

equal. This equality was chosen since the estimates for these two periods were most similar in the time-varying and day-varying models.

To choose between these models, table 4.6 presents model fit statistics, including the model deviance ($-2 \log L$), AIC (Akaike, 1973), and BIC (Read & Cressie, 1988). Lower values on these fit statistics imply better models. As can be seen, AIC would suggest the fourth model as best, whereas BIC points to the Rasch or random-intercepts model. As noted by Rost (1997) BIC penalizes overparameterization more than AIC, and so the results here are not too surprising. Because these are nested models, likelihood-ratio tests can also be used; these are presented in the bottom half of table 4.6. The likelihood-ratio tests support the model with day- and time-varying slopes (i.e., discrimination parameters) over the three simpler models. Thus, there is evidence that the day of week and time of day

Table 4.6 IRT model fit statistics

	Common slopes	Time-varying slopes	Day-varying slopes	Day- and time-varying slopes
$-2 \log L$	2811.79	2803.20	2798.37	2788.27
AIC	2835.79	2835.20	2834.37	2832.27
BIC	2872.08	2883.58	2888.80	2898.80
Parameters q	12	16	18	22
LIKELIHOOD-RATIO TESTS				
Comparisons to common slopes model				
χ^2		8.59	13.42	23.52
df		4	6	10
$p <$		0.072	0.037	0.009
Comparisons to time-varying slopes model				
χ^2			4.83	14.93
df			2	6
$p <$			0.089	0.021
Comparison to day-varying slopes model				
χ^2				10.10
df				4
$p <$				0.039

$\text{AIC} = -2 \log L + 2q$; $\text{BIC} = -2 \log L + q \log N$

are differentially discriminating in relating to adolescents' smoking "ability" or "dependence."

The difficulty estimates in table 4.4 support the notion that adolescent smoking is least frequent on Sunday and Monday, increases during the midweek Tuesday to Thursday period, and is highest on Friday and Saturday. All of the day-of-week indicators, with Monday as the reference cell, are statistically significant, based on Wald statistics, except for Sunday in all models. Figure 4.3 presents the difficulty estimates implied by the final model for the seven days and five time periods. In this figure, the estimates are presented using the IRT parameterization in equation (4.11). Thus, lower values reflect "easier" items, or time periods where smoking behavior is relatively more common. Turning to the time-of-day indicators, where 10 P.M. to 3 A.M. is the reference cell, it is seen that more frequent smoking periods are 2 P.M. to 6 P.M. and 6 P.M. to 10 P.M., whereas 3 A.M. to 9 A.M. is the least frequent period for smoking. The 9 A.M. to 2 P.M. period is intermediate and statistically similar to the 10 P.M. to 3 A.M. reference cell.

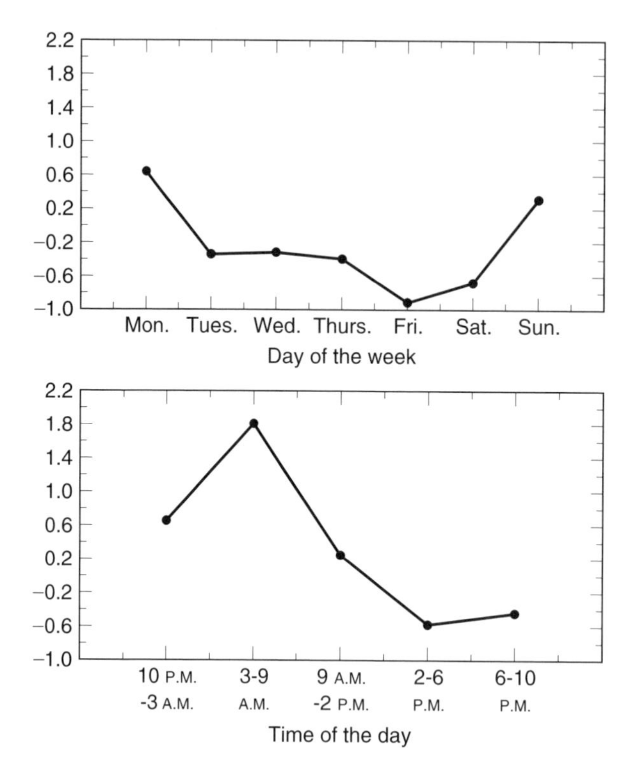

Figure 4.3 Difficulty estimates based on day- and time-varying slopes model.

The results for the difficulty parameters agree with the observed frequency data presented in table 4.3, reflecting diffferences in the frequency of smoking reports. Alternatively, the discrimination parameter estimates reveal the weighting of the days and time periods on the latent smoking level and so can suggest aspects of the data that are not so obvious. For this, figure 4.4 displays the discrimination, or slope, estimates based on the final model for the seven days and five time periods. Again, these are displayed in the IRT representation of equation (4.11). These indicate that the most discriminating days are Monday, Wednesday, and Thursday, and the least discriminating days are Friday, Saturday, Sunday, and Tuesday. This suggests, in combination with the difficulty results, that although weekend smoking is relatively prevalent, it is not as informative as weekday smoking in determining the level of an adolescent's smoking behavior. In terms of time of day, these estimates and their characterization in figure 4.4 clearly point out the time period of 3 A.M. to 9 A.M. as the most discriminating period. This result agrees well with the literature on smoking, because morning smoking and smoking after awakening are important markers of smoking addiction (Heatherton et al., 1991). Interestingly, by comparing figures 4.3 and 4.4, one can see that this discriminating time period is one of very infrequent smoking for adolescents.

Another fundamental aspect of IRT modeling is the estimation of each individual's level of ability θ. In achievement testing, these indicate the relative abilities

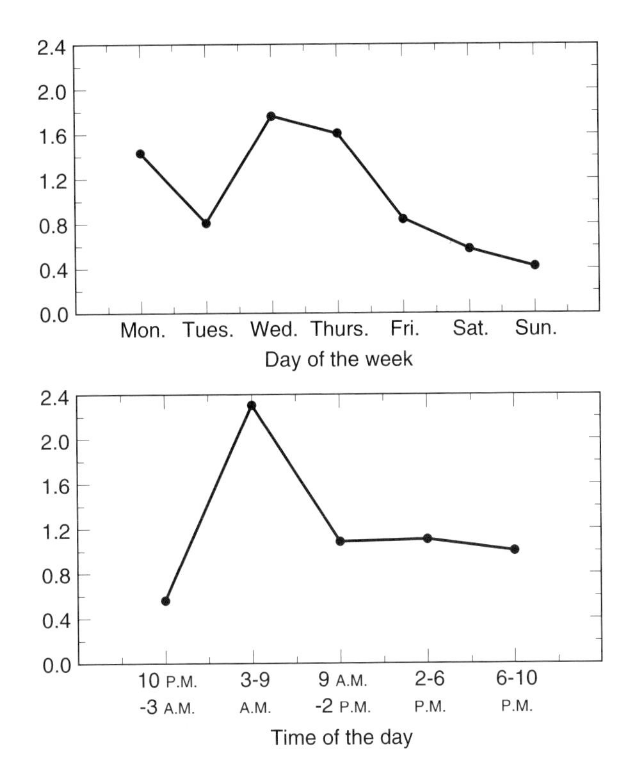

Figure 4.4 Discrimination estimates based on day- and time-varying slopes model.

of a sample of test-takers. Here, these reflect the latent smoking level of the adolescents in this study. As mentioned, estimation of θ is typically done using empirical Bayes methods, whereas estimation of the item parameters is based on maximum likelihood (Bock & Aitkin, 1981). This combination of empirical Bayes and maximum likelihood is also the usual procedure in mixed models (Laird & Ware, 1982). The distribution of ability estimates, based on the day- and time-varying slopes model, is presented in figure 4.5. This plot indicates that many individuals have a low level of smoking, though not all. Given the observed frequencies in table 4.3, this pattern of results for θ estimates is not surprising. Many

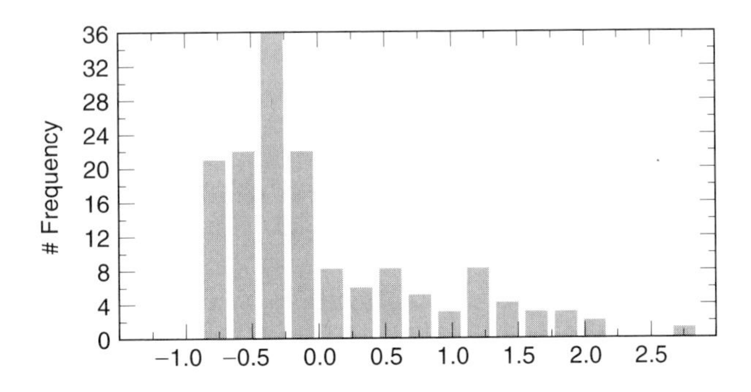

Figure 4.5 Histogram of empirical Bayes ability estimates from the two-parameter model.

subjects provided only one or two smoking reports, and therefore would certainly have low levels on the latent smoking variable. A concern is that the distribution is assumed to be normal in the population; however, this figure indicates a positively skewed distribution. IRT methods that do not assume normality for the underlying distribution of ability are described in Mislevy (1984), who indicates how nonparametric and mixture ability distributions can be estimated in IRT models. Such extensions are also described and further developed within a mixed model framework by Carlin et al. (2001). In the present case, we considered a nonparametric representation and estimated the density at a finite number of points from approximiately -4 to 4, in addition to estimation of the usual item parameters. These results, not shown, gave similar conclusions in terms of the item difficulty and discrimination parameters. Thus, our conclusions seem relatively robust to the distributional assumption for ability.

A natural question to ask is the relationship between the IRT estimate of smoking level and the simple sum of the number of smoking reports an individual provides. Figure 4.6 presents a scatterplot of the number of smoking reports versus the latent smoking ability estimates, based on the day- and time-varying slopes model. These two are, of course, highly correlated ($r = 0.96$), but the IRT estimate is potentially more informative because it includes information about when the smoking events were reported.

To give a sense of this, table 4.7 lists the minimum and maximum IRT ability estimate, stratified by the number of smoking reports, for subjects with 2, 3, 4, or 5 reports. For each, the day of the week and the time of the day are indicated. As can be seen, within each stratum, the lower IRT estimates are more associated with what might be considered "party" smoking, namely, smoking on weekend evenings and nights. Conversely, higher IRT estimates are more associated with weekday and morning smoking reports. Again, this agrees with notions in smoking research that smoking alone or outside of a social event may be more characteristic of the development of smoking dependence (Nichter et al., 2002).

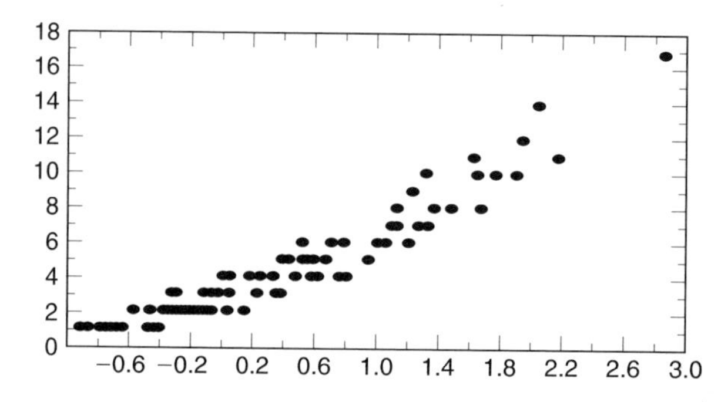

Figure 4.6 Number of smoking reports versus empirical Bayes estimates.

Table 4.7 Minimum and maximum IRT estimate stratified by number of smoking reports: day of week and time of day of smoking reports (1 = report, 0 = no report)

No. of reports	IRT estimate	Day	3 A.M. to 8:59 A.M.	9 A.M. to 1:59 P.M.	2 P.M. to 5:59 P.M.	6 P.M. to 9:59 P.M.	10 P.M. to 2:59 A.M.
2	−0.555	Sat.	0	0	1	0	1
	0.158	Mon.	0	0	0	0	1
		Wed.	1	0	0	0	0
3	−0.315	Fri.	0	0	0	1	1
		Sat.	0	0	0	0	1
	0.383	Tues.	0	0	1	0	0
		Wed.	0	0	1	0	0
		Fri.	1	0	0	0	0
4	0.017	Fri.	0	0	1	1	0
		Sat.	0	0	0	1	1
	0.808	Tues.	0	0	0	1	0
		Wed.	1	0	0	1	0
		Thurs.	0	1	0	0	0
5	0.414	Wed.	0	0	0	1	0
		Fri.	0	0	1	1	0
		Sat.	0	0	0	1	1
	0.947	Thurs.	1	0	1	0	0
		Fri.	1	0	0	0	0
		Sat.	0	1	0	1	0

4.4 Discussion

IRT models have been extensively developed and used in educational and psychological measurement. However, use of IRT models outside of these areas is rather limited. Part of the reason for this is that these methods have not been well understood by nonpsychometricians. With the emergence and growing popularity of mixed models for longitudinal data, formulation of IRT models within this class can help to overcome these obstacles.

In this chapter, we have striven to show the connection between IRT and mixed models, and how standard software for mixed models can be used to estimate basic IRT models. Some translation of the parameter estimates is necessary to properly express the mixed model results in IRT form, and we have illustrated how this is done using a frequently analyzed dataset of LSAT test items.

IRT modeling was illustrated using Ecological Momentary Assessment (EMA) data from a study of adolescent smoking. Here, we analyzed whether or not a smoking report had been made in each of 35 time periods, defined by the crossing of seven days and five time intervals within each day. The IRT model was able to identify which time periods were most associated with smoking reports, and also which time periods were the most informative, in the sense of discriminating underlying levels of smoking behavior. As indicated, weekend and evening hours yielded the most frequent smoking reports; however, morning and, to some extent, midweek reports were most discriminating in separating smoking levels.

IRT modeling provides a useful method for addressing questions about patterning of behavior beyond mere frequency reports. For example, in the case presented here, we examined whether the time of day or day of week that an adolescent smokes discriminates underlying levels of smoking behavior. This is not possible if one only considers smoking quantity alone. Although our data indicate that adolescents are more likely to smoke on weekend evenings (the stereotypic weekend party phenomenon), the data also indicate that those smoking events may be a less important indicator of the underlying level of smoking behavior than smoking episodes that occur either midweek or in the early mornings. Thus, one might consider midweek smoking as an indicator of a more pronounced level of adolescent smoking behavior, and perhaps as an indicator of the subsequent development of smoking dependence. Indeed, these data also lead one to address questions such as whether "binge" smoking episodes are less associated with underlying smoking behavior than is a pattern of smoking that is more evenly distributed over the week. The IRT models presented here are clearly useful in furthering the empirical investigations of a number of behavioral phenomena for which both quantity and patterns of behavior may be important. For example, one could easily apply similar models to addressing questions about patterns of lapses and relapses following abstinence as well as to models of escalation.

In addition, other applications of IRT modeling of EMA data are clearly possible. For example, in this chapter we have focused on dichotomous data, but IRT models for ordinal and nominal outcomes have also been developed. Additionally, we did not include any covariates in our analyses, but these could easily be handled. Thus, we could explore whether the item parameters differed between males and females, or whether the number of friends present during a given time period influenced the probability of a smoking report.

ACKNOWLEDGMENTS

Thanks are due to Siu Chi Wong for statistical analysis. This work was supported by National Institutes of Mental Health Grant MH56146, National Cancer Institute Grant CA80266, and by a grant from the Tobacco Etiology Research Network, funded by the Robert Wood Johnson Foundation.

References

Adams, R.J., Wilson, M., & Wang, W. (1997a). The multidimensional random coefficients multinomial logit model. *Applied Psychological Measurement, 21,* 1–23.

Adams, R.J., Wilson, M., & Wu, M. (1997b). Multilevel item response models: An approach to errors in variable regression. *Journal of Educational and Behavioral Statistics, 22,* 47–76.

Akaike, H. (1973). Information theory and an extension of the maximum likelihood principle. In B.N. Petrov & F. Csaki (Eds.), *Second International Symposium on Information Theory,* (pp. 267–281). Budapest: Academiai Kiado.

Anderson, E.B. (1985). Estimating latent correlations between repeated testings. *Psychometrika, 50,* 3–16.

Bartholomew, D.J., & Knott, M. (1999). *Latent Variable Models and Factor Analysis* (2nd ed.). New York: Oxford University Press.

Birnbaum, A. (1968). Some latent trait models and their use in inferring an examinee's ability. In F.M. Lord & M.R. Novick (Eds.), *Statistical Theories of Mental Test Scores.* Reading, MA: Addison-Wesley.

Bock, R.D. (1972). Estimating item parameters and latent ability when responses are scored in two or more nominal categories. *Psychometrika, 37,* 29–51.

Bock, R.D. (1997). A brief history of item response theory. *Educational Measurement: Issues and Practice, 16,* 21–32.

Bock, R.D., & Aitkin, M. (1981). Marginal maximum likelihood estimation of item parameters: An application of the EM algorithm. *Psychometrika, 46,* 443–459.

Bock, R.D., Gibbons, R.D., & Muraki, E. (1988). Full-information item factor analysis. *Applied Psychological Measurement, 12,* 261–280.

Bock, R.D., & Lieberman, M. (1970). Fitting a response model for *n* dichotomously scored items. *Psychometrika, 35,* 179–197.

Breslow, N.E., & Lin, X. (1995). Bias correction in generalised linear mixed models with a single component of dispersion. *Biometrika, 82,* 81–91.

Bretelera, M.H.M., Hilberinkb, S.R., Zeemanc, G., & Lammersa, S.M.M. (2004). Compulsive smoking: The development of a Rasch homogeneous scale of nicotine dependence. *Addictive Behaviors, 29,* 199–205.

Camilli, G., & Shepard, L.A. (1994). *Methods for Identifying Biased Test Items.* Thousand Oaks, CA: Sage.

Carlin, J.B., Wolfe, R., Brown, C.H., & Gelman, A. (2001). A case study on the choice, interpretation and checking of multilevel models for longitudinal binary outcomes. *Biostatistics, 2,* 397–416.

Clayton, D. (1996). Generalized linear mixed models. In W.R. Gilks, S. Richardson, & D.J. Spiegelhalter (Eds.), *Markov Chain Monte Carlo Methods in Practice* (pp. 275–303). New York: Chapman & Hall.

Corcoran, C., Coull, B., & Patel, A. (1999). *EGRET for Windows User manual.* Cambridge, MA: CYTEL Software Corporation.

Embretson, S.E. (1991). A multidimensional latent trait model for measuring learning and change. *Psychometrika, 56,* 495–516.

Embretson, S.E. (2000). Multidimensional measurement from dynamic tests: Abstract reasoning under stress. *Multivariate Behavioral Research, 35,* 505–542.

Embretson, S.E., & Reise, S.P. (2000). *Item Response Theory for Psychologists.* Mahwah, NJ: Erlbaum.

Fahrmeir, L., & Tutz, G.T. (2001). *Multivariate Statistical Modelling Based on Generalized Linear Models* (2nd ed). New York: Springer.

Fischer, G.H., & Pononcy, I. (1994). An extension of the partial credit model with an application to the measurement of change. *Psychometrika, 59,* 177–192.

Fraser, C. (1988). *NOHARM II: A Fortran Program for Fitting Unidimensional and Multidimensional Normal Ogive Models of Latent Trait Theory.* Armidale, N.S.W.: University of New England, Centre for Behavioral Studies.

Gibbons, R.D., & Hedeker, D. (1992). Full-information item bi-factor analysis. *Psychometrika, 57,* 423–436.

Gilks, W., Richardson, S., & Speigelhalter, D.J. (1997). *Markov Chain Monte Carlo in Practice.* New York: Chapman & Hall.

Goldstein, H. (1995). *Multilevel Statistical Models,* (2nd ed.). New York: Halstead Press.

Goldstein, H., & Rasbash, J. (1996). Improved approximations for multilevel models with binary responses. *Journal of the Royal Statistical Society, Series B, 159,* 505–513.

Greene, W.H. (1998). *LIMDEP Version 7.0 User's Manual* (revised edition). Plainview, NY: Econometric Software, Inc.

Hambleton, R.K., Swaminathan, H., & Rogers, H.J. (1991). *Fundamentals of Item Response Theory.* Newbury Park, CA: Sage.

Heatherton, T.F., Kozlowski, L.T., Frecker, R.C., & Fagerstrom, K.O. (1991). The Fagerstrom test for nicotine dependence: A revision of the Fagerstrom tolerance questionnaire. *British Journal of Addictions, 86,* 1119–1127.

Hedeker, D. (2005). Generalized linear mixed models. In B. Everitt & D. Howell (Eds.), *Encyclopedia of Statistics for the Behavioral Sciences.* London: John Wiley.

Hedeker, D., & Gibbons, R.D. (1996). MIXOR: A computer program for mixed-effects ordinal probit and logistic regression analysis. *Computer Methods and Programs in Biomedicine, 49,* 157–176.

Heinen, T. (1996). *Latent Class and Discrete Latent Trait Models.* Thousand Oaks, CA: Sage.

Hulin, C.L., Drasgow, F., & Parsons, C.K. (1983). *Item Response Theory.* Homewood, IL: Dow Jones-Irwin.

Kim, S.-H. (2001). An evaluation of a Markov chain Monte Carlo method for the Rasch model. *Applied Psychological Measurement, 25,* 163–176.

Laird, N.M., & Ware, J.H. (1982). Random-effects models for longitudinal data. *Biometrics, 38,* 963–974.

Lawley, D.N. (1943). On problems connected with item selection and test construction. *Proceedings of the Royal Society of Edinburgh, 61,* 273–287.

Lord, F.M. (1980). *Applications of Item Response Theory to Practical Testing Problems.* Hillside, NJ: Erlbaum.

Lord, F.M., & Novick, M.R. (1968). *Statistical Theories of Mental Health Scores.* Reading, MA: Addison-Wesley.

McCulloch, C.E., & Searle, S.R. (2001). *Generalized, Linear, and Mixed Models.* New York: John Wiley.

Mellenbergh, G.J. (1995). Conceptual notes on models for discrete polytomous item responses. *Applied Psychological Measurement, 19,* 91–100.

Mermelstein, R., Hedeker, D., Flay, B., & Shiffman, S. (in press). Real-time data capture and adolescent cigarette smoking. In A. Stone, S. Shiffman, & A. Atienza (Eds.), *The Science of Real-Time Data Capture: Self-Report in Health Research.* New York: Oxford University Press.

Mislevy, R.J. (1984). Estimating latent distributions. *Psychometrika, 49,* 359–381.

Mislevy, R.J., & Bock, R.D. (1998). *BILOG-3: Item Analysis and Test Scoring with Binary Logistic Models.* Chicago: Scientific Software International, Inc.

Muthén, B., & Muthén, L. (2001). *Mplus User's Guide.* Los Angeles: Muthén & Muthén.

Nichter, M., Nichter, M., Thompson, P.J., Shiffman, S., & Moscicki, A.B. (2002). Using qualitative research to inform survey development on nicotine dependence among adolescents. *Drug and Alcohol Dependency, 68* (*Suppl. 1*), S41–S56.

Novick, M.R. (1966). The axioms and principal results of classical test theory. *Journal of Mathematical Psychology, 3*, 1–18.

Patz, R.J., & Junker, B.W. (1999). A straightforward approach to Markov chain Monte Carlo methods for item response theory models. *Journal of Educational and Behavioral Statistics, 24*, 146–178.

Rabe-Hesketh, S., Pickles, A., & Skrondal, A. (2001). *GLLAMM Manual*. Technical Report 2001/01. Institute of Psychiatry, King's College, University of London, Department of Biostatistics and Computing.

Rabe-Hesketh, S., Skrondal, A., & Pickles, A. (2002). Reliable estimation of generalized linear mixed models using adaptive quadrature. *The Stata Journal, 2*, 1–21.

Rasbash, J., Steele, F., Browne, W., & Prosser, B. (2004). *A User's Guide to MLwiN Version 2.0*. London: Institute of Education, University of London.

Rasch, G. (1960). *Probabilistic Models for Some Intelligence and Attainment Tests*. Copenhagen: Danish Institute for Educational Research.

Raudenbush, S.W., & Bryk, A.S. (2002). *Hierarchical Linear Models in Social and Behavioral Research: Applications and Data-Analysis Methods* (2nd ed.). Thousand Oaks, CA: Sage.

Raudenbush, S.W., Bryk, A.S., Cheong, Y.F., & Congdon, R. (2004). *HLM 6: Hierarchical Linear and Nonlinear Modeling*. Chicago: Scientific Software International, Inc.

Raudenbush, S.W., Yang, M.-L. & Yosef, M. (2000). Maximum likelihood for generalized linear models with nested random effects via high-order, multivariate Laplace approximation. *Journal of Computational and Graphical Statistics, 9*, 141–157.

Read, T.R.C., & Cressie, N.A.C. (1988). *Goodness-of-Fit Statistics for Discrete Multivariate Data*. New York: Springer.

Reise, S.P. (2000). Using multilevel logistic regression to evaluate person-fit in IRT models. *Multivariate Behavioral Research, 35*, 543–568.

Richardson, M.W. (1936). The relationship between difficulty and the differential validity of a test. *Psychometrika, 1*, 33–49.

Rijmen, F., Tuerlinckx, F., De Boeck, P., & Kuppens, P. (2003). A nonlinear mixed model framework for item response theory. *Psychological Methods, 8*, 185–205.

Rodriguez, D., & Audrain-McGovern, J. (2004). Construct validity analysis of the early smoking experience questionnaire for adolescents. *Addictive Behaviors, 29*, 1053–1057.

Rodríguez, G., & Goldman, N. (1995). An assessment of estimation procedures for multilevel models with binary responses. *Journal of the Royal Statistical Society, Series A, 158*, 73–89.

Rost, J. (1997). Logistic mixture models. In W.J. van der Linden & R.K. Hambleton (Eds.), *Handbook of Modern Item Response Theory* (pp. 449–463). New York: Springer.

Samejima, F. (1969). Estimation of latent ability using a response pattern of graded scores. *Psychometrika, Monograph No. 17*.

Schaeffer, N.C. (1988). An application of item response theory to the measurement of depression. In C. Clogg (Ed.), *Sociological Methodology 1988* (pp. 271–307). Washington, DC: American Sociological Association.

Skrondal, A., & Rabe-Hesketh, S. (2004). *Generalized Latent Variable Modeling: Multilevel, Longitudinal and Structural Equation Models*. Boca Raton, FL: Chapman & Hall/CRC Press.

Spearman, C. (1904). The proof and measurement of association between two things. *American Journal of Psychology, 15*, 72–101.

StataCorp (1999). *Stata: Release 6.0*. College Station, TX: Stata Corporation.

Thissen, D. (1982). Marginal maximum likelihood estimation for the one-parameter logistic model. *Psychometrika, 47*, 175–186.

Thissen, D. (1991). *MULTILOG User's Guide: Multiple Categorical Item Analysis and Test Scoring Using Item Response Theory*. Chicago: Scientific Software International, Inc.

Thissen, D., & Steinberg, L. (1986). A taxonomy of item response models. *Psychometrika, 51*, 567–577.

Thissen, D., & Steinberg, L. (1988). Data analysis using item response theory. *Psychological Bulletin, 104*, 385–395.

Thissen, D., & Wainer, H. (2001). *Test Scoring*. Mahwah, NJ: Erlbaum.

Thurstone, L.L. (1925). A method of scaling psychological and educational tests. *Journal of Educational Psychology, 16*, 433–451.

Thurstone, L.L. (1926). The scoring of individual performance. *Journal of Educational Psychology, 17*, 446–457.

Thurstone, L.L. (1927). Psychophysical analysis. *American Journal of Psychology, 38*, 368–389.

van der Linden, W.J., & Hambleton, R.K. (Eds.) (1997). *Handbook of Modern Item Response Theory*. New York: Springer.

Wright, B.D. (1977). Solving measurement problems with the Rasch model. *Journal of Educational Measurement, 14*, 97–116.

Wu, M.L., Adams, R.J., & Wilson, M. (1998). *ACER ConQuest: Generalized Item Response Modelling Software*. Melbourne: Australian Council for Educational Research.

5

Fitting Curves with Periodic and Nonperiodic Trends and Their Interactions with Intensive Longitudinal Data

Carlotta Ching Ting Fok and James O. Ramsay

In longitudinal series with only a few waves of measurement, growth or change over time can often be well approximated by simple time-graded polynomials (e.g., linear or quadratic trends). With more intensive schedules of measurement, however, patterns of fluctuation tend to be more complex, and modeling them in a flexible and parsimonious way becomes a challenge. Drawing upon ideas from functional data analysis (Ramsay & Silverman, 1997), we show how to build multilevel regression models for intensive longitudinal data that exhibit a combination of periodic and nonperidic trends.

5.1 Periodic and Nonperiodic Trends

A time series is said to be periodic or cyclic if there is some length of time P over which the data values tend to repeat themselves. The period P may be a day, a week, a year, or any other natural unit. Raw, unadjusted time-series data often contain periodic effects as well as long-term trends or drifts. For example, Ramsay and Silverman (1997, p. 219) examined the quarterly values of gross domestic product (GDP) reported by sixteen countries from 1980 to 1994. Each country experienced a marked increase in GDP over the fifteen-year period. Many countries adjusted their quarterly GDP figures to remove the effects of annual business cycles, but some did not. The countries that did not adjust their figures clearly show cyclic trends with a period of one year.

Long-term trends are sometimes seen within periodic effects; that is, periodic and nonperiodic effects may interact. Imagine, for example, a study of hospitalized bipolar patients in which each patient's mood is measured several times per day over a period of weeks. One would probably find that mood tends to swing in a recurrent fashion. As medication and other forms of treatment are applied, however, one would also hope that the patients demonstate improvement over time, experiencing less severe cycles of change as the study progresses. In nonclinical settings, long-term periodic and nonperiodic changes can also be seen in measures of personality and interpersonal behavior.

In this chapter, we apply the multilevel regression models described by Walls et al. (chapter 1, this volume) but in a different way. We account for complex patterns of change in intensive longitudinal data by a combination of periodic effects, nonperiodic effects, and autocorrelated noise. Our motivating application is a set of data on interpersonal behaviors collected and previously analyzed by Brown and Moskowitz (1998).

5.1.1 Overview of the Data

Although psychologists tend to regard personality traits as stable over time, it is clear that situational influences lead to considerable fluctuation in how these traits are expressed and measured. For the most part, these fluctuations have been treated as random noise. Recent analyses have shown, however, that intraperson variability in expression may show stable and reliable patterns. Many types of behavior demonstrate regular cycles, and these cycles may account for a large part of the overall variance. For example, a person might tend to be quarrelsome on Mondays but more agreeable later in the week.

Brown and Moskowitz (1998) selected and measured behaviors from the interpersonal circumplex, a personality model that organizes certain interpersonal characteristics into a circle (Kiesler, 1983; Leary, 1957; Wiggins, 1979, 1982). The circle is composed of two major bipolar axes, agency and communion, which describe the way in which a person deals with the outside world. Agency is the extent to which a person demonstrates free will and dominance in his or her own actions. Hence, an individual who demonstrates a high degree of agency might be involved in more dominant than submissive behaviors. Communion, on the other hand, represents "strivings for intimacy, union, and solidarity with a social or spiritual entity" (Brown & Moskowitz, 1998, p. 108); a high degree of communion would be reflected by agreeable interactions with others and infrequent quarrelsome behavior.

In this study, 89 participants were asked to complete a one-page form after each social interaction of five minutes or longer over a period of 20 consecutive days, immediately recording the details of each just-completed interaction. Participants were supplied with ten forms per day, together with stamped, addressed envelopes, and were requested to return each day's completed forms to the researchers on the next weekday. Based on these responses, we computed daily mean values of summary measures in four dimensions: dominance, submissiveness, agreeableness, and quarrelsomeness. Plots of the mean (across participants) standardized levels of each dimension by day of study is shown in figure 5.1. This figure suggests that, on average, the behaviors exhibit periodic trends; that dominant and submissive behaviors tend to rise and fall in opposite phase; and that agreeableness and quarrelome behavior tend to occur in opposite phase.

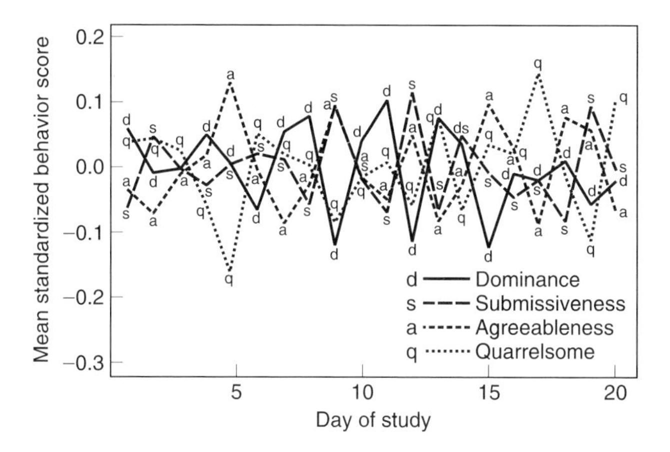

Figure 5.1 Mean levels of the four dimensions by day of the study.

For simplicity, our analyses will focus only on the dimension of agreeableness. Before collecting these data, Brown and Moskowitz (1998) hypothesized that personality measures would exhibit weekly cycles. Although each participant remained in the study for 20 days, they did not all begin on the same day of the week. To align the curves with respect to day of the week, an integer from 0 to 6 was added to each participant's day number, so that day 1 would always correspond to Tuesday. With this shift, a participant who began on Tuesday would have observations for days $1, \ldots, 20$; one who began on Wednesday would have observations for days $2, \ldots, 21$; and so on. The mean of the daily measures for agreeableness after adjusting for different starting days is shown in figure 5.2. We shall return to these data in section 5.3 and, using our extended multilevel model, demonstrate that a weekly cycle is indeed present.

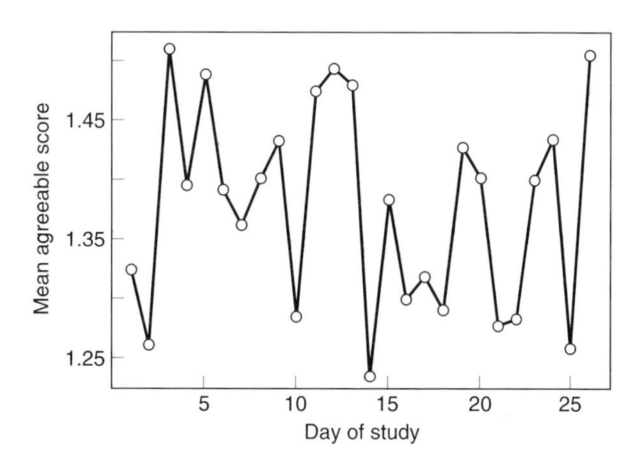

Figure 5.2 Mean agreeable scores across participants.

5.2 The Model

Our model is a multilevel linear regression that combines spline smoothing with Fourier analysis to describe periodic and nonperiodic fluctuations over time. In this section, we briefly review the multilevel model and then discuss how to create basis functions that capture periodic and nonperiodic trends.

5.2.1 Multilevel Models with Autocorrelated Errors

Multilevel models for longitudinal data, also known as linear mixed-effects models, were described at length by Walls et al. (chapter 1, this volume). Reviews of this model are also given by Wallace and Green (2002). Using the notation of Laird and Ware (1982), the basic model is

$$y_i = X_i\beta + Z_i b_i + e_i, \tag{5.1}$$

where y_i is a vector of responses for individual i, X_i and Z_i are matrices of regressors, β is a vector of fixed coefficients to be estimated, b_i is a vector of random effects, and e_i is a vector of residual errors. The latter two vectors are both assumed to have mean zero. Suppose that y_{ij}, the jth measurement for subject i, was taken at time t_{ij}. The mean response for subject i at time t_{ij} is then

$$\mu_i(t_{ij}) = x_{ij}^T\beta + z_{ij}^T b_i, \tag{5.2}$$

where x_{ij}^T and z_{ij}^T denote the jth rows of X_i and Z_i. Averaging across all subjects with these same regressors, the mean response for the population at occasion t_{ij} becomes

$$\mu(t_{ij}) = x_{ij}^T\beta. \tag{5.3}$$

For the most part, our attention will focus on estimating the population-average mean $\mu(t)$ and assessing how it varies with t, both in a periodic and a long-term sense.

In many applications of this model, the elements of e_i are assumed to be independent; that is, e_i is assumed to be normally distributed with covariance matrix $\sigma^2 V_i$, where $V_i = I$ is the identity matrix. When intensive measurements are taken over time, the assumption of independent within-subject residuals may be unrealistic. Many factors can cause residuals for neighboring time points to be correlated with each other. For example, one could imagine a compensatory process that causes a negative residual to be followed by a positive residual, and vice versa. There may be short-term fluctuations spanning only a few consecutive points. As the study progresses, subjects may exhibit carryover effects, practice effects, or fatigue.

Borrowing some ideas from time-series analysis, we can account for relationships among neighboring residuals by changing the form of V_i to allow autocorrelation. Two simple and popular classes of models for autocorrelated series are the autoregressive (AR) and moving average (MA) processes (Pinheiro & Bates, 2000). In an AR process, the residual for individual i at the tth occasion is predicted by a linear combination of the last p residuals,

$$e_{ij} = \alpha_1 e_{i,j-1} + \alpha_2 e_{i,j-2} + \cdots + \alpha_p e_{i,j-p} + \epsilon_{i,j}.$$

This autoregressive process is designated AR(p). In the above equation, ϵ_{ij} represents a white-noise process, and the coefficients $\alpha_1, \ldots, \alpha_p$ weight the history of the process to predict its current state. In most cases, it is assumed that $|\alpha_r| < 1$ for $r = 1, \ldots, p$. An MA(q) process, on the other hand, has the form

$$e_{ij} = \epsilon_{ij} + \alpha_1 \epsilon_{i,j-1} + \alpha_2 \epsilon_{i,j-2} + \cdots + \alpha_q \epsilon_{i,j-q}.$$

Notice that, in this case, the coefficients $\alpha_1, \ldots, \alpha_q$ weight the history, not of the process itself, but of its most recent disturbances. Both types of dependence can be combined into a more general process called ARMA(p, q). Popular software packages for multilevel modeling allow the user to select various correlational forms for V_i. In particular, the S-PLUS function lme(), described by Pinheiro and Bates (2000), now supports the general ARMA(p, q) structure for arbitrary $p \geq 0$ and $q \geq 0$.

In general, inferences about the form of V_i can be sensitive to the choice of regressors included in X_i and Z_i. Periodic trends for the population or individual subjects can sometimes masquerade as autocorrelation among the e_{it}'s. When population-average trends are of primary interest, we ought to be reasonably certain that these trends are adequately described by the model's mean structure before concluding that autocorrelated noise is present. Autocorrelated error structures tend to be needed when the number of measurements is large. With fewer measurements per subject, apparent autocorrelation can sometimes be reduced or eliminated by including random intercepts, as described in section 5.2.7.

5.2.2 Time-Varying Effects

We desire to fit models to intensive longitudinal data in which the mean response fluctuates over time in complicated ways, as illustrated by the plot in figure 5.2. In the multilevel model (5.1), trends in the mean are reflected in the fixed-effects component $X_i\beta$. Let us rewrite the population-average mean function (5.3) as

$$\mu(t) = \beta_0\phi_0(t) + \beta_1\phi_1(t) + \beta_2\phi_2(t) + \cdots,$$

where $\phi_0(t), \phi_1(t), \phi_2(t), \ldots$ are functions of time known as basis functions that are incorporated into the columns of X_i. With short longitudinal series, the average trend may be well approximated by simple time-graded polynomials; a quadratic trend, for example, could be modeled by taking $\phi_0(t) = 1$, $\phi_1(t) = t - t_0$, and $\phi_2(t) = (t - t_0)^2$, where t_0 is a time point chosen as an origin. Simple polynomials can effectively approximate smooth functions in the neighborhood of any particular t_0. To approximate complicated functions over a long time span, however, different kinds of basis functions are usually needed.

Hastie and Tibshirani (1993) introduced a varying-coefficient regression model to describe situations where the effects of predictors on a response vary as smooth functions of another variable T. Let Y be a response variable and $X_1, X_2, \ldots, X_p$ be a set of predictors. In ordinary multiple regression, the mean of Y is assumed to be

$$\mu = \beta_0 + \beta_1 X_1 + \beta_2 X_2 + \cdots + \beta_p X_p.$$

In a varying-coefficient model, the mean of Y at $T = t$ is

$$\mu(t) = \beta_0(t) + X_1\beta_1(t) + \cdots + X_t\beta_p(t),$$

where the regression coefficients now vary with t. In longitudinal applications where T is a measure of time, the varying-coefficient model is often expressed as

$$\mu(t) = \beta_0(t) + X_1(t)\beta_1(t) + \cdots + X_p(t)\beta_p(t),$$

where the covariates may also vary with time. The intercept function $\beta_0(t)$ is decomposed into a linear combination of basis functions to reflect periodic and nonperiodic trends in the overall mean of Y. Temporal changes in the effect of a time-varying covariate $X_j(t)$ can be incorporated by decomposing $\beta_j(t)$ into basis functions.

5.2.3 Regression Splines for Nonperiodic Trends

Spline smoothing is a statistical technique that approximates trends in data by piecewise polynomials. The goal of spline smoothing is to achieve a degree of global smoothness while, at the same time, allowing the fit to capture local features of the data.

Briefly speaking, a polynomial spline (or, simply, a "spline") is a piecewise polynomial function constructed over an interval $t \in [a, b]$ by partitioning it into n subintervals. Within any subinterval, the function is a simple polynomial of degree K, that is, a linear combination of $1, t, t^2, \ldots, t^K$, but the coefficients of these polynomials are constrained to make the first $K - 1$ derivatives agree wherever one subinterval adjoins another. The subintervals of $[a, b]$ are $[\xi_{\ell-1}, \xi_\ell]$, $\ell = 1, \ldots, n$, where

$$a = \xi_0 < \xi_1 < \cdots < \xi_n = b.$$

The points ξ_ℓ are called knots. The endpoints ξ_0 and ξ_n are known as boundary knots, whereas $\xi_1, \xi_2, \ldots, \xi_{n-1}$ are called interior knots. The degree of smoothness is determined by the choice of K and n and by the placement of the interior knots. Small values of n make the function smooth; larger n leads to greater precision for approximating local features.

More formally, a polynomial spline of degree K (also said to be of order $K+1$), which we denote by $s_K(t)$, is a function that satisfies the following three properties (Bojanov et al., 1993; Eubank, 1988; Greville, 1969):

1. $s_K(t)$ is a polynomial of degree K or less in any subinterval $[\xi_{\ell-1}, \xi_\ell]$;
2. $s_K(t)$ is a continuous function with $K-1$ continuous derivatives if $K > 0$; and
3. the Kth derivative of the spline is a piecewise constant or step function, with discontinuities at the knots $\xi_0, \ldots, \xi_n$.

From this definition, one can show (e.g., Eubank, 1988) that a polynomial spline with degree K and knots $\xi_0, \ldots, \xi_n$ can be written as

$$s_K(t) = \sum_{k=0}^{K} \theta_k t^k + \sum_{\ell=1}^{n-1} \delta_\ell (t - \xi_\ell)_+^K, \tag{5.4}$$

where θ_ℓ and δ_ℓ are real coefficients, and

$$(t - \xi_\ell)_+^K = \begin{cases} (t - \xi_\ell)^K, & t > \xi_\ell \\ 0, & t \leq \xi_\ell. \end{cases} \tag{5.5}$$

When written in this form, the spline is said to use the *truncated power basis*. The truncated power basis for a spline of degree K consists of the monomials $1, t, t^2, \ldots, t^K$, plus one additional function $(t - \xi_\ell)_+^K$ for each interior knot. Thus, the number of basis functions required is equal to $A = (K+1) + (n-1) = K + n$.

When $K = 0$, the spline is a step function, a set of horizontal line segments with jumps at the interior knots. When $K = 1$, the spline becomes a polygonal arc function, which is a set of straight line segments joined together at the knots. In that case the function itself is continuous but the first derivative is a step function. A quadratic spline ($K = 2$) has a continuous first derivative, and its second derivative is a step function. A cubic spline ($K = 3$), perhaps the most common choice in functional data analysis, is

$$s_K(t) = \theta_0 + \theta_1 t + \theta_2 t^2 + \theta_3 t^3 + \sum_{\ell=1}^{n} \delta_\ell (t - \xi_\ell)_+^3. \tag{5.6}$$

When fitting a spline to a longitudinal series for a single subject, the coefficients $\theta_0, \ldots, \theta_K$ and $\delta_1, \ldots, \delta_n$ can be estimated using multiple regression. With multiple subjects, the $(K+1) + (n-1)$ basis functions can be placed into the columns of the matrix X_i in the multilevel regression model (5.1).

When using splines, we encounter a tradeoff between fit and smoothness. Increasing the number of knots gives the spline greater flexibility to match the anomalies of the current data. More knots make the curves less smooth, however, which may diminish the validity and accuracy of prediction for future datasets. Because the number of coefficients to be estimated increases linearly with the number of knots, too many knots may also lead to numerical instability in the estimation of the coefficients. The subjectivity in choosing the number and position of knots is sometimes considered a drawback of regression splines. In practice, however, the placement of the knots is rarely arbitrary. For a longitudinal study with equally spaced measurements, one would typically spread the knots uniformly across the study interval. When measurements are not uniformly distributed in time, the knot positions are usually determined by quantiles of the measurement times, so that more knots are placed in the regions of higher measurement intensity.

5.2.4 The B-Spline Basis

Although the truncated power basis is the simplest way to describe splines, it is often not the best way to incorporate splines into a regression model. The reason is that these basis functions $(t - \xi_\ell)_+^K$ tend to be highly intercorrelated, especially when the number of knots is large, producing cross-product matrices that are nearly singular (Ramsay & Silverman, 1997). For computational purposes, a more practical and attractive alternative is the so-called B-spline basis.

B-spline basis functions can be algebraically defined as normalized divided differences (Eubank, 1988). To begin, consider a set of knots $\xi_0 < \xi_1 < \ldots < \xi_n$. The rth-order divided difference of a function f at ξ_ℓ is defined recursively as

$$[\xi_\ell, \ldots, \xi_{\ell+r}]f = \frac{[\xi_{\ell+1}, \ldots, \xi_{\ell+r}]f - [\xi_\ell, \ldots, \xi_{\ell+r-1}]f}{\xi_{\ell+r} - \xi_\ell},$$

where

$$[\xi_\ell]f = f(\xi_\ell).$$

For example, the first-order divided difference is

$$[\xi_\ell, \xi_{\ell+1}]f = \frac{f(\xi_{\ell+1}) - f(\xi_\ell)}{\xi_{\ell+1} - \xi_\ell},$$

a numerical approximation to the first derivative of f between ξ_ℓ and $\xi_{\ell+1}$. The second-order divided difference is

$$\begin{aligned}[\xi_\ell, \xi_{\ell+1}, \xi_{\ell+2}]f &= \frac{[\xi_{\ell+1}, \xi_{\ell+2}]f - [\xi_\ell, \xi_{\ell+1}]f}{\xi_{\ell+2} - \xi_\ell} \\ &= \frac{\dfrac{f(\xi_{\ell+2}) - f(\xi_{\ell+1})}{\xi_{\ell+2} - \xi_{\ell+1}} - \dfrac{f(\xi_{\ell+1}) - f(\xi_\ell)}{\xi_{\ell+1} - \xi_\ell}}{\xi_{\ell+2} - \xi_\ell},\end{aligned}$$

which approximates the second derivative. The rth normalized divided difference is defined as

$$(\xi_{\ell+r} - \xi_\ell)[\xi_\ell, \ldots, \xi_{\ell+r}]f = [\xi_{\ell+1}, \ldots, \xi_{\ell+r}]f - [\xi_\ell, \ldots, \xi_{\ell+r-1}]f. \tag{5.7}$$

An important property of divided differences is the following: If f is a polynomial of order r, then $[\xi_\ell, \ldots, \xi_{\ell+r}]f = 0$ (Eubank, 1988).

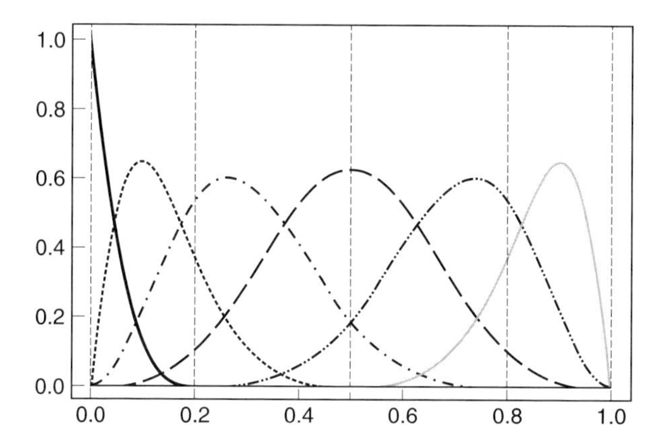

Figure 5.3 An example of a cubic B-spline basis with three interior knots.

A B-spline basis function can be generated by computing normalized divided differences with f's equal to the functions $(t - \xi_\ell)_+^K$ from the truncated power basis. For a precise statement and proof, refer to Eubank (1988). Using this result, it is a fairly easy matter to compute a B-spline basis of any degree with any set of knots. For the purposes of this chapter, the details of this computation are not important, because routines for generating B-spline bases are available in many statistical packages. In S-PLUS, for example, B-splines can be quickly generated by the function bs(). An attractive property of B-spline basis functions is that they have compact support, that is, they are nonzero over at most four subintervals. When used as regressors, the resulting cross-product matrices are banded and easy to invert (Hastie & Tibshirani, 1993). A plot of the seven B-spline basis functions for a cubic spline with three interior knots is shown in figure 5.3. As this plot shows, the individual B-spline basis functions and their respective coefficients have little or no substantive meaning; the most important summary is the smooth curve produced by the particular linear combination of B-splines that fits the data well.

5.2.5 Fourier Basis for Periodic Trends

Splines can effectively represent smooth long-term trends, but they are not an ideal tool for capturing periodic effects. A function f is said to be periodic with period P if $f(t) = f(t + P)$ for all t. Most periodic functions of interest can be represented by an infinite Fourier series, a weighted sum of sine–cosine pairs with increasing frequency,

$$f(t) = c_0 + c_1 \sin(\omega t) + c_2 \cos(\omega t) + c_3 \sin(2\omega t) + c_4 \cos(2\omega t) + \cdots . \qquad (5.8)$$

If a function f with known period P is smooth, it may be well approximated by the first few terms,

$$f(t) \approx c_0 + \sum_{m=1}^{M} c_{2m-1} \sin(m\omega t) + \sum_{m=1}^{M} c_{2m} \cos(m\omega t), \tag{5.9}$$

where $\omega = 2\pi/P$. The sum of the squared coefficients for the mth sine–cosine pair,

$$s_m = c_{2m-1}^2 + c_{2m},$$

is called the power in the sequence at period P/m or at frequency m. A larger number of basis functions (i.e., a higher value of M) is needed to approximate periodic functions with strong local features such as sharp turns or discontinuities.

The simplest kind of periodic function is a sinusoidal curve,

$$f(t) = c_0 + c_1 \sin(2\pi t/P),$$

which takes the baseline value c_0 at the origin ($t = 0$), rises to $c_0 + c_1$ at $t = P/4$, drops to $c_0 - c_1$ at $t = 3P/4$, and returns to baseline at $t = P$. If the data oscillate in a sinusoidal fashion, and if the time scale is arranged so that the phase happens to begin in the vicinity of $t = 0$ (or $t = P/2$), then the periodic effect may be well captured by including the single basis function $\sin(2\pi t/P)$ among the regressors. A phase that begins at an arbitrary known value t_0 may be handled by reexpressing time as $t' = t - t_0$. Alternatively, the shift parameter t_0 can be freely estimated by including a cosine term,

$$f(t) = c_0 + c_1 \sin(2\pi t/P) + c_2 \cos(2\pi t/P),$$

because the trigonometric identity

$$a \cos(\theta) + b \sin(\theta) = \sqrt{a^2 + b^2} \sin\left(\theta + \tan^{-1} \frac{a}{b}\right)$$

shows that a sine function with a phase shift can be written as a linear combination of sine and cosine functions.

5.2.6 Combining Periodic and Nonperiodic Trends

Periodic trends in longitudinal data are often accompanied by long-term nonperiodic variation. The periodic and nonperiodic effects may also interact—for example, the cyclic variation may gradually increase or decrease over time. A mean response with evolving periodic trends may be written as

$$\mu(t) = \alpha(t) + \sum_{m=1}^{M} \beta_m(t)\phi_m(t), \tag{5.10}$$

where $\phi_m(t)$ is equal to $\sin(2\pi t/P)$ if m is odd and $\cos(2\pi t/P)$ if m is even. (Notice that we have now redefined M to be the total number of sine and cosine functions rather than the number of sine–cosine pairs.) The function $\alpha(t)$, which captures the long-term nonperiodic trend, is decomposed as

$$\alpha(t) = \sum_{a=1}^{A} c_a B_a(t),$$

where the B_a's are B-spline basis functions. The long-term evolution in periodic trend is decomposed as

$$\beta_m(t) = \sum_{b=1}^{B} d_{mb} B_b(t).$$

The total number of coefficients to be estimated in the mean function (5.10) is $A + BM$. In any given application, we shall want to choose A, B, and M to be just large enough to capture the apparent signal in the data. Notice that the value of A or B may be changed either by changing the degree of the splines or by changing the number of knots. Each additional degree or each additional knot adds another term to A or B. Taking $B = 1$ and $B_1(t) = 1$ produces a stable periodic trend that does not change over time. An empirical strategy for choosing A, B, and M will be described in section 5.3.

In our decomposition (5.10), the estimated coefficients $\hat{c}_1, \ldots, \hat{c}_A$ and $\hat{d}_{11}, \ldots, \hat{d}_{MB}$ of the individual basis functions may be difficult to interpret. In functional data analysis, substantive meaning is usually attached not to the coefficients but to the function itself. In practice, one would usually interpet the model by plotting the estimated function,

$$\hat{\mu}(t) = \sum_{a=1}^{A} \hat{c}_a B_a(t) + \sum_{m=1}^{M} \sum_{b=1}^{B} \hat{d}_{mb} B_b(t),$$

over the domain of time represented by the study. Additional insight may be gleaned by separately plotting the periodic part $\sum_m \sum_b \hat{d}_{mb} B_b(t)$ and the nonperiodic part $\sum_a \hat{c}_a B_a(t)$.

5.2.7 Including Random Effects

The basis functions in (5.10) will define the fixed components $X_i\beta$ of our multilevel regression model (5.1). Thus far, however, we have said little about the random components $Z_i b_i$. If variation is seen in the mean levels of response across subjects, then it may be reasonable to include random intercepts, in which case the model becomes

$$y_{ij} = \alpha(t_{ij}) + \sum_{m=1}^{M} \beta_m(t_{ij}) \phi_m(t_{ij}) + b_i + e_{ij}, \tag{5.11}$$

where y_{ij} is the response score for subject i at time t_{ij}, and b_i is the random disturbance to the intercept for subject i. Note that this corresponds to a multilevel model (5.1) in which Z_i consists of a single column of ones.

In practice, the dependence among responses within an individual may be more elaborate than a random-intercepts structure. For example, individuals' trends over time may appear to depart from the average trend (5.10) in complicated ways. With a large number of individuals and many occasions per individual, it may be possible to allow some of the coefficients for the basis functions in $\alpha(t)$ or even $\beta_m(t)$ to vary by subject. A simpler and more parsimonious way to allow individual variation in trends is to allow autodependence among the residuals e_{ij} through an AR or MA process, as we described earlier in this section.

5.3 Application: Personality Data

5.3.1 Preparing the Data

Recall that in the personality study of Brown and Moskowitz (1998), 89 participants were asked to record details of their social interactions each day for a period of 20 consecutive days. On average, participants completed about six forms per day. In particular, an average of 2.1 forms were completed each morning (6 A.M. to noon), 2.8 each afternoon (noon to 6 P.M.), and 1.6 each evening (6 P.M. to 6 A.M.). Each form had a roughly equal number of items measuring the four personality dimensions (dominance, submissiveness, agreeableness, and quarrelsomeness). For our analysis of agreeableness, we computed daily raw mean scores, the average number of agreeableness items checked by each participant on each day. The responses for each participant were thus reduced to a single vector of length 20. As mentioned in section 5.1.1, the measure of time, which in this case corresponds to study day, was shifted to align the participants' curves with respect to day of the week. One who began on Tuesday would have day coded as $1, \ldots, 20$; one who began on Wednesday would have day coded as $2, \ldots, 21$; and so on. Coding Tuesday as day 1 was reasonable, because when we introduced a sine–cosine pair with a period of $P = 7$ as a Fourier basis (as described below), the coefficient for the cosine term was small and insignificant, suggesting that the weekly sinusoidal cycle essentially begins on Monday (day 0).

5.3.2 Model Selection

Multilevel models were fitted to the daily raw mean scores using the function lme() in the statistical package S-PLUS. B-spline bases were generated using the function bs(). Recall from section 5.2.6 that the total number of basis functions in the model is $A + BM$, where A is the number of B-spline basis functions describing the long-term trend, M is the number of Fourier basis functions describing the periodic trend, and B is the number of B-spline basis functions describing the long-term evolution of the periodic trend. With day now coded as $1, 2, \ldots, 26$, these data can support a maximum total of $A + BM = 26$ basis functions before the matrix of stacked regressors X_i becomes rank-deficient. We decided to limit our attention to models with $A + BM \leq 10$. This choice, which is admittedly somewhat arbitrary, represents our attempt to prevent overfitting. A random intercept was included in each model to capture the intersubject variation in the mean agreeableness scores. Attempts to include additional random effects (i.e., random slopes) to allow the long-term trends to vary by subject did not improve the model fit. Autoregressive or moving-average error structures for the e_{ij}'s did not improve the fit either, so the covariance matrix for e_i was taken to be σ^2 times the identity matrix.

The simplest model that we considered had a linear long-term trend with no knots ($A = 2$), one sine function with a period of $P = 7$ for the Fourier basis ($M = 1$), and a constant term indicating no change in the periodic trend over time ($B = 1$). In this model, the coefficient for the long-term trend was highly significant ($p < 0.0001$) and the coefficient for the sine term was also highly significant ($p < 0.005$). Starting from this model, we tried all possible combinations of $A \geq 2$, $B \geq 1$, and $M = 1, 2$ such that $A + BM = 4, 5, \ldots, 10$ with 0, 1, or 2 equally spaced interior knots. For example, as we increased A from 2 to 3, we attempted to represent the long-term trend by a quadratic function with no knots and by a piecewise linear function with one knot at day $(26 - 1)/2$. In each of these models, we found no significant long-term effects beyond the simple linear trend. The cosine term in the Fourier basis was not significant, and no long-term changes in the periodic effect could be found. The simplest model was thus taken as the final model. Using the notation of equation (5.3), the population-average mean response at day t is

$$\mu(t) = \beta_0 + \beta_1(t) + \beta_2 \sin(2\pi t/7).$$

5.3.3 Results

The estimated coefficient for the long-term linear trend was negative, indicating a gradual decline in the mean agreeableness over the study period. The presence of a sine term without a cosine indicates sinusoidal fluctuation with a phase beginning at Monday ($t \approx 0$) or late Thursday ($t \approx 3.5$). A plot of the fitted trend over the entire study period is shown in figure 5.4. To make the long-term trend apparent, this plot also has a dashed line showing the fit with the long-term linear trend removed. The weekly cycle achieves its minimum predicted value on Wednesday ($t \approx 1.75$) and rises to its maximum on Saturday ($t \approx 5.25$) before declining again.

The declining linear trend, which we have dubbed "long-term" because it spans the entire study period, represents only a tiny fraction of any participant's lifespan. Because this was a simple observational study with no intervention, we are inclined to believe that the downward trend does not represent any natural development but merely a reaction to the measurement process itself; more discussion of reactivity appears in the introduction to this volume. We shall not attempt to explain or interpret the periodic effect. We do note, however, that this is rather

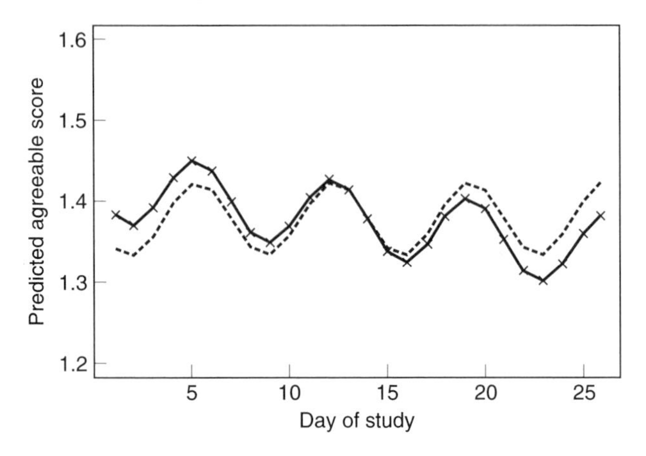

Figure 5.4 Predicted agreeable level by day of the study. The dotted line represents the periodic model obtained by a constant and a sine term with no long-term trend.

consistent with the findings of Brown and Moskowitz (1998), who also reported a weekly cycle.

As with any application of multilevel modeling, it is important to check the adequacy of the distributional assumptions by examining relevant diagnostic plots. Residual plots from this model (not shown) did not reveal any heteroscedasticity or serious departures from normality. Residual plots for multilevel models are discussed at some length by Walls et al. (chapter 1, this volume).

5.4 Discussion

In this chapter, we have shown how a set of carefully chosen basis functions may capture both regular periodic effects and irregular long-term trends in a mean response over time. The existing literature on varying-coefficient models tends to emphasize modeling of a single series. As we have shown, however, series for multiple subjects are easily handled by treating the basis functions as time-varying covariates in a multilevel model. With many occasions per subject, regression splines tend to fit long-term trends better than ordinary time-graded polynomials, and interacting a spline basis with a Fourier basis allows periodic effects to gradually evolve. As an alternative to regression splines, Li et al. (chapter 3, this volume) fit smooth curves to intensive longitudinal data by local polynomial regression. Local polynomial regression does an excellent job of fitting irregular curves, but it is not an ideal tool for describing regular periodic effects.

Because we focused attention on the population-average mean structure, similar results could have been obtained by fitting a marginal mean model using generalized estimating equations (GEE) (Schafer, chapter 2, this volume).

In fact, a GEE analysis with the same regressors and an exchangeable or compound symmetry working covariance assumption would have led to nearly identical estimates. An advantage of GEE is that it can handle nonnormal continuous outcomes and discrete outcomes such as binary indicators or frequencies. One of its disadvantages is that a larger sample (i.e., more subjects) may be required for the sandwich-based standard errors to be appropriate. Alternatively, binary or count data could be described parametrically by a generalized linear mixed model (GLMM); a brief overview of GLMMs is given by Hedeker et al. (chapter 4, this volume).

Throughout this chapter, we have assumed that the base period P of the cyclic trend is known and identical for all subjects. In many studies involving human participants, the hypothesized period will be some obvious natural unit of time (day, week, or year). If the period were unknown, one could apply a sine–cosine pair with different periods and look for a value of P for which the power (i.e., the sum of the two squared coefficients) is high. Oscillatory behavior that varies by subject can be described by the dynamical systems models (Boker & Laurenceau, chapter 9, this volume).

References

Bojanov, B.D., Hakopian, H.A., & Sahakian, A.A. (1993). *Spline Functions and Multivariate Interpolations*. London: Kluwer.

Brown, K.W., & Moskowitz, D.S. (1998). Dynamic stability of behavior: The rhythms of our interpersonal lives. *Journal of Personality, 66*, 105–134.

Eubank, R.L. (1988). *Spline Smoothing and Nonparametric Regression*. New York: Marcel Dekker.

Greville, T.N.E. (1968). *Theory and Applications of Spline Functions*. New York: Academic Press.

Hastie, T., & Tibshirani, R. (1993). Varying-coefficient models. *Journal of the Royal Statistical Society, Series B, 55*, 757–796.

Kiesler, D.J. (1983). The 1982 interpersonal circle: A taxonomy for complementarity in human transactions. *Psychological Review, 90*, 185–214.

Laird, N.M., & Ware, J.H. (1982). Random-effects models for longitudinal data. *Biometrics, 38*, 963–974.

Leary, T. (1957). *Interpersonal Diagnosis of Personality*. New York: Ronald Press.

Pinheiro, J.C., & Bates, M.D. (2000). *Mixed-Effects Models in S and S-PLUS*. New York: Springer.

Ramsay, J.O. & Silverman, B.W. (1997). *Functional Data Analysis*. New York: Springer.

Wallace, D., & Green, S.B. (2002). Analysis of repeated measures designs with linear mixed models. In D.S. Moskowitz & S.L. Hershberger (Eds.), *Modeling Intraindividual Variability with Repeated Measures Data: Methods and Applications*. Mahwah, NJ: Lawrence Erlbaum.

Wiggins, J.S. (1979). A psychological taxonomy of trait-descriptive terms: The interpersonal domain. *Journal of Personality and Social Psychology, 37*, 395–412.

Wiggins, J.S. (1982). Circumplex models of interpersonal behavior in clinical psychology. In P.C. Kendall & J. N. Butcher (Eds.), *Handbook of Research Methods in Clinical Psychology* (pp. 183–221). New York: John Wiley.

6

Multilevel Autoregressive Modeling of Interindividual Differences in the Stability of a Process

Michael J. Rovine and Theodore A. Walls

Most psychological phenomena have been studied through the consideration of data collected at relatively few occasions of measurement over time. However, a recent wave of diary-based studies has produced many databases with more intensively collected longitudinal data. The number of occasions can be larger than those collected in a typical panel study, but less than the very high numbers of occasions typically found in the time-series domain (Brockwell & Davis, 1991; Shumway & Stoffer, 2000). The same kinds of questions that could be answered based on a fewer number of occasions could also be proposed for these longer time series. Traditionally, scientific questions for which longitudinal data have been collected have been concerned with stability and change (Baltes et al., 1988; Belsky & Rovine, 1990; Belsky et al., 1983; Brim & Kagan, 1980).

Questions related to stability can be framed as follows: to what extent can we expect to predict the next occasion from the current occasion? In panel studies, stability has been typically assessed separately for each pair of adjacent occasions either using correlation or regression coefficients. Models of this type are in the class of lagged regression panel models suggested by Kenny (1973), Kenny and Campbell (1989), and Shingles (1985). In this model, the value of a variable at each occasion is treated as a separate dependent variable in a regression model, with the previous value(s) and possibly some additional covariates acting as predictors. This yields a set of coefficients which can be compared to determine whether the stability between adjacent occasions changes across the series. In terms of regression models, stability on a variable is high when individuals tend to occupy the same location in the distribution at different occasions. With many occasions of data, this method yields a large number of coefficients and, as a result, becomes quite unwieldy.

Questions related to change can be framed as follows: to what extent can we identify discernible patterns in the means across all occasions of measurement?

In panel studies, mean level change is typically assessed using repeated measures ANOVA or growth curve models.

With more measurement occasions, we can still ask the following questions. Does the series have an overall increase? Does the series follow some known developmental function? We can still approach these questions using a repeated measures or growth curve approach in which one looks for increasing or decreasing trends in the data (Cook & Campbell, 1979). However, to answer questions related to stability in these longer series, we shall suggest a time-series approach based on an autoregressive model (Box & Jenkins, 1976) which will allow us to define a single prediction equation that we shall apply across the complete range of the series. What we refer to as stability in a panel study will be indicated by the autoregressive relationships that describe the process. These relationships reflect regularities in the sequential dependences across the length of the series (Wohlwill, 1973). Questions related to the existence of these regularities could include: is the behavior today a good indicator of the behavior tomorrow? Or, more genenerally, is there some set of previous occasions (i.e., lags) that can predict future behavior across the whole length of the series?

In this chapter, we propose a multilevel time-series approach to model individual differences in the regularities of behaviors. Because this approach is not part of the standard multilevel modeling repertoire, we shall first show how this model represents a simple extension of more typical approaches such as repeated measures ANOVA and growth curve modeling. Time-series models are typically used for a single series. Here, we shall use time-series parameters estimated separately for each individual as an index of interindividual variability in the regularity of a process. Under certain strict assumptions, this multilevel model can be described as an instance of the linear mixed model (Laird & Ware, 1982). As such, we shall next show how to use a simple autoregressive model (AR(1)) to define the first level of a multilevel-type model. We shall then show how to express this model both as a hierarchical linear or multilevel model (Bryk & Raudenbush, 1992; Goldstein, 1995, 2003) and as a linear mixed model (Laird & Ware, 1982). We shall then implement the linear mixed model (LMM) form of the equations in SAS PROC MIXED (Littell et al., 1996; SAS Institute Inc., 1995). With this approach, we can use other variables (e.g., characteristics of the individual) to explain variability in these random coefficients. We shall give a practical example of this method using data from a daily diary study containing 60 nightly reports of alcohol consumption in ounces as reported by 93 heavy drinkers (Armeli et al., 2000; Carney et al., 2000).

6.1 Defining Stability as Regularity in a Time Series

In a panel design, stability has been represented by the degree of correlation between adjacent occasions of measurement. Looking at these coefficients, one could assess the degree of stability for a particular pair of occasions or determine whether the degree of stability changes across all occasions. Using a structural equations approach (Bollen, 1989), the equivalence of the stability coefficients could be tested.

For more intensively collected data, the large number of stability coefficients may not be as useful as a more parsimonious set of coefficients that can adequately describe the complete series. Wohlwill (1973) used the term *regularity* to refer to any lawfulness in the moment-to-moment measurements. Specifically, he proposed that stable "regularity of occurrence" is achieved when measurement is available at each occasion, and unstable when some occurrences are missing. By extension, lawful characteristics revealed by analyses of the nature of measurements over many occurrences provide richer description of regularity of occurrence; the goal of time-series analysis is to provide such a rich description (Gottman, 1981; Shumway & Stoffer, 2000). By contrast, Wohlwill also described "regularity of form of change," that is, when measurements are stable a smooth developmental function results and when unstable this function takes on a complex shape. Many chapters in this volume excel at characterizing this kind of regularity, in particular, Li et al. (chapter 3) and Fok and Ramsay (chapter 5).

These regularities represent systematic relationships among the observations in the series that serve to describe the complete series. In terms of stability coefficients, we can think of the simplest regularity, the first-order autoregressive ((AR(1)) coefficient, as equivalent to a model in which all of the stability coefficients between adjacent occasions are required to be equal. Under this assumption, we can estimate a single coefficient that functions as a stability between any two occasions in the series.

Although we shall consider mainly simple autoregressive models (AR(1) and AR(2)) in this chapter, we shall describe the more general autoregressive model of order, p (AR(p)). The multilevel AR(p) model that we propose is a stability model. It does not describe the stability between a single pair of adjacent occasions; rather the AR(p) coefficients represent a parsimonious set of coefficients that summarize the regularity of the complete range of the series. To properly estimate the AR(p) coefficients, we need to first identify any trends in the series and remove them. Trends in the series would induce an additional autocorrelation that would be confounded with the regularities of interest. We could first model the trends using, for example, a polynomial function and then extract the residuals. We would then use the autocorrelation function of the resulting residuals to estimate the AR(p) coefficients. If we were not explicitly interested in modeling the trends, we could detrend the series by differencing (Chatfield, 1996).

6.2 Multilevel Models

For correlated data based on longitudinal replications of a repeated measures factor, models based on the linear mixed model (Laird & Ware, 1982) have become more prevalent in the social sciences. Software programs for different implementations of this general model include SAS PROC MIXED (SAS Institute Inc., 1995) and many others enumerated in Walls et al. (chapter 1, this volume). Two conceptually different ways of formulating instances of the model are available and related to specific software. In programs like SAS PROC MIXED, a single model equation can be written that includes fixed and random effects. These effects are then separated and identified in different parts of the program. In programs like HLM a separate model is written for each level of the analysis. These two different conceptualizations are interchangeable (Singer, 1998; Walls et al., chapter 1, this volume). Both general conceptualizations allow for the estimation of statistical models for correlated data that result from both longitudinal data collection and from clustering due to nested research designs.

A nested research design could result from data collected in multiple classrooms in different schools. As we would expect that students in the same classroom may be more alike than students from different classrooms, we would model those data as correlated. In this model, students would be nested in classrooms which would, in turn, be nested within schools. If we were interested in specifically testing mean differences between schools, school would be treated as a fixed effect. If we were interested in, for example, testing a treatment/control difference in a study taking place in a number of different schools, we could treat school as an additional random effect.

For correlated longitudinal data we can imagine time as a factor with individuals measured on multiple occasions. With a single repeatedly measured dependent variable and time treated as a categorical predictor, we would be estimating a repeated measures ANOVA model. With time as a continuous predictor, we would be estimating an individual curve model. In this chapter, we shall consider more intensively measured data with time as a discrete factor.

6.2.1 The Linear Mixed Model

The linear mixed model (LMM) proposed by Laird and Ware (1982) and based on the work of Harville (1974, 1976, 1977) is

$$y_i = X_i\beta + Z_i\gamma_i + \varepsilon_i, \tag{6.1}$$

where, for the *ith* case, y_i is an $n_i \times 1$ vector of response values, X_i is an $n_i \times b$ design matrix for the fixed effects, Z_i is an $n_i \times g$ design matrix for the random effects, γ_i is a $g \times 1$ vector of random effects, and β is a $b \times 1$ vector of fixed effect parameters. The coefficients $\beta = (\beta_1, \ldots, \beta_b)$ are common to all individuals, while the $\gamma_i = (\gamma_{i1}, \ldots, \gamma_{ig})$ vary from individual to individual. The γ_i are assumed to be independently distributed across subjects with a distribution $\gamma_i \sim N(0, \sigma^2 D)$, where D is an arbitrary "between" covariance matrix, and the ε_i, the within-subject errors, are assumed to have the distribution $\varepsilon_i \sim N(0, \sigma_\varepsilon^2 W_i)$, where W_i is a weight matrix. This model allows us to estimate a wide variety of submodels including repeated measures analysis of variance and growth curve models. Jennrich and Schlucter (1986) described a number of different models based on equation (6.1). While many software products can estimate these models, certain of these products represent a more direct implementation of LMM, including SAS PROC MIXED.

For longitudinal models in which time is a repeated measures factor, the LMM can be used to estimate different models depending on the level of measurement of time. When we test a hypothesis related to whether the means of some dependent variable differ from occasion to occasion, we treat time as a categorical variable and typically perform a repeated measures ANOVA. In terms of the LMM, we drop the $Z_i\gamma_i$ and the model becomes

$$y_i = X_i\beta + \varepsilon_i. \tag{6.2}$$

y_i represents a vector of the occasion values; ε_i represents a vector of occasion errors which can be patterned to reflect a hypothetical error structure (Jennrich & Schlucter, 1986; Rovine & Molenaar, 1998).

When we treat time as a continuous regression-type predictor, we estimate what is typically called a growth curve model (Goldstein, 1987; Grizzle & Allen, 1969). For this model, the term $Z_i\gamma_i$ is added back into the model and represents the individual intercepts and slopes that we could imagine are fitted through the individual set of repeatedly measured observations. For this model, then, β represents the fixed effects estimate of the population model, which here would be the population growth model, and γ_i represents random effects, indicating the degree to which the individual's curve deviates from the group curve. The covariance matrix of these random effects is estimated and the solution is used to predict the individual growth curves (Robinson, 1991).

6.2.2 A Common Conceptualization

Each of the above models represents a prototypical use of the LMM (Rovine & Molenaar, 1998). There is a conceptual way to tie them together. In each case, new dependent variables are created either explicitly (as in the case of the repeated measures ANOVA) or implicitly (as in the case of the growth curve analysis).

Consider a 2 group $\times$ 3 time repeated measures ANOVA where group is a between effect and the variables y_1, y_2, and y_3 represent the three repeated measures. We can perform an analysis equivalent to the repeated measures ANOVA by first transforming the dependent variables using orthogonal polynomial coding coefficents (Myers, 1979) as follows:

$$\begin{aligned} dep_{\text{linear}} &= (-1 * y_1) + (0 * y_2) + (1 * y_3), \\ dep_{\text{quadratic}} &= (1 * y_1) - (2 * y_2) + (1 * y_3), \\ dep_{\text{average}} &= ((1 * y_1) + (1 * y_2) + (1 * y_3))/3, \end{aligned} \tag{6.3}$$

and then treat each of these variables as a dependent variable in a one-way ANOVA with group as the factor. In this manner, we have explicitly created dependent variable values that capture characteristics of change. In fact, we have decomposed the time effect into its linear and quadratic components (McCall & Appelbaum, 1973). We can imagine these variables being "passed along" to a second set of equations where they become the dependent variables predicted by, in this case, the between-group variable. This results in the set of regressions:

$$\begin{aligned} y_{\text{linear}} &= \beta_{0\text{linear}} + \beta_{1\text{linear}} group + \varepsilon_{\text{linear}}, \\ y_{\text{quadratic}} &= \beta_{0\text{quadratic}} + \beta_{1\text{quadratic}} group + \varepsilon_{\text{quadratic}}, \\ y_{\text{average}} &= \beta_{0\text{average}} + \beta_{1\text{average}} group + \varepsilon_{\text{average}}. \end{aligned} \tag{6.4}$$

$\beta_{1\text{average}}$ tests the between-*group* effect. $\beta_{0\text{linear}}$ and $\beta_{0\text{quadratic}}$ test linear and quadratic contrasts on the main effect time. $\beta_{1\text{linear}}$ and $\beta_{1\text{quadratic}}$ test the $time_{\text{linear}} * group$ and the $time_{\text{quadratic}} * group$ interaction effects (Hertzog & Rovine, 1985). The sums of squares related to the first two regression equations can be pooled to estimate the sums of squares related to the *time* and *time* $*$ *group* effects and the corresponding error sums of squares.

Consider the same design except that now we treat time as an interval-level variable. This is a simple example of a two-group growth curve model which we model with a random level and slope. In this model, we write the equation for individual i as:

$$y_{ti} = \pi_{0i} + \pi_{1i} time_{ti} + r_{ti}, \qquad r_{ti} \sim N(0, \sigma^2), \tag{6.5}$$

where π_{0i} and π_{1i} are the intercept and slope for subject i, and r_{ti} is the residual around the individual line at occasion t for subject i. The π_{0i} and π_{1i} are then "passed along" to the set of regression equations where they become the dependent variables:

$$\begin{aligned} \pi_{0i} &= \beta_{00} + \beta_{01} group + u_{0i}, \\ \pi_{1i} &= \beta_{10} + \beta_{11} group + u_{1i}, \end{aligned} \tag{6.6}$$

where

$$\begin{pmatrix} u_{0i} \\ u_{1i} \end{pmatrix} \sim N\left(\begin{pmatrix} 0 \\ 0 \end{pmatrix}, \begin{pmatrix} \tau_{00} & \tau_{01} \\ \tau_{10} & \tau_{11} \end{pmatrix} \right) \tag{6.7}$$

indicates that the covariance matrix of the random effects is normally distributed with zero mean vector and an unstructured covariance matrix.[1]

We see that for both the repeated measures ANOVA and the growth curve model, there is a *level 1* where the dependent variable capturing some characteristic of change is defined, and a *level 2* where that characteristic is treated as a dependent variable in a regression analysis. This conceptualization generalizes to more complex models.

For the case of more intensively collected longitudinal data, implementation of either of these models without accounting for diverse patterns of change over time in new and appropriate ways leads to conceptual and computational challenges (Walls et al., chapter 1, this volume). In addition, these models consider characteristics of means. Consistently, if the degree to which the regularities related to occasion-to-occasion stability is of interest, these models will not address the proper questions.

In this chapter, we suggest a new regression-based model as a way of creating the dependent variable that does address the notion of stability. This variable will be based on a first-order autoregressive (AR(1)) model.

6.2.3 The AR(p) Model

The AR(p) model, a submodel of the more general autoregressive moving average (ARMA) model (Box & Jenkins, 1976), is typically used to describe sequential dependences with the minimum number of parameters. The model is often applied to a single set of repeated measures obtained on a single subject and, as a result, describes within-subject variation.

The model can be written as

$$y_t = \beta_{t,t-1} y_{t-1} + \beta_{t,t-2} y_{t-2} + \cdots + \beta_{t,t-p} y_{t-p} + Z_t, \tag{6.8}$$

where the autoregressive relationships are modeled through the p structural regression coefficients ($\beta_{t,t-k}$), which represent the coefficient relating two occasions separated by lag k. Z_t represents an uncorrelated innovation. The order of the process, p, is often determined heuristically. One of the requirements for estimating the parameters is that the process must be stationary. Broadly defined, a stochastic process is weakly stationary if there is no systematic change in the mean across the length of the series and if there is no systematic change in variance across the length of the series. For the models we shall consider, the sufficient assumption of weak stationarity is adopted, formally defined as:

$$\begin{aligned} E[Y_t] &= \mu, \\ Cov([Y_t, Y_{t+\tau}]) &= \gamma(\tau). \end{aligned} \tag{6.9}$$

As such, the mean of the series, μ, will be constant, and the autocovariance function, γ, will depend only on the lag, t. In this case a regression weight (e.g., $\beta_{t,t-1}$) depends only on the lag and is, thus, identical regardless of t. We can think of the $\beta_{t,t-1}$ coefficient as describing how well the next observation is predicted by current observation. Under weak stationarity this prediction holds for any observation. Weak stationarity is critical to the interpretation of the time series. Since a trend in the data will dominate the autocovariance function, the series must be detrended before the parameters of the series can be properly estimated.

While AR(p) and other ARMA models are most typically used to estimate the parameters of a single time series, Molenaar's (1985; Wood & Brown, 1994) dynamic factor model marked an innovative shift in the field toward the proper use of time-series models for the study of intraindividual differences in behavior and development (Gregson, 1983; Nesselroade & Molenaar, 2003). The question of how these single subject methods can be applied to multiple individuals has been addressed for small samples (Nesselroade & Molenaar, 1999). Structural equation modeling approaches to ARMA modeling have been particularly fruitful (Molenaar, 1999; van Buuren, 1997), and have resulted in multivarate

ARMA models (DuToit & Browne, 2001) and multiple indicator ARMA models (Sivo, 2001). For multiple individuals, these methods have been applied to univariate and multivariate time series (Hamaker et al., 2003, 2005).

6.3 A Multilevel AR(1) Model

In this chapter, we shall model a process that we hypothesize is described by the simplest possible autoregressive model, the AR(1). As a result, we shall not consider the problem of estimating the order of the process based on each individual time series.

The estimation of the coefficients for each individual time series is the *level 1* of our model. In the multilevel formulation, we shall implicitly estimate the individual AR(1) coefficients and pass them along to a next step where they become the values of a dependent variable in a regression or ANOVA model.

To show that the individual parameter estimates of this multilevel approach yield results equivalent to those we would obtain from a more standard time-series approach, we shall compare the coefficients predicted by the multilevel approach with the ordinary least squares (OLS) results obtained by estimating each individual's model separately.

To estimate these models, we shall need to create the appropriate dataset based on the original time series. For the AR(p) model, we must lag the dependent variable, y_{ti}, which is indexed by t representing the occasion of measurement. We can add the lagged variable by creating a second variable, $y_{t-1,i}$, which includes the value of the same variable on the previous occasion. We could add the second lag by creating the variable $y_{t-2,i}$ in the same fashion. Table 6.1 shows a small dataset in which a variable and its first- and second-order lags are included.

Table 6.1 Raw score and first two lagged variables

Raw score	Lag 1	Lag 2
0.40	–	–
0.50	0.40	–
0.50	0.50	0.40
0.00	0.50	0.50
0.00	0.00	0.50
0.36	0.00	0.00
0.00	0.36	0.00
0.00	0.00	0.36
1.01	0.00	0.00
1.20	1.01	0.00
...	...	...

Typically, when one is interested in determining the order of the process, one sets the lag window, the number of lags considered necessary to be able to include all terms necessary to identify the process, at some number larger than the maximum number of variables that will eventually be included. If we specify the order of the process as either AR(1), we include the first lag. If the order of the process is AR(2), we include the first and second lags of the variable.

For the comparison analysis, we shall analyze each separate time series using OLS regression. Such procedures are described in any text on time series (cf. Brockwell & Davis 1991; Shumway & Stoffer, 2000). The results of this analysis will include a set of parameter values that will describe the interindividual variability in the regularity of a process over the range of a moderate span dataset.

6.3.1 OLS Estimates of the Individual Series

The pattern of correlations represented by the autocorrelation function of a non-stationary series has contributions from at least two sources. For a stationary series, the autocorrelation function indicates which lagged variables are sufficient to predict the current value of the dependent variable. The addition of a trend induces additional correlations among the lagged variables. With a trend in the time series, these two sources are confounded. For purposes of both estimation and interpretation, it is necessary to separate them, typically by first removing the trend, and then estimating the order of the process.

Assuming the condition that each series has been detrended, we estimate the AR(1) model for each series as

$$y_{ti} = \beta_{0i} + \beta_{t,t-1,i} y_{t-1,i} + \varepsilon_{ti}, \tag{6.10}$$

where $\beta_{t,t-1,i}$ is the first-order autoregression coefficient for subject i, and ε_{ti} is residual for subject i at time t. In time-series modeling the residuals are often referred to as innovations. The variance of the innovations for subject i is $\sigma^2_{\varepsilon i}$, and is assumed equal for each subject. We include the intercept β_{0i} to allow for an uncentered series.

The equation used to account for the variability in the $\beta_{t,t-1,i}$ coefficients is

$$\beta_{t,t-1,i} = \gamma_0 + \gamma_1 x_1 + \gamma_2 x_2 + \cdots + \gamma_k x_k + u_{t,t-1,i}, \tag{6.11}$$

where the γ_k represent the fixed effect regression coefficient, and $u_{t,t-1,i}$ represents the equation residual with $u_{t,t-1,i} \sim N(0, \sigma^2_u)$.

These estimates are, of course, OLS. The problems with OLS estimates have been well documented (Kreft & de Leeuw, 1998; Snijders & Bosker, 1999). We use them here to compare with the estimates to be obtained through the multilevel formulation of the same model.

6.4 Application: Daily Alcohol Use

To exemplify the use of AR(p) coefficients in a slopes-as-outcomes type analysis, we reviewed data from Armeli and colleagues (2000), which considered daily alcohol drinking behavior in adults who were moderate to heavy social drinkers. The sample was gender-balanced and was composed of adults working in administrative or professional careers at diverse low–moderate to high income levels. The participants were trained to report their daily consumption of alcohol in ounces, stress levels, desire to drink, and related psychosocial states each evening for 60 days. The question we are interested in regards the regularity of alcohol amounts consumed by participants in this study from day to day. In particular, is the next day's amount consumed predicted by the prior day's consumption?

The model that we assume to account for the process is an AR(1) model. We first estimated the AR(1) coefficients separately for each subject. The model estimated is

$$y_{ti} = \beta_{0i} + \beta_{1i}\, y_{t-1,i} + r_{ti}, \tag{6.12}$$

where β_{0i} represents the intercept of the uncentered series for subject i, and β_{1i} represents the AR(1) regression slope. We include in table 6.2 the regression and intercept coefficients for the individual series for the first ten cases. The AR(1) coefficients are variable ranging from −0.35 to 0.55. Summary statistics including the mean vector and covariance matrix of the coefficients are reported in table 6.3. Note that the mean level (intercept) of drinking is 1.65 ounces per day, just over the NIAAA recommended level for adult daily consumption of 1.5 ounces for men.

The AR coefficients reflect a range of magnitudes, indicating a heterogeneous and generally moderate strength of the relationship among consecutive measurements (see Velicer and Fava, 2003, or Chatfield, 1996, for interpretational

Table 6.2 ARIMA-based individual μ and AR(1) parameters

Intercept	AR(1) estimate
0.17	0.00
0.77	0.02
0.51	−0.04
0.80	0.17
0.98	0.17
1.75	−0.21
0.97	0.14
1.16	−0.04
1.18	0.39
1.41	0.14
...	...

Table 6.3 Means and covariance of the ARIMA estimates

	Intercept	AR(1) coefficient
Intercept	0.54	0.01
AR(1) estimate	0.01	0.03
Mean	1.65	0.10

guidelines). This means that the prior day of drinking does a pretty good job of predicting for the next day of drinking. In terms of direction, only 20 (21%) of the AR coefficients were negative, indicating that for most subjects, drinking on a given day predicted for a higher level of drinking the next day. When the AR coefficient was negative, magnitudes were generally low, indicating that only a few subjects drank notably less the next day.

Both the reasons for the responses on the outcome variable and the explanation of the variability are most likely more complex than the simple model we have selected. For example, contextual reasons may exist for a particular amount of alcohol consumption. These contextual reasons could come as concomitantly measured time series (e.g., mood) or as time-invariant characteristics (e.g., social network or socioeconomic status). The initial model could easily be expanded to include either of these contingencies. For the concomitant time series, we would consider the cross-covariance function (Box & Jenkins, 1976) as the basis for the regression coefficients. As before, we would not determine the order of the process for each series, but hypothesize an order and then describe the regression equation. For the time-invariant factors, we would add them as additional regression predictors, as we demonstrate later in the chapter. The regression coefficients for all of these effects would be passed along to the set of equations which would attempt to account for the variability in these coefficients.

6.5 Estimating This Model in SAS PROC MIXED

To take advantage of the estimation properties of the linear mixed (multilevel) model (Laird & Ware, 1982), we now consider how to estimate the AR(1) model using SAS PROC MIXED. For these models it will be easiest to determine the correct expression of the model by beginning with a level 1/level 2 formulation (Bryk & Raudenbush, 1992; Goldstein, 1985) and then substituting to create a single equation (Singer, 1998). We begin with a model designed to test the variability in the random slopes related to the AR(1) model. *Model 1* can be written as:

Level 1:

$$y_{ti} = \pi_{0i} + \pi_{1i} y_{t-1,i} + r_{ti}, \qquad r_{ti} \sim N(0, \sigma^2). \tag{6.13}$$

Level 2:

$$\pi_{0i} = \beta_{00} + u_{0i},$$
$$\pi_{1i} = \beta_{10} + u_{1i} \tag{6.14}$$

where

$$\begin{pmatrix} u_{0i} \\ u_{1i} \end{pmatrix} \sim N\left(\begin{pmatrix} 0 \\ 0 \end{pmatrix}, \begin{pmatrix} \tau_{00} & \tau_{01} \\ \tau_{10} & \tau_{11} \end{pmatrix}\right). \tag{6.15}$$

For the level 1 model, the residuals, r_{ti}, are assumed independent and normally distributed. Since the level 1 model is AR(1), we can justify this assumption if (i) the series has been detrended, and (ii) the correct order of process is, indeed, AR(1). The autocorrelation function (ACF) functions as a good diagnostic of the residuals for each case. It allows us to determine which series are adequately modeled by the specified process. If (i) and (ii) hold, then the residuals of the series would be a white noise process. Recognizing that the level 1 in this model can be pictured as a $y_t \times y_{t-1}$ scatterplot, we could check the distribution of the residuals for each case. Unlike the $y \times$ time individual scatterplot of a growth curve model based on just a few occasions of measurement, we have here a considerable number of residuals for each case and can take a more considered look at the assumptions of the model.

As mentioned previously, the individual series must be detrended to avoid a confounding autocorrelation. The series does not have to be demeaned if the individual intercepts are estimated as part of the model. Since we estimate the AR(1) model by regressing $y(t)$ onto $y(t-1)$, we center the predictor variable, $y(t-1)$, separately for each case. We do not center the dependent variable, $y(t)$. This results in an intercept estimating the mean of the series. This also "lines up" the different series and makes the intercepts for the different series comparable. Using this case-by-case centering, the intercepts for the individual series will match the estimates obtained using a dedicated ARIMA modeling program such as SAS PROC ARIMA.

It is important to note that the model we estimate differs from the SAS PROC MIXED (AR(1)) option. SAS PROC MIXED allows one to structure the covariance matrix of the residuals in a number of different ways, including some autoregressive structures (AR(1), AR(2), AR(p)). The selected structure is used to create a block diagonal matrix in which the parameter values are identical for each block (which is equivalent to a case). As a result, this autoregressive structure would only yield results comparable to the models considered in this chapter when the parameter values are identical for each case. The approach we present here allows the values of the AR coefficients to vary from case to case.

To implement this model in SAS PROC MIXED, we first substitute level 2 into level 1:

$$y_{ti} = [\beta_{00} + u_{0i}] + [\beta_{10} + u_{1i}]y_{t-1,i} + r_{ti} \tag{6.16}$$

Table 6.4 GLMM means and covariance

	Intercept	Slope
Intercept	0.52	0.01
Slope	0.01	0.01
Mean	1.65	0.08

and rearrange terms:

$$y_{ti} = [\beta_{00} + \beta_{10}\, y_{t-1,i}] + [u_{0i} + u_{1i}\, y_{t-1,i}] + r_{ti}, \tag{6.17}$$

where the first bracket represents the fixed effects of the model and the second bracket represents a random intercept and slope related to the AR(1) coefficient. β_{00} and β_{10} test the average intercept and slope for the AR(1) model. The random coefficients, u_{0i} and u_{1i}, represent the individual deviations around the respective means for the intercepts and slopes. To line up the individual random intercepts, we center the lagged variable which appears along the abcissa of the $y_t \times y_{t-1}$ scatterplot. The variability of the u_{1i} thus indicates the variability of the individual AR(1) regression coefficients.

As we can see, the fixed effect regression coefficient related to the AR(1) is almost identical to the average of the individual separately estimated AR(1) coefficients. One advantage of the multilevel formulation is that we obtain a significance test for the pooled AR(1) coefficient that takes into consideration the correlated structure of the data. For this model, the correlation structure of the residuals includes a random slope and intercept. The covariance matrix of these random slopes and intercepts is included in table 6.4. We can compare this matrix to the covariance matrix of the individual AR(1) coefficients that appears in table 6.3. The values are nearly identical, demonstrating the near equivalence of these models. As one would expect, the variances of the random slopes and intercepts is smaller than those estimated separately by OLS regression.

The model we present here is relatively simple. We can extend the model in a number of different ways. For example, if a dichotomous *Group* variable were expected to account for the variability in the random coefficients, we would write model 2 as:

Level 1:

$$y_{ti} = \pi_{0i} + \pi_{1i}\, y_{t-1,i} + r_{ti}, \qquad r_{ti} \sim N(0, \sigma^2). \tag{6.18}$$

Level 2:

$$\begin{aligned} \pi_{0i} &= \beta_{00} + \beta_{01} * Group + u_{0i}, \\ \pi_{1i} &= \beta_{10} + \beta_{11} * Group + u_{1i}, \end{aligned} \tag{6.19}$$

where

$$\begin{pmatrix} u_{0i} \\ u_{1i} \end{pmatrix} \sim N\left(\begin{pmatrix} 0 \\ 0 \end{pmatrix}, \begin{pmatrix} \tau_{00} & \tau_{01} \\ \tau_{10} & \tau_{11} \end{pmatrix}\right). \tag{6.20}$$

Substituting level 2 into level 1:

$$y_{ti} = [\beta_{00} + \beta_{01}Group + \mu_{0i}] + [\beta_{10} + \beta_{11}Group + u_{1i}] * y_{t-1,i} + r_{ti} \tag{6.21}$$

and rearranging terms:

$$y_{ti} = [\beta_{00} + \beta_{01}Group + \beta_{10}y_{t-1,i} + \beta_{11}Group * y_{t-1,i}] + [u_{0i} + u_{1i}y_{t-1,i}] + r_{ti}, \tag{6.22}$$

where the first bracket represents the fixed effects of the model and the second bracket represents a random intercept and slope related to the AR(1) coefficient. To the previous model, we have added coefficients for the fixed *Group* and *Group* $\times$ y_{t-1} interaction. We are still modeling the variability in the individual slopes and intercepts of the AR(1) model through the random coefficients, u_{0i} and u_{1i}. If we considered an AR(2) process as our level 1 model, we would write model 2 as:

Level 1:

$$y_{ti} = \pi_{0i} + \pi_{1i}y_{t-1,i} + \pi_{2i}y_{t-2,i} + r_{ti}, \qquad r_{ti} \sim N(0, \sigma^2). \tag{6.23}$$

Level 2:

$$\begin{aligned} \pi_{0i} &= \beta_{01}Group + u_{0i}, \\ \pi_{10} &= \beta_{11}Group + u_{1i}, \\ \pi_{2i} &= \beta_{21}Group + u_{2i}, \end{aligned} \tag{6.24}$$

where

$$\begin{pmatrix} u_{0i} \\ u_{1i} \\ u_{2i} \end{pmatrix} \sim N\left(\begin{pmatrix} 0 \\ 0 \\ 0 \end{pmatrix}, \begin{pmatrix} \tau_{00} & \tau_{01} & \tau_{02} \\ \tau_{10} & \tau_{11} & \tau_{12} \\ \tau_{20} & \tau_{21} & \tau_{22} \end{pmatrix}\right).$$

The multilevel AR(1) presents some interpretational challenges. The fixed effect related to each lagged variable represents the average of a set of individual AR coefficients. This average coefficient does not represent the "process" of the "average" individual. It is nothing more than the central tendency of the distribution of AR coefficients. This fixed effect would have a strong interpretation only if it could be assumed both that the order of the process is identical for each individual and that the magnitude of the process is identical. The fixed effects regression coefficient related to the product (or interaction) between the lagged variable and a presumed covariate is accounting for interindividual differences in the variability of the individual AR coefficients, that is, variability in the regularity of the process.

The individual AR coefficients are deviations around the average value. The range of the lagged variable can differ for each individual and will depend on the individual's characteristics. If one thinks of a y_t by y_{t-1} scatterplot, the data points representing each individual will only be properly aligned with the other individuals if each individual time series is centered on the lagged variable.

6.6 Predicting the Individual AR(1) Coefficients

SAS PROC MIXED delivers Best Linear Unbiased Prediction (BLUP) values for the individual regression coefficient values. We can estimate the individual random slopes and intercepts as BLUP. Given the estimates of σ^2 and D, one can predict the individual random effects (Harville, 1977; Jones, 1993; Robinson, 1991) as the deviation of each individual's slope and intercept from the group values, which are given by

$$\hat{\gamma} = DZ_i^T \hat{V}^{-1}(y_i - X_i\hat{\beta}), \tag{6.25}$$

where

$$\hat{\beta} = \left(\sum_i X_i^T \hat{V}^{-1} X_i\right)^{-1} \left(\sum_i X_i^T \hat{V}^{-1} y_i\right) \tag{6.26}$$

are the fixed effect parameter estimates. Given γ and β, we can now predict the y_i as

$$\hat{y}_i = X_i\hat{\beta} + Z_i\hat{\gamma}_i \tag{6.27}$$

and similarly estimate the innovations.

To generate the estimated individual AR(1) coefficients, we add these deviations to the fixed effect regression values. The random slopes and intercepts for the first ten cases are included in table 6.5. We correlated the AR(1) coefficients generated in this manner with the individually estimated OLS coefficients. The correlation between the two sets of coefficients was 0.90.

Figure 6.1 shows a histogram of all of the regression coefficients for these data. That is, figure 6.1 is a histogram of slopes of the centered lag 1 ounces consumed on ounces consumed (TOTOZ). Note that the magnitudes of these

Table 6.5 GLMM individual intercepts and slopes

Intercepts	Slopes
−1.41	−0.03
−0.82	−0.03
−1.08	−0.03
−0.79	0.01
−0.66	0.02
0.11	−0.17
−0.64	0.00
−0.47	−0.03
−0.43	0.04
−0.22	0.01
...	...

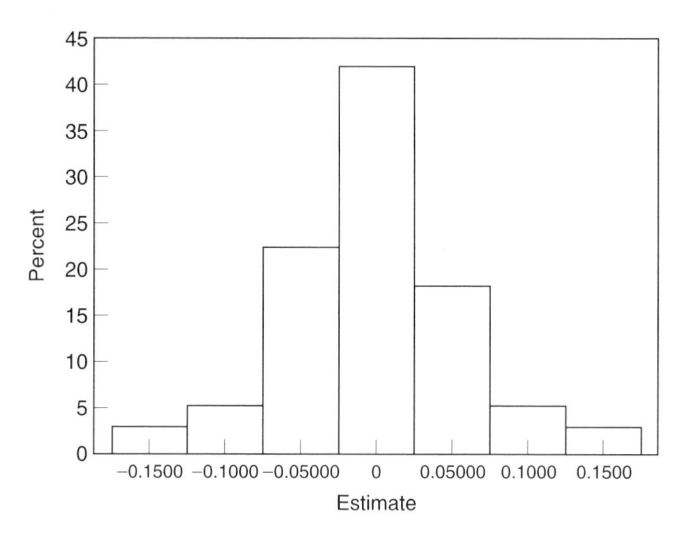

Figure 6.1 Histogram of baseline estimates of ounces from centered, lagged ounces.

slopes are normally distributed and assume both negative and positive values, reflecting different kinds of regularity. These slopes reflect the same variability as described earlier in the case of individual ARIMA estimates and as documented by the earlier comparisons of the covariance matrices. The SAS PROC MIXED code for this baseline model is included in Appendix A, and we include parts of the output from a SAS PROC MIXED run in Appendix B (both these items can be found in the companion website to this book).

6.6.1 Extending the Model to Include Covariates

Because this model will provide greatest applied utility when covariates are hypothesized to explain the variability of the AR-based coefficients, we provide an extension in two models, first with a single covariate for simplicity and then with two covariates for greater explication. In the first model, the reported level of desire to drink at each occasion is used as a continuous covariate. The general equation for the model is:

$$Y_{ti\ TOTOZ} = \beta_0 + \beta_1 LAGOZ + \beta_2 DESIRE + \beta_3 LAGOZ * DESIRE + r_{ti} \tag{6.28}$$

and for our estimated model the coefficients are:

$$Y_{ti\ TOTOZ} = -0.68_{\text{intercept}} + 0.05_{LAGOZ} + 0.54_{DESIRE} + 0.02_{LAGOZ*DESIRE} + r_{ti}. \tag{6.29}$$

In this model, the effect of the first-order lagged reported *desire* to drink variable on actual daily drinking level (/textitTOTOZ) is large. The cross-level effect of desire on the random slopes of lagged drinking desire with ounces consumed daily is significant, indicating that their drinking desire accounts for at least

Table 6.6 Alcohol consumption model with drinking desire as a covariate

Effect	Estimate	SE	DF	t-value	Prob.
Intercept	−0.69	0.09	92	−7.92	<0.0001
LAGOZ	0.05	0.03	5367	1.41	0.1587
DESIRE	0.54	0.01	5367	49.17	<0.0001
LAGOZ * DESIRE	0.02	0.01	5367	2.57	0.0102

some of the variability in desire by drinking slopes. In typical multilevel longitudinal models, this effect would be time * desire. Here, the interpretation is different; time is replaced by drinking level at the previous day. So, drinking regularity is accounted for by desire. This indicates that the regularity of drinking behavior can be explained by the covariate desire to drink. The regression coefficients appear in table 6.6. Figure 6.2 shows the plots of these coefficients for prototypical values one standard deviation above and below the mean of drinking desire (Singer & Willett, 2003).

However, there may be further differences to be explored by using both the continuous drinking desire variable and a categorical grouping variable. Because this may frequently be of interest in treatment-control designs, we demonstrate how to conduct this model here. We begin by plotting specific subjects' time-series plots for prototypical male subjects one standard deviation above and

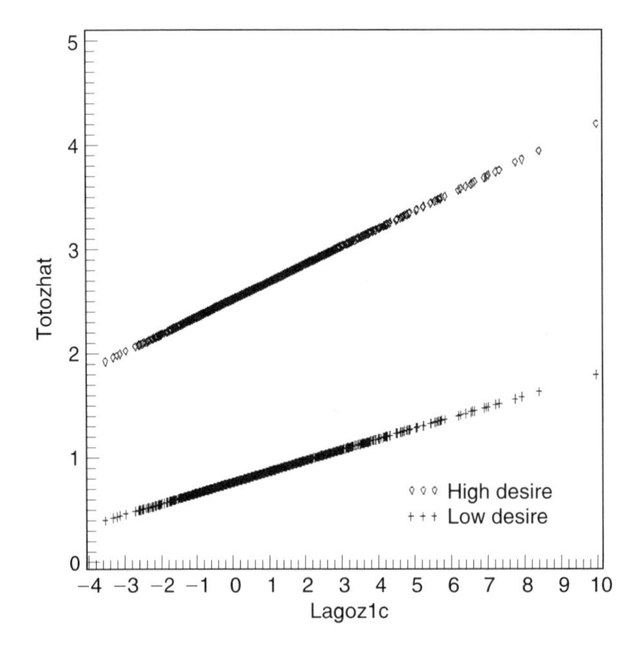

Figure 6.2 Controlled effects of sex for two levels of desire.

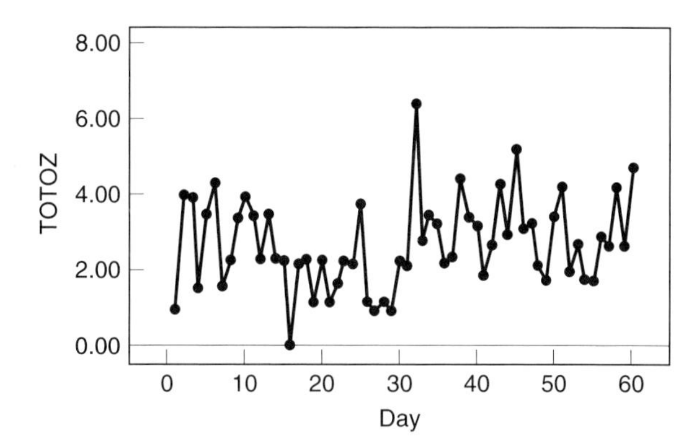

Figure 6.3 Ounces per day for a selected male, high drinking desire.

below the mean of drinking desire in figures 6.3 and 6.4. The plot in figure 6.3 shows ounces of alcohol consumed each day for a man with high mean reported drinking desire (one standard deviation above the mean of all subjects' desire). This plot reflects a fairly erratic pattern of drinking regularity. By contrast, a male with low reported desire to drink (one standard deviation below the mean) frequently has no drinks and appears more stable (regular) (figure 6.4). We see that, at least for men, the prototypical plots differ in their regularity. In order to explore the predominance of this kind of difference across subjects for both men and women, we specify a new model.

In the second model, the reported level of desire to drink at each occasion is used as a continuous covariate just as in the first model and a grouping variable, sex, is added to the model. All possible interaction terms of the covariates were assessed and sequentially removed if not significant. The general equation for the

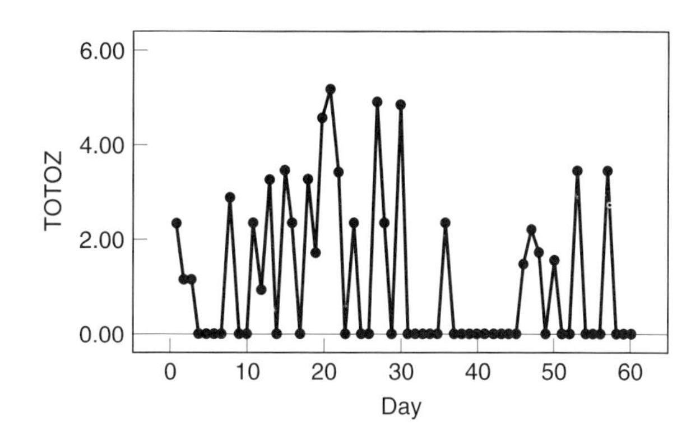

Figure 6.4 Ounces per day for a selected male, low drinking desire.

second model extension is

$$Y_{ti}\ TOTOZ = \beta_0 + \beta_1 LAGOZ + \beta_2 DESIRE + \beta_3 SEX + \beta_4 DESIRE * SEX + \beta_4 LAG * DESIRE + r_{ti}, \tag{6.30}$$

$$Y_{ti\ TOTOZ} = 1.24 + 0.06\ LAGOZ + 0.69\ DESIRE + 0.87\ SEX - 0.25\ DESIRE * SEX + 0.02\ LAGOZ * DESIRE + r_{ti},$$

where sex is coded 0 for men and 1 for women. The equations can be rewritten for each sex as:

$$\text{Men}: Y_{ti\ TOTOZ} = -1.24 + 0.06\ LAGOZ + 0.69\ DESIRE + 0.02\ LAG * DESIRE$$
$$\text{Women}: Y_{ti\ TOTOZ} = -1.24 + 0.44\ LAGOZ + 0.02\ LAG * DESIRE. \tag{6.31}$$

This model indicates that desire to drink influences the regularity of men's and women's drinking for the two sexes. Specifically, the effect is 0.69 for men and 0.44 for women for a one-unit increase in the level of the lagged total ounces. All three-way interactions, as well as the interaction of lagged drinking by sex, were nonsignificant and were removed from the model. The plot in figure 6.5 shows the slopes for TOTOZ for prototypical levels of desire for each sex, based on the last equation. This shows that the regularity of day-to-day drinking differs for our prototypical values of sex and drinking desire. A positive coefficient represents the level of day-to-day prediction with higher coefficients meaning more consistency. A participant with a negative coefficient would have

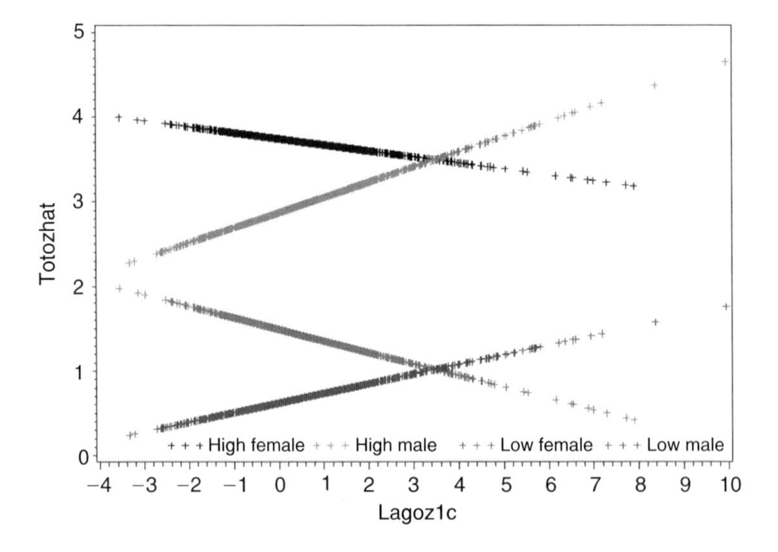

Figure 6.5 Controlled effects of sex for two levels of desire.

Table 6.7 Alcohol consumption model with sex as a covariate

Effect	Estimate	SE	DF	t-value	Prob.
Intercept	−1.24	0.13	91	−9.53	<0.0001
lagoz1c	0.06	0.03	5366	1.77	0.0771
DRDESIRE	0.69	0.02	5366	40.06	<0.0001
sexREC	0.87	0.17	91	5.07	<0.0001
DRDESIRE*sexREC	−0.25	0.02	5366	−11.28	<0.0001
lagoz1c*DRDESIRE	0.02	0.01	5366	2.18	0.0295

a more intermittent drinking pattern. For example, a large negative coefficient would represent a sawtooth pattern of an individual. Table 6.7 shows all of the estimates from this model.

6.7 Discussion

In this chapter, we have presented an atypical use of a multilevel model, one in which the level 1 model is an AR(p) model. The regression coefficients represent the separate estimation of an AR(p) process for each individual's time series. Variability in these coefficients is explained in a level 2 model. For the Armeli and colleagues (2002) alcohol use dataset, we modeled this multilevel AR(1) model as a general linear mixed model using SAS PROC MIXED. We further demonstrated the application potential of this model by developing two models with covariates, one with a continuous covariate and the other with a continuous covariate, a categorical covariate, and their interactions.

This work represents a first step in the modeling of multilevel AR data. As such, a number of limitations of this approach must be considered. Most important, we are not identifying the order of the AR process; rather, we are requiring that each series conform to the specified process. If the process is properly specified, we can assume that the residuals for each series represent a white noise process and can thus be modeled by the random coefficients error structure. In this case, the individual residuals are hypothesized to be normally distributed with equal variances.

We are also assuming that the random coefficients, in this case the AR(p) coefficients, are normally distributed. We can only realistically expect this if the AR process is the correct process for each individual. We can indicate the kind of problem that could occur by imagining a situation in which an AR(1) process is correct for one half of our sample and an AR(2) process is correct for the other half. In this case the distribution of AR(2) coefficients could be a mixture of a normal distribution for those properly specified and a uniform distribution (all 0's) for those cases improperly specified. In this case we would more appropriately estimate two different models: one for those appropriately identified by an AR(1) process and one for those appropriately identified by an AR(2) process. Using the autocorrelation functions as diagnostic values would allow us to more appropriately group individuals according to the correct order of the process.

In situations in which we need to identify the process for each individual, we could imagine having to a prior set of analyses. For moderate span data which still has relatively few occasions for proper time-series identification, this could be problematic (Jones, 1991). A next step would involve identifying the complete group model by comparing AR(1) to AR(2) to AR(3) by means of AIC, BIC, and so on, checking the distribution of the random coefficients for each subject's model.

A simpler way to approach the problem of a mixture of model types would involve estimating the more complex model and checking the variance component for the higher-order term to see if it is necessary. If not, that term could be dropped, and the simpler model should suffice for that sample. With significant variability in the AR(2) coefficients, the resultant distribution of the predicted AR(2) coefficients is normal. However, if the true distribution of the AR(2) coefficients is a mixture of a uniform (all values 0) and a normal, the variability of the coefficients related to the uniform part of the distribution will be slightly overestimated while the variability related to the normal part of the distribution will be slightly underestimated.

We also note that the models we consider here are for equally spaced data. In this model, time need not appear explicitly as a variable. The individual regression coefficient represents the change in $y(t)$ per unit change in $y(t-1)$. To adjust the coefficient to represent a longer or shorter span of time could be most easily accomplished by moving to a state-space representation. Multilevel models can be reformulated as state-space models in which the state and observation equations form the levels. Icaza and Jones (1999) compare SAS PROC MIXED and state-space approaches to handling time-series models. The models discussed here also assume weak stationarity. Nonstationary models would also be approached using the state-space framework. Chapter 7 by Ho et al. (this volume) covers the state-space form as it could apply to this and other models.

Another particularly unusual aspect of this form of a multilevel model is related to the fact that each series must be detrended. We could still have level differences related to the overall average of the series. These levels could be estimated by the intercept of the model, which would be the constant of the AR(p) process. We would typically assume that any variability in the level of the process would be modeled in some complementary consideration of the trends in the data. As a result, it would probably be a good strategy to center the dependent variable of each series and estimate a model with no intercept.

In summary, the models described in this chapter address questions relating to the regularity of a process across the complete range of a series. The series is summarized by means of a set of AR coefficients. The general question to be answered is whether the current behavior predicts the future behavior across the whole series. Variability in these coefficients could be explained by other covariates. Under an assumption of stationarity, this set of prediction coefficients describes the whole series for a prespecified process of order AR(p). There are a number of data collection techniques for which such methods would seem particularly appropriate. Studies making use of diary data would seem particularly amenable to questions related to whether a current behavior predicts a future behavior across the observed series. Among these, variables likely to reflect regularities over the term of the study and appropriate at the time scale to reflect these regularities are most viable (Walls et al., 2005). If such a prediction holds true, the individual autoregressive coefficient would represent a single parsimonious description of the regularity of sequential stability for each participant. The distribution of this kind of coefficient could represent a good indication that individual differences exist for the process studied, one that could be further explained with selected design-based or theoretically meaningful covariates.

NOTE

1. If we treat *time* as a categorical variable and replace it with a set of dummy-coded variables, we have a common multilevel formulation of the repeated measures ANOVA. We can model the covariance structure of the errors setting u_{0i} and u_{1i} to *zero* and patterning r_{ti} (Rovine & Molenaar, 1998).

References

Armeli, S., Carney, M.A., O'Neil, T.P., Tennen, H., & Affleck, G. (2000). Stress and alcohol use: A daily process examination of the stressor-vulnerability model. *Journal of Personality and Social Psychology, 78*, 979–994.

Baltes, P.B., Reese, H.W., & Nesselroade, J.R. (1988). *Life-Span Developmental Psychology: Introduction to Research Methods*. Hillsdale, NJ: Erlbaum.

Belsky, J., & Rovine, M. (1990). Patterns of marital change across the transition to parenthood. *Journal of Marriage and the Family, 52*, 5–19.

Belsky, J., Spanier, G.B., & Rovine, M.J. (1983). Stability and change in marriage across the transition to parenthood. *Journal of Marriage and the Family, 45*, 567–577.

Bollen, K.A. (1989). *Structural Equations with Latent Variables*. New York: John Wiley.

Box, G.E.P., & Jenkins, G.M. (1976). *Time Series Analysis, Forecasting, and Control.* Oakland, CA: Holden-Day.

Brim, O.G., & Kagan, J. (1980). Constancy and change: A view of the issues. In O.G. Brim & J. Kagan (Eds.), *Constancy and Change in Human Development* (pp. 1–25). Cambridge, MA: Harvard University Press.

Brockwell, P.J., & Davis, R.A. (1991). *Time Series: Data Analysis and Theory*. New York: Springer.

Bryk, A.S., & Raudenbush, S.W. (1992). *Hierarchical Linear Models*. Newbury Park, CA: Sage.

Carney, M.A., Armeli, S., Tennen, H., Affleck, G., & O'Neil, T.P. (2000). Positive and negative daily events, perceived stress, and alcohol use: A diary study. *Journal of Consulting and Clinical Psychology, 68*, 788–798.

Chatfield, C. (1996). *The Analysis of Time Series: An Introduction*. Boca Raton, FL: Chapman & Hall/CRC Press.

Cook, T.D., & Campbell, D.T. (1979). *Quasi-experimentation: Design and Analysis Issues for Field Settings*. Chicago: Rand McNally.

DuToit, S., & Browne, M.W. (2001). The covariance structure of vector ARMA time series. In R. Cudeck, S. Du Toit, & D. Sorbom (Eds.), *Structural Equation Modeling: Present and Future* (pp. 279–314). Chicago: Scientific Software International.

Goldstein, H.I. (1987). Multilevel mixed linear model analysis using iterative generalized least squares. *Biometrika, 73*, 43–56.

Goldstein, H.I. (1995). *Multilevel Statistical Modeling*. London: Edward Arnold.

Goldstein, H.I. (2003). *Multilevel Statistical Models* (3rd ed.). New York: Oxford University Press.

Gottman, J.M. (1981). *Time Series Analysis*. New York: Cambridge University Press.

Gregson, R.A.M. (1983). *Time Series in Psychology*. Hillsdale, NJ: Erlbaum.

Grizzle, J.E., & Allen, D.M. (1969). Analysis of growth and dose response curves. *Biometrics, 25*, 357–382.

Hamaker, E.L., Dolan, C.V., & Molenaar, P.C.M. (2003). ARMA-based SEM when the number of time points T exceeds the number of cases N: Raw data maximum likelihood. *Structural Equation Modeling, 10*, 352–379.

Hamaker, E.L., Dolan, C.V., & Molenaar, P.C.M. (2005). Statistical modeling of the individual: Rationale and application of multivariate time series analysis. *Multivariate Behavioral Research*, in press.

Harville, D.A. (1974). Bayesian inference for variance components using only error contrasts. *Biometrika, 61*, 383–385.

Harville, D.A. (1976). Extensions of the Gauss–Markov theorem to include the estimation of random effects. *Annals of Statistics, 4*, 384–395.

Harville, D.A. (1977). Maximum likelihood approaches to variance component estimation and to related problems. *Journal of the American Statistical Association, 72*, 320–340.

Hertzog, C.H., & Rovine, M.J. (1985). Repeated-measures analysis of variance in development research: Selected issues. *Child Development, 56*, 787–809.

Icaza, G., & Jones, R.H. (1999). A state space EM algorithm for longitudinal data. *Journal of Time Series Analysis, 20*, 537–550.

Jennrich, R.I., & Schlucter, M.D. (1986). Unbalanced repeated measures models with structured covariance matrices. *Biometrics, 42*, 805–820.

Jones, K. (1991). The application of time series methods to moderate span longitudinal data. In L.M. Collins & J.L. Horn (Eds.), *Best Methods for the Analysis of Change*. Washington, DC: American Psychological Association.

Jones, R.H. (1993). *Longitudinal Data with Serial Correlation: A State-Space Approach*. London: Chapman & Hall.

Kenny, D.A. (1973). Cross-lagged and common forms in panel data. In A.E. Goldberger & O.D. Duncan (Eds.), *Structural Equation Models in the Social Sciences*. New York: Seminar Press.

Kenny, D.A., & Campbell, D.T. (1989). On the measurement of stability in overtime data. *Journal of Personality, 57*, 445–581.

Kreft, I., & de Leeuw, J. (1998). *Introducing Multilevel Modeling*. London: Sage.

Laird, N.M., & Ware, H. (1982). Random-effects models for longitudinal data. *Biometrics, 38*, 963–974.

Littell, R.C., Milliken, G.A., Stroup, W.W., & Wolfinger, R.D. (1996). *SAS System for Mixed Models*. Cary, NC: SAS Institute, Inc.

McCall, R.B., & Appelbaum, M.I. (1973). Bias in the analysis of repeated-measures designs: Some alternative approaches. *Child Development, 44*, 401–415.

Molenaar, P.C.M. (1985). A dynamic factor model for the analysis of multivariate time series. *Psychometrika, 50*, 181–202.

Molenaar, P.C.M. (1999). Comment on fitting MA time series by structural equation models. *Psychometrika, 64*, 91–94.

Myers, J. (1979). *Fundamentals of Experimental Design*. New York, Allyn-Bacon.

Nesselroade, J.R., & Molenaar, P.C.M. (1999). Pooled lagged covariance structures based on short, multivariate time series for dynamic factor analysis. In R.H. Hoyle (Ed.), *Statistical Strategies for Small Sample Research* (pp. 223–250). Thousand Oaks, CA: Sage.

Nesselroade, J.R., & Molenaar, P.C.M. (2003). Quantitative methods for developmental processes. In J. Valsiner & K. Connolly (Eds.), *Handbook of Developmental Psychology* (pp. 622–639). London: Sage.

Robinson, G.K. (1991). That BLUP is a good thing: The estimation of random effects. *Statistical Science, 6*, 15–51.

Rovine, M.J., & Molenaar, P.C.M. (1998). A LISREL model for the analysis of repeated measures with a patterned covariance matrix. *Structural Equation Modeling, 5*, 318–343.

SAS Institute Inc. (1995). *Introduction to the Mixed Procedure*. Cary, NC: SAS Institute, Inc.

Shingles, R.D. (1985). Causal inference in cross-lagged panel analysis. In H.M. Blalock, Jr. (Ed.), *Causal Models in Panel and Experimental Designs*. New York: Aldine.

Shumway, R.H., & Stoffer, D.S. (2000). *Time Series Analysis and its Applications*. New York: Springer.

Singer, J.D. (1998). Using SAS PROC MIXED to fit multilevel models, hierarchical models, and individual growth models. *Journal of Educational and Behavioral Statistics, 24*, 323–355.

Singer, J.D., & Willett, J.B. (2003). *Applied Longitudinal Data Analysis: Modeling Change and Event Occurrence*. New York: Oxford University Press.

Sivo, S.A. (2001). Multiple indicator stationary time series models. *Structural Equation Modeling, 8*, 599–612.

Snijders, T.A.B., & Bosker, R.J. (1999). *Multilevel Analysis*. London: Sage.

van Buuren, S. (1997). Fitting ARMA time series by structural equation models. *Psychometrika, 62*, 215–236.

Velicer, W.F., & Fava, J.L. (2003). Time series analysis. In J.A. Schinka and W.F. Velicer (Eds.) [I.B. Weiner (Editor-in-Chief)], *Handbook of Psychology: Research Methods in Psychology* (Vol. 2, pp. 589–606). New York: John Wiley.

Walls, T.A., Höppner, B.B., & Goodwin, M.S. (2005). Statistical issues in intensive longitudinal data analysis. In A. Stone, S. Shiffman, A. Atienza, & L. Nebelling (Eds.), *The Science of Real-Time Data Capture*. New York: Oxford University Press (forthcoming).

Wohlwill, J.F. (1973). *The Study of Behavioral Development*. New York: Academic Press.

Wood, P., & Brown, D. (1994). The study of intraindividual differences by means of dynamic factor models: Rationale, implementation, and interpretation. *Psychological Bulletin, 116*, 166–186.

7

The State-Space Approach to Modeling Dynamic Processes

Moon-Ho Ringo Ho, Robert Shumway, and Hernando Ombao

In this chapter we seek to present a self-contained treatment of state-space modeling and attempt to make the exposition accessible to those who have relatively little prior knowledge of the subject. We focus on issues of modeling and show how state-space models offer a rich and flexible class of structures that accommodate both the static and dynamic nature of intensive longitudinal data.

Longitudinal data collected from a group or groups of subjects following over time often exhibit within-subject serial correlations, involving random subject effects and the presence of observational errors. Researchers are usually interested in describing the trend over time (e.g., is it linear or curvilinear?, is it increasing or decreasing?), whether there are significant differences in the trend across groups of subjects, and what factors can account for this trend and the differences. Longitudinal data present opportunities for exploiting state-space methods for multivariate longitudinal models that can be expressed in state-space form. Such representation opens new opportunities for modeling intensive longitudinal data by extending the usual mixed model to allow dynamic random effects and time-varying covariates with stochastic coefficients in a parametric or nonparametric manner, and to estimate long-term and short-term covariate effects. Furthermore, state-space models provide a convenient methodology for treating incomplete and unequally spaced data.

Parameters in the state-space models are estimated using the Kalman filter and smoother. When used in conjunction with the expectation-maximization (EM) algorithm, they offer an elegant approach to handling incompletely observed multivariate vectors. The computational burden is much less in state-space models than in mixed models ($p \times p$ vs. $pN \times pN$, where p is the total number of variables measured per subject and N is the total number of subjects). We illustrate the application of state-space models in nonstandard situations for analyzing intensive longitudinal data collected in neuroscientific and traffic network studies. The applications of state-space models reported here are new.

The use of state-space models in the social sciences, except in economics, is uncommon. Through the two case studies presented in this chapter, we hope readers will be convinced that state-space models provide an effective means for practical analysis of intensive longitudinal data and will consider the use of state-space models in their own work.

Because of space limitations, a complete coverage of state-space models is not possible. We list here a number of books that contain treatment of state-space methods for readers' reference. Two early texts written from an engineering point of view are Jazwinski (1970) and Anderson and Moore (1979). Harvey (1989) gives a comprehensive treatment of the state-state approach to the structural time-series models that are widely used in economics. An up-to-date treatment with detailed discussion on estimation of non-Gaussian and nonlinear state-space models by simulation methods is given by Durbin and Koopman (2001). Jones (1993) discusses the application of state-space models in the analysis of longitudinal data. More general books on time-series analysis with partial treatment on state-space modeling include Fahrmeir and Tutz (1994) and Shumway and Stoffer (2000). West and Harrison (1997) present a Bayesian perspective on state-space models with a focus on forecasting. Most of the discussions in these references emphasize the state-space approach to univariate time-series analysis without replication. Our chapter, which is supplementary to these texts, emphasizes the application of state-space models for multiple time-series analysis and replicated univariate time-series analysis.

The rest of this chapter is structured as follows. We begin with an introduction to the linear Gaussian state-space models. In section 7.2, we demonstrate that some familiar models in the time-series and longitudinal data analysis literature can be reformulated as special cases of the state-space models. The parameter estimation and statistical inference procedures for state-space models are described in section 7.3. Applications of state-space models to analyzing datasets from a neuroscientific experiment and a traffic study are discussed in sections 7.4 and 7.5, respectively. Finally, we conclude this chapter in section 7.6 with some discussion on handling missing data and nonlinear non-Gaussian data in state-space models.

7.1 Gaussian State-Space Models

A linear Gaussian state-space model is characterized by an unobserved series of vectors $\theta_1, \ldots, \theta_n$ (called *states*), which are associated with a series of observations $y_1, \ldots, y_n$. The relation between the states and the observations is specified through the observation and state equations. In longitudinal settings with N subjects, the observation equation and state equations are defined as

$$y_{t,j} = \Gamma x_{t,j} + A\theta_{t,j} + e_{t,j}, \qquad e_{t,j} \sim N_p(0, R), \tag{7.1}$$

$$\theta_{t,j} = \Phi\theta_{t-1,j} + w_{t,j}, \qquad w_{t,j} \sim N_k(0, Q), \qquad t = 1, \ldots, n; \; j = 1, \ldots, N, \tag{7.2}$$

where $y_{t,j}$ is a $p \times 1$ vector of observations, and $x_{t,j}$ is a $q \times 1$ input vector of fixed functions which can be used to incorporate fixed trends or treatment effects. Deterministic (e.g., polynomial or sinusoidal) and stochastic (e.g., cubic spline) trends can be included in a straightforward manner under state-space models (see case study 2 for an illustration) so as to accommodate the nonstationarity in the time-series data. Γ is a $p \times q$ regression coefficient matrix relating the input vector $x_{t,j}$ to the output time series $y_{t,j}$. Equation (7.1) is called the observation equation, and has the structure of a linear regression relating the state vector to the observed time series. Equation (7.2) is called the state equation, and describes the dynamics of the states, denoted by a $k \times 1$ vector $\theta_{t,j}$, in terms of a first-order Markov process. The relation between the state vector to the observed values is characterized by the matrix A of size $p \times k$. The dependence of the current state on the past is determined by the transition matrix, Φ.

The unobserved Markovian process of the states might be of interest in its own right or as a technical tool for formulating a specific correlation structure. The classical autoregressive integrated moving average (ARIMA) models, for instance, can be put into the state-space form, with the latent process of the states containing the lagged observations of the observed process that is not of separate interest. In contrast, in the original application of the state-space models for navigation systems of a space rocket's position, the observation equation (7.1) describes how the radar observations, disturbed by noise, depend on the state vector (such as position and velocity) of a spacecraft, and the state equation (7.2) describes the spacecraft motion in space.

Another example is in financial time-series analysis, where the variance of the observations, known as volatility, which usually changes over time, can be represented as the latent state. The process of the latent states measures the stability of a stock market. The model for the latent state, $\theta_{t,j}$, is quite general and can easily be specialized to the classical mixed effects model (see section 7.2 for more details). The specialization involves setting the error in (7.2) to zero and the transition matrix to $\Phi = I$. Then the distribution of the initial state vector, $\theta_{0,j}$, can be taken as normal, with zero mean and covariance matrix $\sigma_0^2 I$, and the sequence of state vectors become conventional random effects that do not change over time ($\theta_{t,j} = \theta_{t-1,j}, j = 1, 2, \ldots, N$). More random effects can be added by stacking the state vectors. The error terms $e_{t,j}$ and $w_{t,j}$ are assumed to be independent of each other and are also independent of the state vector $\theta_{t,j}$. The measurement errors ($e_{t,j}$) and the state errors ($w_{t,j}$) will be assumed in this chapter to follow multivariate normal distributions with zero means and covariance matrices R and Q. The state vector ($\theta_{t,j}$) will be assumed to have normally distributed initial state vector θ_{0j} with mean μ_{0j} and covariance Σ_0. In practice, the values for μ_{0j} and Σ_0 are usually unknown and are set to some preset numbers (e.g., zero for μ_{0j} and a large value such as 10^6 for Σ_0). The impact of these chosen values become negligible when the length of the time series is long. We comment further on this in section 7.3.

The matrices $\Gamma, \Phi, R,$ and Q may depend on an unknown parameter vector, which can be estimated using a maximum-likelihood method via the use of the Kalman filter and smoother (see section 7.2). Note that the matrices Γ, A, Φ, R, and Q in (7.1) and (7.2) do not change over time. However, these matrices can be generalized to vary over time if necessary as in case study 1. Another situation for allowing A and Γ to be time-varying (i.e., A_t and Γ_t) is to account for missing data. Under a state-space modeling framework, time-series observations do not need to be equally spaced or measured at a common set of occasions for multiple time series. The latter situation can be regarded as a form of missing data and can be handled easily in state-space models through the use of an expectation-maximization (EM) algorithm. In this chapter, our applications focus on intensive longitudinal data that are equally spaced and collected at a common set of occasions. In the conclusion section, we discuss how to handle missing data in a state-space modeling framework. For a univariate response, interested readers can refer to Icaza and Jones (1999) for the analysis of unequally spaced longitudinal data by the state-space model. Notice that the formulation given in (7.1) covers both the single and multiple time-series analysis, although most of the state-space applications are primarily for single time-series analysis. One way to capture the dependence between the time series is to have correlated disturbances (i.e., the matrix R is nondiagonal). An alternative approach is via a factor analysis model (see section 7.2).

We can further extend the state-space model by including covariates in the state equation. In other words, the model allows two types of covariate, namely, long-term covariates, entering via the state equation, and short-term covariates, entering via the observation equation (Jørgensen et al., 1996). This opens up a new opportunity for modeling intensive longitudinal data which may not be easily achieved in the mixed modeling framework. The terminology "long-term" is used because the Markov structure of the latent process creates a carryover effect on the observed values. In contrast, short-term covariates have an immediate effect on the observed values.

Consider, for example, an application to analysis of the impact of air pollution and seasonal variables on the number of emergency room visits due to respiratory diseases. We may include air pollution covariates such as daily measurements of an air pollutant in the state equation to model the day-to-day carryover effect of air pollution. The state equation can then be interpreted as a latent morbidity process representing the air pollution's potential for causing respiratory disease. The effect of seasonal variables such as temperature or humidity are usually expected to be short-term and may be included in the observation equation to represent modulating factors that have an acute effect on the emergency visit counts relative to the value of the latent state. In this way, we can handle substantive questions that cannot be easily addressed by the classical mixed model for longitudinal data analysis.

The state-space modeling approach provides a unified way of describing a wide class of linear stochastic processes. Examples are Box–Jenkins ARIMA models, conditional heteroscedastic models, spline smoothing, structural time-series models, transfer function models, vector autoregressive moving average models, and many others, which all have a state-space representation. We review some of these models and discuss their connection with the state-space model in the next section. These statistical models are commonly used to capture specific features in intensive longitudinal data. The state-space approach, however, can analyze these features simultaneously in the same model.

7.2 Some Special Cases of State-Space Models

7.2.1 Linear Regression Models with Time-Varying Coefficients

Suppose that we observe data pairs (x_{tj}, y_{tj}) at occasions $t = 1, 2, \ldots, n$ from subjects $j = 1, 2, \ldots, N$. A simple linear regression of $y_{1j}, y_{2j}, \ldots, y_{nj}$ on $x_{1j}, x_{2j}, \ldots, x_{nj}$ would assume that the effect of x on y is invariant over time. However, this may be too restrictive in examining, for example, the neural control on speech production by regressing the lip movement data y_{tj} on the electromyogram x_{tj}, as a measure of neural activity. The influence of the brain on the lip movement β can vary over time. To capture this feature, we can extend the usual linear regression model as $y_{tj} = \beta_{0tj} + x_{tj}\beta_{1tj} + e_{tj}, e_{tj} \sim N(0, \sigma_e^2)$, where the regression coefficients vary over time as $\beta_{0tj} = \beta_{0,t-1,j} + w_{0tj}, w_{0tj} \sim N(0, \sigma_{w0}^2)$ and $\beta_{1tj} = \beta_{1,t-1,j} + w_{1tj}, w_{1tj} \sim N(0, \sigma_{w1}^2)$. Let $\mathbf{X}_{tj} = (1, x_{tj})'$ and $\boldsymbol{\beta}_{tj} = (\beta_{0tj}, \beta_{1tj})'$. Then the corresponding state-space form is

$$y_{tj} = \mathbf{X}_{tj}'\boldsymbol{\beta}_{tj} + e_{tj},$$

$$\begin{pmatrix}\beta_{0tj}\\ \beta_{1tj}\end{pmatrix} = \begin{pmatrix}1 & 0\\ 0 & 1\end{pmatrix}\begin{pmatrix}\beta_{0,t-1,j}\\ \beta_{1,t-1,j}\end{pmatrix} + \begin{pmatrix}w_{0tj}\\ w_{1tj}\end{pmatrix}.$$

The evolution of regression coefficients over time is described by a simple random walk process which means the current value of β is the sum of its immediate past value and a random white noise. Note that the regression coefficients do not have to be varying but can also be "fixed" by setting the state errors (w_{0tj}, w_{1tj}) to zero. More complex mechanisms to characterize the changing pattern of the mean (intercept) and the effect of x_t over time can be considered. For example, we can impose additional constraints on the mean trend and the influence of x_t to guarantee that their changes vary over time in a smooth manner. This can be achieved by imposing second difference constraints: $\nabla^2\beta_{0t} = w_{0t}$ and $\nabla^2\beta_{1t} = w_{1t}$. The second difference operation means $\nabla^2\beta_{0t} = \nabla\beta_{0t} - \nabla\beta_{0,t-1}$, where $\nabla\beta_{0t} = \beta_{0tj} - \beta_{0,t-1,j}$. Thus, $\nabla^2\beta_{0t} = (\beta_{0tj} - \beta_{0,t-1,j}) - (\beta_{0,t-1,j} - \beta_{0,t-2,j}) = \beta_{0tj} - 2\beta_{0,t-1,j} + \beta_{0,t-2,j}$. Similarly, $\nabla^2\beta_{1t} = (\beta_{1tj} - \beta_{1,t-1,j}) - (\beta_{1,t-1,j} + \beta_{1,t-2,j}) = \beta_{1tj} - 2\beta_{1,t-1,j} + \beta_{1,t-2,j}$. In state-space form, the observation equation is same as the above and the state equation becomes:

$$\begin{pmatrix} \beta_{0tj} \\ \beta_{1tj} \\ \beta_{0,t-1,j} \\ \beta_{1,t-1,j} \end{pmatrix} = \begin{pmatrix} 2 & 0 & -1 & 0 \\ 0 & 2 & 0 & -1 \\ 1 & 0 & 0 & 0 \\ 0 & 1 & 0 & 0 \end{pmatrix} \begin{pmatrix} \beta_{0,t-1,j} \\ \beta_{1,t-1,j} \\ \beta_{0,t-2,j} \\ \beta_{1,t-2,j} \end{pmatrix} + \begin{pmatrix} w_{0tj} \\ w_{1tj} \\ 0 \\ 0 \end{pmatrix}.$$

The time-varying function β_{0t} and β_{1t} are called the spline function in nonparametric statistics literature (Green & Silverman, 1994). This above formulation is equivalent to modeling the effect of x_t on y_t by the cubic splines (Wecker & Ansley, 1983). We illustrate the application of such smoothing constraints in case study 2.

7.2.2 Autoregressive Moving Average Models

Box and Jenkins assumed that a univariate time series is made up of trends, seasonal and irregular components. Instead of modeling these components directly, they suggested removing the trend and seasonal components by differencing prior to the analysis, which applies successive operations of the form $\nabla x_t = x_t - x_{t-1}$ to achieve a stationary time series. The resulting differenced series, denoted as y_t, is then modeled by an autoregressive moving average (ARMA) process: $y_t = \phi_1 y_{t-1} + \cdots + \phi_p y_{t-p} + \epsilon_t + \theta_1 \epsilon_{t-1} + \cdots + \theta_q \epsilon_{t-q}$. The constants $\phi_1, \phi_2, \ldots, \phi_p$ are called autoregressive coefficients and relate the value of y at time t to its p past values, $y_{t-1}, y_{t-2}, \ldots, y_{t-p}$. The constants $\theta_1, \theta_2, \ldots, \theta_q$ are called moving average coefficients and relate the present white noise, ϵ_t, to its q past values, $\epsilon_{t-1}, \epsilon_{t-2}, \ldots, \epsilon_{t-q}$, where $\epsilon_t \sim N(0, \sigma_\epsilon^2)$. This model is usually referred to as the ARMA(p, q) model. Consider the ARMA(2,1) model $y_t = \phi_1 y_{t-1} + \phi_2 y_{t-2} + w_t + \theta_1 w_{t-1}$, which has the state-space representation:

$$y_t = \begin{pmatrix} 1 & 0 \end{pmatrix} \begin{pmatrix} \theta_{1t} \\ \theta_{2t} \end{pmatrix},$$

$$\begin{pmatrix} \theta_{1t} \\ \theta_{2t} \end{pmatrix} = \begin{pmatrix} \phi_1 & 1 \\ \phi_2 & 0 \end{pmatrix} \begin{pmatrix} \theta_{1,t-1} \\ \theta_{2,t-1} \end{pmatrix} + \begin{pmatrix} w_t \\ \theta_1 w_t \end{pmatrix}.$$

In this state-space form, the state vector contains the lagged value of the observed process and the observation noise is set to zero. In this application, the state vector serves merely as a technical tool for formulating the correlation structure corresponding to the ARMA(2,1) model and does not have any substantive meaning. An alternative state-space representation for the ARMA model can be found, for example, in Shumway and Stoffer (2000, p. 338).

7.2.3 Linear Mixed Model

The linear mixed model, introduced by Laird and Ware (1982), is one of the most widely used methods for analyzing longitudinal data. A key feature of this model is the use of random effects (γ_j) to account for the unspecified source of variability across subjects. In a typical longitudinal study, a series of measurements is collected from N subjects, denoted as $\mathbf{y_j} = (y_{1j}, \ldots, y_{n_j,j})'$, $j = 1, \ldots, N$, and $t = 1, \ldots, n_j$. Each subject's response vector is modeled as $\mathbf{y_j} = X_j\mathbf{b} + Z_j\gamma_j + u_j$, where X_j is an $n_j \times b$ design matrix, $\mathbf{b}$ is a $b \times 1$ vector of fixed parameters, and Z_j is an $n_j \times g$ design matrix corresponding to the $g \times 1$ vector of random effects, γ_j, which is assumed to be independent across subjects and distributed as $N(0, \Sigma_\gamma)$. The within-subject errors, u_j, are independently distributed as $N(0, \Sigma_u)$.

Again, the Laird–Ware model has a state-space representation (Icaza & Jones, 1998; Jones, 1993). Without loss of generality, we consider a mixed model with a subject-specific random intercept and slope effects and first-order autoregressive within-subject error. For longitudinal observations, y_{tj}, taken over time $t = 1, 2, \ldots, n_j$ from a subject j: $y_{tj} = \beta_{0j} + x_{tj}\beta_{1j} + u_{tj}$, where $u_{tj} = \rho u_{t-1,j} + w_{tj}$, $w_{tj} \sim N(0, \sigma_w^2)$; β_{0j} and β_{1j} represent the subject-specific intercept and slope, $(\beta_{0j}, \beta_{1j})' \sim N((\overline{\beta_0}, \overline{\beta_1})', \Sigma_\beta)$, and $\Sigma_\beta = \begin{pmatrix} \sigma_0^2 & \sigma_{01} \\ \sigma_{01} & \sigma_1^2 \end{pmatrix}$. They can be reparameterized as $\beta_{0j} = \overline{\beta_0} + \gamma_{0j}$ and $\beta_{1j} = \overline{\beta_1} + \gamma_{1j}$, where $(\gamma_{0j}, \gamma_{1j})' \sim N((0, 0)', \Sigma_\beta)$. Therefore, $y_{tj} = (\overline{\beta_0} + x_{tj}\overline{\beta_1}) + (\gamma_{0j} + x_{tj}\gamma_{1j}) + u_{tj}$, $u_{tj} = \rho u_{t-1,j} + w_{tj}$. Expressed in state-space form, the observation equation is given by:

$$y_{tj} = \begin{pmatrix} 1 & x_{tj} \end{pmatrix} \begin{pmatrix} \overline{\beta_0} \\ \overline{\beta_1} \end{pmatrix} + \begin{pmatrix} 1 & x_{tj} & 1 \end{pmatrix} \begin{pmatrix} \gamma_{0tj} \\ \gamma_{1tj} \\ u_{tj} \end{pmatrix} + e_{tj},$$

where e_{tj}, the observational errors, have a normal distribution with zero mean and variance $R = \sigma_e^2$, and are uncorrelated at different times and uncorrelated with the state vector. The state equation is given by

$$\begin{pmatrix} \gamma_{0tj} \\ \gamma_{1tj} \\ u_{tj} \end{pmatrix} = \begin{pmatrix} 1 & 0 & 0 \\ 0 & 1 & 0 \\ 0 & 0 & \rho \end{pmatrix} \begin{pmatrix} \gamma_{0tj} \\ \gamma_{1tj} \\ u_{t-1,j} \end{pmatrix} + \begin{pmatrix} 0 \\ 0 \\ w_{tj} \end{pmatrix},$$

where w_{tj} is a sequence of uncorrelated identically distributed random variables with zero mean and variance σ_w^2 which are uncorrelated with the past of the u_{tj} process. The variance of the first-order autoregressive error, u_{tj}, is equal to $\sigma_w^2/(1 - \rho^2)$. The variances of the subject random intercept and slope effects are expressed as the variances of the initial state for γ_{0tj} and γ_{1tj}, and are estimated as the unknown parameters in the state-space models. Subject random effects are usually assumed to be static in linear mixed models (i.e., the errors are zero in the state equation). In the state-space modeling framework, they can be extended to be dynamic such as $\gamma_{1,t,j} = \varphi\gamma_{1,t-1,j} + \varepsilon_{tj}$ in a straightforward manner. Another possible extension is the inclusion of the covariates in the state equation to account for the random subject effects.

7.2.4 Dynamic Factor Analysis Model

When there are massive amounts of time series collected, such as in an EEG (electroencephalogram) experiment, some of them may show common behavior over time and it may be useful to discover the latent structure that drives multiple time series simultaneously on a common system. One way to capture this common dependence among multiple time series is to use a factor analysis model. In classical factor analysis, it is assumed that each of p variables can be expressed in terms of a linear combination of k ($<p$) common factors and random errors. Factor analysis is usually applied to the cross-sectional data. Generalization of factor analytic theory to time-series data was presented by Brillinger (1975), Geweke (1977), and Stock and Watson (1988), among many others.

For illustration, we consider two time series y_{1tj} and y_{2tj} collected from a subject j, with the noise components obeying a first-order autoregressive process. These two series share a common component, c_t, which also follows a first-order autoregressive process:

$$
\begin{aligned}
y_{1tj} &= \gamma_1 c_{tj} + z_{1tj}, \\
y_{2tj} &= \gamma_2 c_{tj} + z_{2tj}, \\
c_{tj} &= \phi c_{t-1,j} + \nu_{tj}, \qquad \nu_{tj} \sim N(0,1), \\
z_{1tj} &= \rho_1 z_{1,t-1,j} + w_{1tj}, \qquad w_{1tj} \sim N(0, \sigma^2_{w1}), \\
z_{2tj} &= \rho_2 z_{2,t-1,j} + w_{2tj}, \qquad w_{2tj} \sim N(0, \sigma^2_{w2}).
\end{aligned}
$$

The coefficients γ_1 and γ_2 are the factor loadings, which are assumed to be the same across all the subjects, and their values do not change over time. ϕ, ρ_1, and ρ_2 are the autoregressive coefficients. ν_{tj}, w_{1tj}, and w_{2tj} are independent of one another. This model incorporates lagged effects of factors on variables by allowing the factor scores c_{tj} to manifest time-related dependences in the form of autocorrelations. This can be interpreted, for instance, as today's factor scores being directly influenced by yesterday's factor scores. The strength of the influence is determined by the value of ϕ. The corresponding state-space representation is given by

$$
\begin{pmatrix} y_{1t} \\ y_{2t} \end{pmatrix} = \begin{pmatrix} \gamma_1 & 1 & 0 \\ \gamma_2 & 0 & 1 \end{pmatrix} \begin{pmatrix} c_t \\ z_{1t} \\ z_{2t} \end{pmatrix},
$$

$$
\begin{pmatrix} c_t \\ z_{1t} \\ z_{2t} \end{pmatrix} = \begin{pmatrix} \phi_1 & 0 & 0 \\ 0 & \rho_1 & 0 \\ 0 & 0 & \rho_2 \end{pmatrix} \begin{pmatrix} c_{t-1} \\ z_{1,t-1} \\ z_{2,t-1} \end{pmatrix} + \begin{pmatrix} \nu_t \\ w_{1t} \\ w_{2t} \end{pmatrix}.
$$

Note that the observational errors are set to zero and the covariance matrix for the state errors are Q, which is a diagonal matrix with the variance for ν_t, w_{1t}, and w_{2t} on the diagonal. Dynamic factor models are usually fitted by structural equation modeling (SEM) software such as LISREL (Jöreskog & Sörbom, 1996). A lagged covariance matrix, formed by shifting the time series by 0 up to ℓ time points for each subject, has to be constructed first before the LISREL analysis. This matrix has dimension $p(\ell+1) \times p(\ell+1)$, where p is the number of variables and ℓ is the total number of lags (including lag 0). It requires a large memory space to manipulate and store this matrix during the computation if the number of variables or lags is large. For this reason, simultaneous analysis of multiple subjects' lagged covariance matrix by SEM is uncommon. Instead, it is common to pool the lagged covariance matrices from multiple subjects into a single one before the analysis is performed (see, e.g., Nesselroade & Molenaar, 1999). Moreover, users need to specify many parameter constraints in SEM softwares which are easily prone to error. Interested readers can refer to Nesselroade et al. (2002) for examples of fitting the dynamic factor analysis model by LISREL.

Fitting the dynamic factor model as a state-space model, on the other hand, does not require the creation and storage of such a large-dimension matrix. Pooling of multiple subjects' data is also not necessary. In contrast, under the state-space model, the size of the matrix needed to be stored or manipulated is only $p \times p$. This is due to the possibility of running a Kalman filter separately for each subject in calculating the likelihood function. We discuss the details of the Kalman filter in the next section. This example also suggests that the widely used structural equation model in social sciences can be expressed in state-space form. Interested readers can refer to Otter (1986) and MacCallum and Ashby (1986) for more details.

7.3 Parameter Estimation

In state-space models, there are two types of unknown quantities, namely, the state vector, θ_t, and the parameters embedded in the matrices, Γ, A, Φ, R, and Q. The unknown parameter, collectively denoted as Ω, can be obtained by maximizing the likelihood function via the Newton–Raphson or quasi-Newton-type algorithm. These types of algorithm have some disadvantages. First, the corrections in the successive iterations generally involve calculating the inverse of the matrix of second-order partial derivatives, which can be quite large if there are a large number of parameters. Moreover, the successive steps involved in a Newton–Raphson may not necessarily increase the size of the likelihood, or one may encounter extremely large steps which actually decrease the likelihood.

Another common method of estimating these parameters is the expectation-maximization (EM) algorithm of Dempster et al. (1977), as proposed by Shumway and Stoffer (1982). The unattractive features of the Newton–Raphson algorithm can be circumvented using EM. The EM steps always increase the likelihood and guarantee convergence to a stationary point for an exponential family (Wu, 1983). The EM equations usually take on a simple heuristically appealing form, in contrast to the highly nonlinear appearance of the Newton–Raphson or scoring corrections. The EM algorithm may converge slowly in the latter stages of the iterative procedure; one may consider switching to another algorithm such as quasi-Newton at this stage. Since the matrix of second partial derivatives is not computed in the EM algorithm, standard errors of the parameter estimates cannot be obtained directly. However, these partial derivatives can be approximated by perturbing the likelihood function in the neighborhood of the maximum. Computation of the information matrix via recursions is possible as in Harvey (1989) or Cavanaugh and Shumway (1986). Versions of the information matrix, obtained from outputs arising naturally in the EM algorithm, such as in Meng and Rubin (1991) or Oakes (1999), are either hard to compute, as in the former, or will involve relatively untractable derivatives as in the latter. A compromise that is easy to apply and will be robust toward distributional assumptions is the bootstrap, as derived in Stoffer and Wall (1991).

All these methods requires the computation of the estimate for θ_{tj} by Kalman filter and the Kalman smoother procedures. In the two case studies presented in the next section, the parameter estimates were estimated by the EM algorithm and the standard errors were obtained by bootstrapping. Because of space limitations, we cannot discuss the EM algorithm and bootstrapping in state-space models here. Interested readers can refer to Shumway and Stoffer (2000, pp. 321–344) for the details. A technical appendix describing the EM algorithm and bootstrap procedures used in the case studies can be requested from the first author.

In what follows, we discuss the Kalman filter and the Kalman smoother procedures for the state estimation. The following notation will be used. The conditional expectation of the state vector at time t given the data $Y_s = \{y_1, \cdots, y_s\}$, from time 1 up to time s, is denoted as $\theta_t^s = E(\theta_t | y_1, \ldots, y_s)$. The error between the actual value of the state vector at time t and its conditional expectation given Y_s is $(\theta_t - \theta_t^s)$. The covariance matrix between the errors at time t and u is expressed as $P_{t,u}^s = E\{(\theta_t - \theta_t^s)(\theta_u - \theta_u^s)'\}$. Moreover, given the observations, $Y_1, \ldots, Y_n$, the problem of estimating θ_t based on observations *before* time t is called forecasting or prediction in time-series literature. The problem of estimating θ_t based on observations *up to* time t $(t \leq n)$ is called filtering. The problem of estimating θ_t based on *all* of the n observations is called smoothing. Assuming Ω are known, we can estimate the unknown state vector by the Kalman filter procedure. The Kalman filter procedure consists of the two steps, namely, prediction and filtering:

1. *Prediction*. At the beginning of time t, based on all the available information up to time $(t-1)$ (i.e., only observations $y_1, \ldots, y_{t-1}$ have been observed, but not y_t), we can obtain an optimal estimator (in the sense that it minimizes the mean square error) for the state vector, θ_t^{t-1}, and the corresponding error variance–covariance matrix, P_t^{t-1}, can be computed using (7.A.1) and (7.A.2) in the Appendix.
2. *Filtering*. Once y_t is realized at the end of time t, we can compute the prediction error, $e_t = y_t - y_t^{t-1} = y_t - \Gamma z_t - A\theta_t^{t-1}$, with its covariance equal to $\Sigma_t = AP_t^{t-1}A' + R$. This error contains new information for θ_t beyond that contained in θ_t^{t-1}. The object of filtering is to update our knowledge of the system each time a new observation y_t is brought in.

In other words, we can make more accurate inference on θ_t based on the information at time t using (7.A.3) and (7.A.4). The filtered value for the state, θ_t^t, is a combination of predicted value, θ_t^{t-1}, and the prediction error, e_t, weighted by the Kalman gain, K_t, defined in (7.A.5). The Kalman gain, K_t, determines the weight assigned to new information about θ_t contained in the prediction error (v_t). The estimator for the unknown states obtained from the Kalman filter is the best *linear* unbiased estimator for every distribution of the noises, e_t and w_t, and initial distribution of θ_0—best in the sense that the error covariance of any other linear estimator exceeds the error covariance of the Kalman filter estimator by a positive semidefinite matrix. It is the best among all linear and nonlinear estimators if the noises, e_t and w_t, and initial distribution of θ_0 are assumed to be normal (Harvey, 1989). The mean (μ_0) and variance (Σ_0) of the initial distribution are usually unknown in empirical studies. It can be shown, however, that the Kalman filter estimates become independent of the estimates of μ_0 and Σ_0 after sufficient time points. Moreover, the Kalman filter estimator is asymptotically unbiased in the limit (as $n \to \infty$; see Otter, 1985).

Given the Kalman filter estimates for the unknown states, we can compute the log-likelihood of the model based on the prediction error decomposition given by Schweppe (1965), and estimate the unknown parameters by the maximum likelihood method. The log-likelihood for the state-space model (excluding the constant) for longitudinal data is given by

$$-2 \log L(Y_{n1}, \ldots, Y_{nN}; \Omega) \propto \sum_{j=1}^{N} \sum_{t=1}^{n} \log |\Sigma_t| + \sum_{j=1}^{N} \sum_{t=1}^{n} (v_{tj})' \Sigma_t^{-1} (v_{tj}), \tag{7.3}$$

where $\Sigma_t = Cov(v_{tj}) = AP_t^{t-1}A' + R$, and $v_{tj} = y_{tj} - \Gamma z_{tj} - A\theta_{tj}^{t-1}$.

Assuming statistical independence between time series from different subjects, we can compute the Kalman filter estimates separately for each subject with the same initial values and the same parameter estimates. The matrix dimension that needs to be manipulated is only $p \times p$. In other words, after running the Kalman filter for one subject, we restart the filter in the sense of using the same initial conditions and the same parameter values before applying the Kalman filter on another subject's data. Running the Kalman filter from $t = 1, \ldots, n$ for a subject gives us the value of the log-likelihood for that subject. Because of the statistical independence assumption, the joint log-likelihood for the state-space models (7.1) and (7.2) is just the sum of all the individual log-likelihood values.

The EM algorithm for state-space models also requires the computation of the Kalman smoother for θ_t. The Kalman smoother is another estimator for the state (θ_t). At any time t, the Kalman smoother estimator for the state is based on the entire sample (in contrast, the Kalman filter estimator for the state at time t is based on the samples up to t only), and we denote it as $\theta_t^n = E(\theta_t | y_1, \ldots, y_n)$ with the corresponding mean squared covariance estimator as $P_t^n = E\{(\theta_t - \theta_t^n)(\theta_t - \theta_t^n)'|y_1, \ldots, y_n\}$. The recursion formulae to compute these quantities are summarized in the Appendix from (7A.6) to (7A.10). The smoothing procedure provides us with more accurate inference on θ_t, since it uses the entire information from $t = 1, \ldots, n$.

Though state-space modeling has become widespread over the last decade in economics and statistics, there has not been much flexible software for the statistical analysis of general models in state-space form. A modern set of state-space modeling tools are available in SsfPack, which is a suite of C routines for performing computations involving the statistical analysis of univariate and multivariate models in state-space form. The routines can fit a wide variety of state-space forms from standard autoregressive moving average to complicated time-varying models. Interested readers can refer to the papers by Koopman et al. (1999, 2001) for the technical details of the algorithms used in the SsfPack package. The SsfPack routines are implemented in Ox (Doornik, 2001) and S-Plus module S+FinMetrics (Insightful Corporation, 2002). An extension of SsfPack called the SsfNong.ox package was developed to fit some specific non-Gaussian state-space models. More details can be found in http://www.ssfpack.com. There are other softwares such as ASTSA (Shumway and Stoffer, 2000; see http://www.stat.pitt.edu/stoffer/tsa.html for download details), the SAS PROC STATESPACE procedure (SAS Institute Inc., 1999), and the SSATS and TSM modules in the Gauss package (Aptech Systems Inc., 2003) that can fit linear state-space models with time-invariant A, Γ, and Φ matrices.

In addition, the high-performance numeric computation and visualization software package MATLAB (MathWorks, 2004) provides a flexible environment for straightforward implementation of the Kalman filter and smoother. Estimation of state-space models can be done through the nonlinear optimization built-in functions such as "fminunc" and "fminsearch" available in MATLAB Optimization toolbox. There is also a MATLAB toolbox for state-space models analysis called E4 that can be downloaded for free from http://www.ucm.es/info/icae/e4/e4download.htm#downe4. The software is not designed for analysis of replicated time series and user's modification is needed. The analyses reported in the next section were performed by MATLAB. West and Harrison (1997) presented the formulation and estimation of Gaussian linear state-space models from a Bayesian point of view, and a free shareware called BATS can be downloaded from ftp.stat.duke.edu/pub/bats/ to perform the Bayesian computation for linear state-space models.

7.4 Application 1: Connectivity Analysis with fMRI Data

7.4.1 Model Formulation

In this section, we present an empirical application of the state-space models for investigating the mechanism of attentional control network from a single-subject functional magnetic resonance imaging (fMRI) experiment.

Prior to the experiment, the subject learned how to name three unfamiliar shapes with a unique color word ("blue," "yellow," and "green"). During the experiment, the subjects were required to name each shape subvocally with the corresponding color labels they had learned and to ignore the ink color in which the shape was presented. Some of these shapes were printed in white (control stimulus) and some of them were printed in an ink color that did not match the color label (experimental stimulus). The ink color information in the stimulus is processed by the lingual gyrus (LG) while the shape information is processed by the middle occipital gyrus (MOG) in the brain. Our brains have limited processing capacity and it is important to have mechanisms to filter out task-irrelevant information (i.e., the ink color in this experiment) and select task-relevant information (i.e., the shape in this experiment). It has been proposed (Banich et al., 2000) that this is coordinated by an attentional control system (located at the dorsolateral prefrontal cortex, DLPFC). We examine this neuroscientific hypothesis from the fMRI data through a state-space model. Four hundred and forty brain scans ($n = 440$) were collected during the experiment. The time series from LG, MOG, and DLPFC were extracted and the fMRI data are shown in figure 7.1. We denote the fMRI time series from LG, MOG, and DLPFC as y_{1t}, y_{2t}, and y_{3t}, respectively.

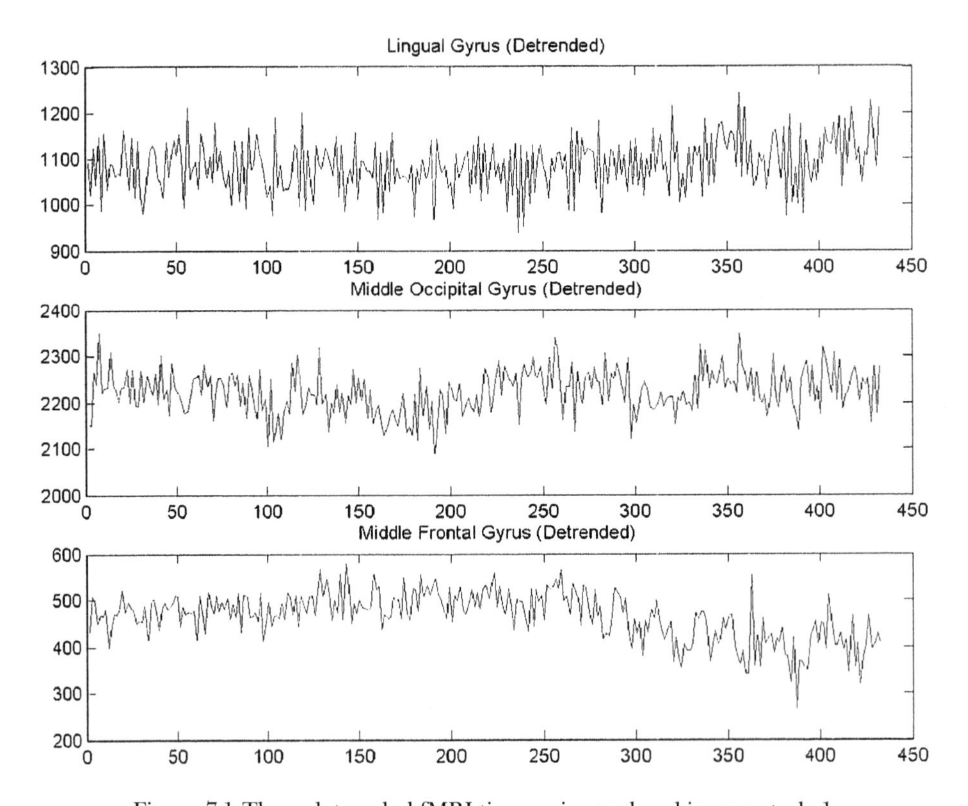

Figure 7.1 Three detrended fMRI time series analyzed in case study 1.

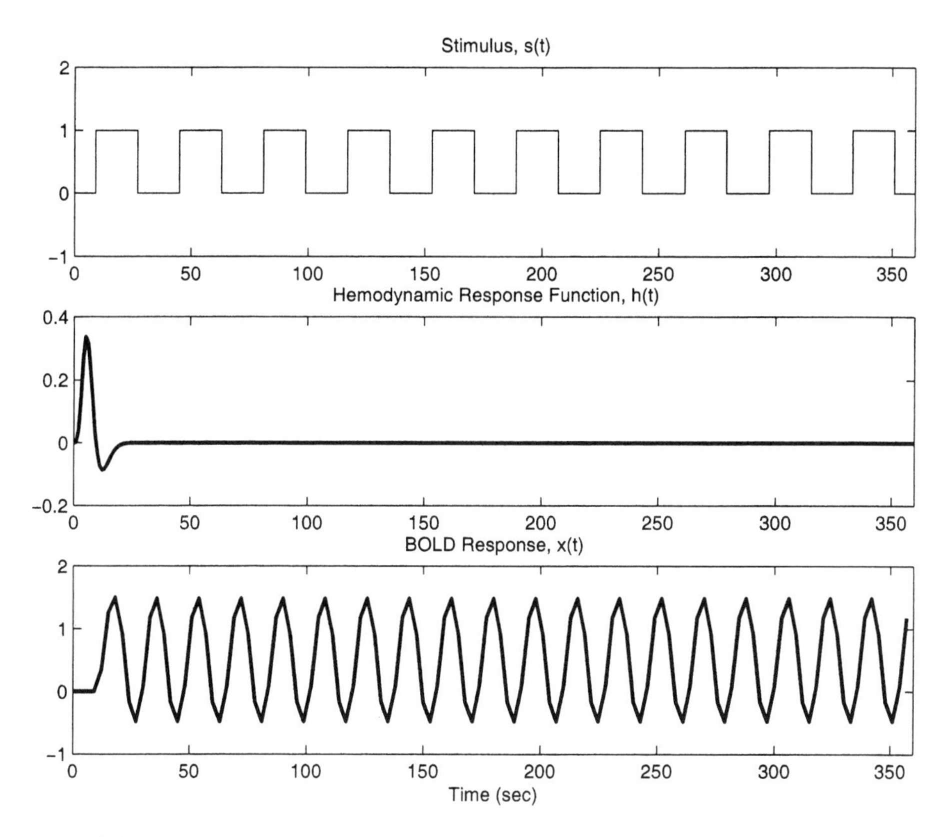

Figure 7.2 Top panel: Stimulus indicator function. Middle panel: Hemodynamic response function. Bottom panel: Convolution of hemodynamic response function with the stimulus indicator function.

The data measured in fMRI experiments are primarily generated by the changes in oxygenation level in the blood. The first step in the statistical analysis of fMRI data is to build a model of how the data are generated by an experimental stimulus. Suppose the external stimulus is given by s_t and a noise-free fMRI response at time t is given by a_t. A fMRI response in the brain is generated after an external stimulus is presented. This response is not instantaneous but usually begins roughly 2 seconds later and rises to a peak at about 6 seconds after stimulus onset, and then returns to baseline. This phenomenon is represented by a hemodynamic response function, h_u (see the second panel in figure 7.2), which is usually approximated by a Poisson, Gaussian, or gamma density function with the parameters chosen a priori by the researchers.

In a typical fMRI experiments, stimuli are often presented repeatedly. When the stimuli are presented close in time (i.e., before a fMRI response returns to baseline), the fMRI responses to successive stimuli superpose in a linear fashion and are commonly captured by the convolution operation as $a_t = \int_0^\infty h_u s_{t-u}\, du$, where s_t takes the value of '1' when the stimulus is presented and '0' when there is no stimulus presented at time t. The top panel in figure 7.2 shows a stimulus, s_t, presented periodically in a fMRI experiment and the bottom panel shows what the fMRI response looks like after the convolution with the periodic stimulus function in the top panel. Notice that the value a_t is a quantity primarily determined by the design of the experiment and can be regarded as known.

Since different parts of the brain are responsible for different functions, the magnitude of the fMRI signal varies over brain regions and experimental conditions. This unknown magnitude (β_{it}), referred to as the *activation* in fMRI literature, is usually estimated by the general linear model:

$$y_{it} = \alpha_i + \beta_{it} a_{it} + e_{it}, \tag{7.4}$$

where y_{it} is the observed fMRI signal recorded by the fMRI scanner, and e_{it} is measurement error at region i at time t. The intercept term α_i represents the baseline activity or activity at the control condition.

Functionally specialized brain areas do not operate on their own but interact with one another depending on the context. We therefore augment the above model to estimate the influence of one brain region on another, referred to as the *effective connectivity* in brain-imaging literature (Friston, 2004). We proposed to model the time-varying "activation" in (7.4) in terms of the history of the error-free fMRI signal from itself and another region as

$$\beta_{it} = \lambda_{i \cdot i} Z_{i,t-1} + \lambda_{i \cdot j} Z_{j,t-1} + w_{i,t}, \tag{7.5}$$

where $Z_{i,t-1} = x_{i,t-1}\beta_{i,t-1}$ and $Z_{j,t-1} = x_{j,t-1}\beta_{j,t-1}$ represent the error-free fMRI signal from a previous time point ($t - 1$). Higher-order effects of the BOLD history (from $t - 2$, $t - 3$, and so on) may also be considered. In this model, the coefficients $\lambda_{i \cdot i}$ and $\lambda_{i \cdot j}$ measure the corresponding impact on region i from itself and region j. The first coefficient reflects the self-feedback and the second coefficient characterizes the coupling relationship between the two regions i and j. The models (7.4) and (7.5) have a state-space representation which will be given below.

The flexibility of the state-space approach allows us to explore many possible alternative mechanisms of attentional control, for example, if the DLPFC suppresses the activation of the LG ($\lambda_{1 \cdot 3} < 0$) and facilitate the activation of the MOG ($\lambda_{2 \cdot 3} > 0$); if there is reciprocal suppression between LG and MOG ($\lambda_{1 \cdot 2} < 0$ and $\lambda_{2 \cdot 1} < 0$) (as our brain's processing capacity is limited, the LG and the MOG may need to compete for the "resources"); or if there is feedback from the LG and the MOG on the DLPFC ($\lambda_{3 \cdot 1} \neq 0$ and $\lambda_{3 \cdot 2} \neq 0$). These mechanisms can be tested separately or simultaneously. We present below the state-space formulation of one possible model which implies that the LG, the MOG, and the DLPFC reciprocally influence on each other (M1).

Observation equations:

$$\begin{pmatrix} y_{1t} \\ y_{2t} \\ y_{3t} \end{pmatrix} = \begin{pmatrix} \alpha_1 & 0 & 0 \\ 0 & \alpha_2 & 0 \\ 0 & 0 & \alpha_3 \end{pmatrix} \begin{pmatrix} 1 \\ 1 \\ 1 \end{pmatrix} + \begin{pmatrix} a_{1t} & 0 & 0 \\ 0 & a_{2t} & 0 \\ 0 & 0 & a_{3t} \end{pmatrix} \begin{pmatrix} \beta_{1t} \\ \beta_{2t} \\ \beta_{3t} \end{pmatrix} + \begin{pmatrix} e_{1t} \\ e_{2t} \\ e_{3t} \end{pmatrix},$$

or compactly as $Y_t = \Pi \underline{1}_3 + A_t B_t + E_t$.

State equations:

$$\begin{pmatrix} \beta_{1t} \\ \beta_{2t} \\ \beta_{3t} \end{pmatrix} = \begin{pmatrix} \lambda_{1\cdot 1} a_{1,t-1} & \lambda_{1\cdot 2} a_{2,t-1} & \lambda_{1\cdot 3} a_{3,t-1} \\ \lambda_{2\cdot 1} a_{1,t-1} & \lambda_{2\cdot 2} a_{2,t-1} & \lambda_{2\cdot 3} a_{3,t-1} \\ \lambda_{3\cdot 1} a_{1,t-1} & \lambda_{3\cdot 2} a_{2,t-1} & \lambda_{3\cdot 3} a_{3,t-1} \end{pmatrix} \begin{pmatrix} \beta_{1,t-1} \\ \beta_{2,t-1} \\ \beta_{3,t-1} \end{pmatrix} + \begin{pmatrix} w_{1t} \\ w_{2t} \\ w_{3t} \end{pmatrix},$$

or compactly as $B_t = (\Lambda A_{t-1})B_{t-1} + W_t$, where Λ is a 3×3 matrix containing the 9 $\lambda_{i\cdot j}$ coefficients.

Note that, in this example, the coefficient matrix Π in (7.1) contains the activations at the control conditions, $\alpha_1, \alpha_2, \alpha_3$ on its diagonal, and the coefficient matrix A contains the convoluted hemodynamic response function which is time-varying but the values of which are known. For convenience, we write $\Pi_v = (\alpha_1, \alpha_2, \alpha_3)'$. The coefficient matrix Φ in (7.2) is also time-varying now and can be written as $\Phi_t = \Lambda A_{t-1}$. The state vector consists of the time-varying magnitude of the fMRI response, β_{it}, induced by the experimental stimuli. Assuming both the measurement errors ($R = cov(e_{1t}, e_{2t}, e_{3t})'$) and process errors ($Q = cov(w_{1t}, w_{2t}, w_{3t})'$) to be normally distributed and independent of each other, we apply the EM algorithm to obtain the unknown parameters Π_v, Λ, R, and Q. The results are shown in table 7.1.

7.4.2 Results

The mechanism implied by model M1 is quite complex. We fitted five other models to see if all these connections are necessary. The fit summary is shown in table 7.1. We reported $-2*$log-likelihood and BIC (Bayesian Information Criterion) for model comparison purposes. For those models with nested relationship, we performed a likelihood-ratio test (LRT) to check if the model can be simplified. The results are summarized in table 7.2. From both the nested model comparisons and BIC, among the six candidate models, M2 fits the best in terms of goodness-of-fit and parsimony.

We also checked the normality and homoscedasticity assumptions based on the standardized one-step forecast errors, $r_t = v_t/\sqrt{var(v_t)}$, where $v_t = y_t - \Gamma z_t - A\theta_t^{t-1}$ can be obtained from the Kalman filter. Under the assumption of a Gaussian model, the forecast errors should be normally distributed and serially independent with unit variance. The normality assumption can be checked by a standard normal quantile–quantile (QQ) plot, which is a graphical display of ordered residuals against their theoretical (i.e., normal) quantiles with the 45° line taken as the reference. If the normality assumption is valid, the residuals should be very close to this line.

Table 7.1 Model fitting summary

Model	Λ	−2*log-likelihood	No. of parameters	BIC
M1	$\begin{pmatrix} \lambda_{1\cdot1} & \lambda_{1\cdot2} & \lambda_{1\cdot3} \\ \lambda_{2\cdot1} & \lambda_{2\cdot2} & \lambda_{2\cdot3} \\ \lambda_{3\cdot1} & \lambda_{3\cdot2} & \lambda_{3\cdot3} \end{pmatrix}$	7021.34	36	25.41
M2	$\begin{pmatrix} \lambda_{1\cdot1} & 0 & \lambda_{1\cdot3} \\ 0 & \lambda_{2\cdot2} & \lambda_{2\cdot3} \\ \lambda_{3\cdot1} & \lambda_{3\cdot2} & \lambda_{3\cdot3} \end{pmatrix}$	7023.62	34	25.38
M3	$\begin{pmatrix} \lambda_{1\cdot1} & 0 & \lambda_{1\cdot3} \\ 0 & \lambda_{2\cdot2} & \lambda_{2\cdot3} \\ 0 & 0 & \lambda_{3\cdot3} \end{pmatrix}$	7068.95	32	25.50
M4	$\begin{pmatrix} \lambda_{1\cdot1} & \lambda_{1\cdot2} & 0 \\ \lambda_{2\cdot1} & \lambda_{2\cdot2} & 0 \\ 0 & 0 & \lambda_{3\cdot3} \end{pmatrix}$	7081.01	32	25.54
M5	$\begin{pmatrix} \lambda_{1\cdot1} & 0 & 0 \\ 0 & \lambda_{2\cdot2} & 0 \\ \lambda_{3\cdot1} & \lambda_{3\cdot2} & \lambda_{3\cdot3} \end{pmatrix}$	7056.60	32	25.45
M6	$\begin{pmatrix} \lambda_{1\cdot1} & 0 & 0 \\ 0 & \lambda_{2\cdot2} & 0 \\ 0 & 0 & \lambda_{3\cdot3} \end{pmatrix}$	7145.28	30	25.73

A simple test for heteroscedasticity (constant variance) is to compare the sum of squares of two exclusive subsets of the sample with sizes h and $(n-h)$ (Harvey, 1989). The test statistic is defined as $H(h) = \sum_{t=n-h+1}^{n} r_t^2 / \sum_{t=1}^{h} r_t^2$, which is $F_{h;h}$-distributed for some preset positive integer h, under the null hypothesis of homoscedasticity. A standard test statistic for testing the significance of the first k residual autocorrelations ($H0 : \rho_1 = \cdots = \rho_k = 0$) is based on the Box–Ljung statistic (Ljung & Box, 1978) and is given by $Q(k) = T(T+2)\sum_{j=1}^{k}(\epsilon_j^2/T-j)$, for some preset positive integer k, where $\epsilon_j = (1/Tm_2)\sum_{t=j+1}^{T}(r_t - m_1)(r_{t-j} - m_1)$, and

Table 7.2 Model comparison summary

Model comparisons	LRT	DF	p-value
M1 vs. M2	2.38	2	0.32
M1 vs. M4	59.73	4	<0.0001
M2 vs. M3	45.33	2	<0.0001
M2 vs. M5	32.98	2	<0.0001

Table 7.3 Maximum likelihood estimators and standard errors for M2

Parameters	Maximum likelihood estimators	SE
$\begin{pmatrix}\gamma_1\\ \gamma_2\\ \gamma_3\end{pmatrix}$	$\begin{pmatrix}342.39\\ -22.13\\ 76.34\end{pmatrix}$	$\begin{pmatrix}3.41\\ 3.29\\ 2.78\end{pmatrix}$
$\begin{pmatrix}\lambda_{1\cdot1} & 0 & \lambda_{1\cdot3}\\ 0 & \lambda_{2\cdot2} & \lambda_{2\cdot3}\\ \lambda_{3\cdot1} & \lambda_{3\cdot2} & \lambda_{3\cdot3}\end{pmatrix}$	$\begin{pmatrix}8.18 & - & -11.23\\ - & 1.88 & 0.39\\ 5.77 & 0.15 & -8.39\end{pmatrix}$	$\begin{pmatrix}0.066 & - & 0.078\\ - & 0.20 & 0.18\\ 0.061 & 0.065 & 0.068\end{pmatrix}$
$\begin{pmatrix}\sigma_{w1}^2\\ \sigma_{w2}^2\\ \sigma_{w3}^2\end{pmatrix}$	$\begin{pmatrix}10.66\\ 84.37\\ 2.89\end{pmatrix}$	$\begin{pmatrix}6.01\\ 246.58\\ 0.95\end{pmatrix}$
$\begin{pmatrix}\sigma_{e1}^2\\ \sigma_{e2}^2\\ \sigma_{e3}^2\end{pmatrix}$	$\begin{pmatrix}1020\\ 1720.1\\ 957.74\end{pmatrix}$	$\begin{pmatrix}122.98\\ 180.47\\ 82.03\end{pmatrix}$

m_1 and m_2 are the mean and variance of r_t. The distribution of Q is approximately χ^2 with (k minus number of parameters) degrees of freedom. No significant violation of assumptions was found and we therefore skipped the details here.

The maximum likelihood estimates and the standard errors computed by the bootstrap (250 bootstrapped samples) are shown in table 7.3. From this, we can see that there is a significant suppression from DLPFC on LG ($\lambda_{1\cdot3}$) and relatively weak facilitation from DLPFC on MOG ($\lambda_{2\cdot3}$). These results are consistent with the proposal of Banich et al. (2000). We also find that there is positive feedback from the LG and MOG on DLPFC ($\lambda_{3\cdot1}$ and $\lambda_{3\cdot2}$). Moreover, DLPFC shows negative self-feedback control on itself ($\lambda_{3\cdot3}$). Our results provide support for the attentional control network hypothesis.

7.5 Application 2: Testing the Induced Demand Hypothesis from Matched Traffic Profiles

7.5.1 Model Formulation

In this application,[1] we consider another application of the state-space model for analysis of traffic volumes on 18 California highways for matched pairs of improved and unimproved sections of highways. The sections whose capacities were improved in the 1970s were paired with unimproved (control) segments, matching on facility type, region, approximate size, initial traffic volumes, and congestion. Details of the construction of the matched pairs and a repeated measures analysis can be found in Mokhtarian et al. (2002). The data are from the

California Department of Transportation (CALTRANS) and represent yearly levels of average daily traffic for 22 years spanning the period from 1976 to 1997. The basic question is whether simply improving the road induces increased demand through demographic changes in either or both members of the pair. Hence, the data can be regarded as panel time-series data consisting of 18 paired unimproved and improved time series, measured over the 22-year period. A complicating factor for which the state-space model is ideally suited is that 14 out of the 36 values for 1997 are missing for various unknown reasons.

Figure 7.3 shows all profiles in the left panels and the means of the improved and unimproved flows in the right panels. Because average daily traffic (ADT) increases almost monotonically from year to year, it is convenient to work with the log-transformed variables and to focus on year-to-year differences. The logarithmic transformation also tends to stabilize the variance, as can be noted from the left panels of figure 7.3. In addition, the year-to-year differences in the log-transformed ADT values express the yearly percentage increases and form the natural basis for comparing the unimproved and improved segments. We use

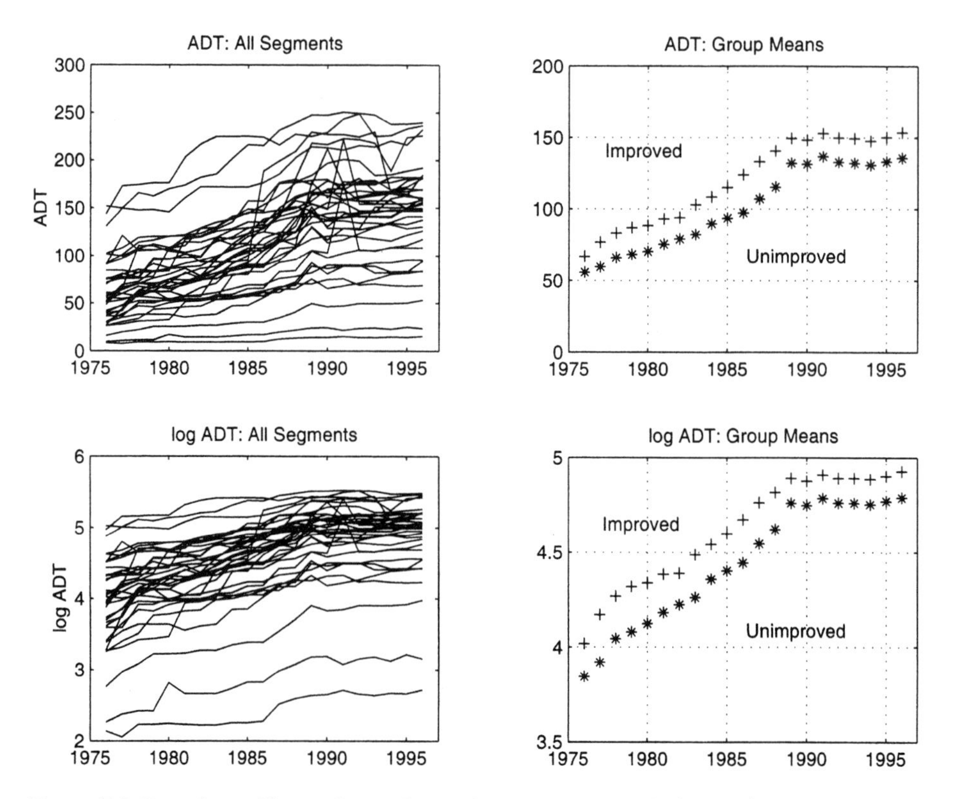

Figure 7.3 Growth profiles and transformed growth profiles (1976–1997). Mean profiles for the unimproved and improved raw and transformed segments are shown on the right side, respectively.

the notation $y_{1\ell t}$ and $y_{2\ell t}$ for the logarithmically transformed unimproved and improved segments, respectively, measured in year $t = 1, \ldots, n$ for the $\ell = 1, \ldots, N$ segments, where, in this case, $n = 22$ and $N = 18$. The yearly differences $y_{i\ell t} - y_{i\ell,t-1}, i = 1, 2$, can be interpreted as percentage increases in going from year $t - 1$ to year t and are also convenient for cases such as this where unimproved and improved segments may have different scales.

To be more specific, assume that the unimproved and improved segment vectors, say $\mathbf{y}_{t\ell} = (y_{1t\ell}, y_{2t\ell})'$, $t = 1, \ldots, 22, \ell = 1, \ldots, N$, can be modeled as the sum of a stochastic (i.e., time-varying) mean vector function $\boldsymbol{\mu}_{t\ell} = (\mu_{1t\ell}, \mu_{2t\ell})'$ and error, say

$$\mathbf{y}_{t\ell} = \boldsymbol{\mu}_{t\ell} + \boldsymbol{e}_{t\ell}. \tag{7.6}$$

The stochastic mean function is usually called the spline function in nonparametric statistics literature. In this case, Mokhtarian et al. (2002) have used this model as a complement to the repeated measures analysis, where the mean value functions are assumed to satisfy a random walk. We propose here an analysis that imposes additional smoothness on fitted profiles by replacing the first difference in the random walk model with a second difference, as discussed in section 7.2, obtained by imposing the constraints

$$\nabla^2 \boldsymbol{\mu}_{t\ell} = w_{t\ell}. \tag{7.7}$$

$\ell = 1, 2, \ldots, N$ where ∇^2 denotes the second difference, that is, $\nabla \boldsymbol{\mu}_{t\ell} = \boldsymbol{\mu}_{t\ell} - \boldsymbol{\mu}_{t-1,\ell}$, and $\nabla^2 \boldsymbol{\mu}_{t\ell} = \nabla(\nabla \boldsymbol{\mu}_{t\ell}) = \nabla \boldsymbol{\mu}_{t\ell} - \nabla \boldsymbol{\mu}_{t-1,\ell} = \boldsymbol{\mu}_{t\ell} - 2\boldsymbol{\mu}_{t-1,\ell} + \boldsymbol{\mu}_{t-2,\ell}$. The stochastic mean vectors are assumed initially to be bivariate normal with unknown mean vectors $\boldsymbol{\mu}_{0\ell}, \ell = 1, \ldots, N$, and known covariance matrix. The first difference, for example $\boldsymbol{\mu}_{t\ell} - \boldsymbol{\mu}_{t-1,\ell}$, can be interpreted as the transformed mean percentage differences between year $t - 1$ and year t. Note here that the observation equation is (7.6) and that (7.7) corresponds to the state equation discussed earlier. We see below that it can be expressed in exactly the state-space form. We assume here that the two errors $\boldsymbol{e}_{t\ell}$ and $w_{t\ell}$ are independent of each other for all ℓ and t, with common covariance matrices R and Q_{11}. The problems of interest for this model are (i) estimating the stochastic mean profiles, $\boldsymbol{\mu}_{t\ell}$, using the Kalman smoother, and (ii) estimating the initial mean and covariance parameters R and Q_{11} by maximum likelihood.

In order to use the state-space model (7.1) and (7.2), we define the state vector as $x_t = (\mu_{1t\ell}, \mu_{2t\ell}, \mu_{1,t-1,\ell}, \mu_{2,t-1,\ell})'$ and write the observation and state equations for (7.6) and (7.7) as

$$\begin{pmatrix} y_{1t\ell} \\ y_{2t\ell} \end{pmatrix} = \begin{pmatrix} 1 & 0 & 0 & 0 \\ 0 & 1 & 0 & 0 \end{pmatrix} \begin{pmatrix} \mu_{1t\ell} \\ \mu_{2t\ell} \\ \mu_{1,t-1,\ell} \\ \mu_{2,t-1,\ell} \end{pmatrix} + \begin{pmatrix} e_{1t\ell} \\ e_{2t\ell} \end{pmatrix}$$

and

$$\begin{pmatrix} \mu_{1t\ell} \\ \mu_{2t\ell} \\ \mu_{1,t-1,\ell} \\ \mu_{2,t-1,\ell} \end{pmatrix} = \begin{pmatrix} 2 & 0 & -1 & 0 \\ 0 & 2 & 0 & -1 \\ 1 & 0 & 0 & 0 \\ 0 & 1 & 0 & 0 \end{pmatrix} \begin{pmatrix} \mu_{1,t-1,\ell} \\ \mu_{2,t-1,\ell} \\ \mu_{1,t-2,\ell} \\ \mu_{2,t-2,\ell} \end{pmatrix} + \begin{pmatrix} w_{t1} \\ w_{t2} \\ 0 \\ 0 \end{pmatrix}.$$

The covariance matrix of $(e_{t1}, e_{t2})'$ will be R and the covariance matrix Q will have the covariance matrix of $(w_{t1}, w_{t2})'$ in the upper left-hand corner and will be zero elsewhere. Under the above specifications, the Kalman smoother results in the estimated means as the first two components of $\boldsymbol{x}_{t\ell}^{n}$ and the estimated covariance matrix of these smoothers as the upper 2×2 matrix of $P_{t\ell}^{n}$.

7.5.2 Results

Figure 7.4 shows partial results for fitting the 18 profiles. The upper panels show the smoothed estimators compared with the observed profiles for transformed average daily traffic in the unimproved and improved segments, numbers $\ell = 6, 11$. The segments are typical in that they both exhibit nonlinear growth patterns with roughly parallel behavior. The 18 segments, in general, tended to be fairly consistent and it was not always the case that the unimproved volumes were lower than the improved profiles. The fitted mean profiles again show roughly parallel behavior. Note that we used the EM algorithm to estimate elements of the observation covariance matrix R and the state covariance matrix Q_{11}. The estimated standard errors were about 0.06 and 0.05 for the observation matrix, with a smallish correlation of 0.17. The standard errors for the states were about 0.04 and 0.05 with a negligible correlation that was constrained to zero in later runs.

We are also interested in evaluating whether induced demand occurs in this context, interpreted here as determining whether there are differential percentage growth rates. This can be evaluated by estimating the difference

$$\boldsymbol{c}'\boldsymbol{\mu}_{t\ell} = [\mu_{2t\ell} - \mu_{2,t-1,\ell}] - [\mu_{1t\ell} - \mu_{1,t-1,\ell}], \tag{7.8}$$

which compares the percentage rates of increase for the improved and unimproved segments on a year-by-year basis using the contrast vector $\boldsymbol{c} = (-1, 1, 1, -1)'$. Here, we note that a test of the form $\boldsymbol{c}'\boldsymbol{\mu}_{t\ell} = 0$ would indicate whether or not the profile ℓ is parallel, that is, the parallelism test often used on group means in sociological applications. The test will be applied later to the mean profiles

$$\bar{\delta}_t = \frac{1}{N}\sum_{\ell=1}^{N} \boldsymbol{c}'\boldsymbol{\mu}_{t\ell}, \tag{7.9}$$

providing a direct test of whether the mean growth profiles in figure 7.3 are parallel.

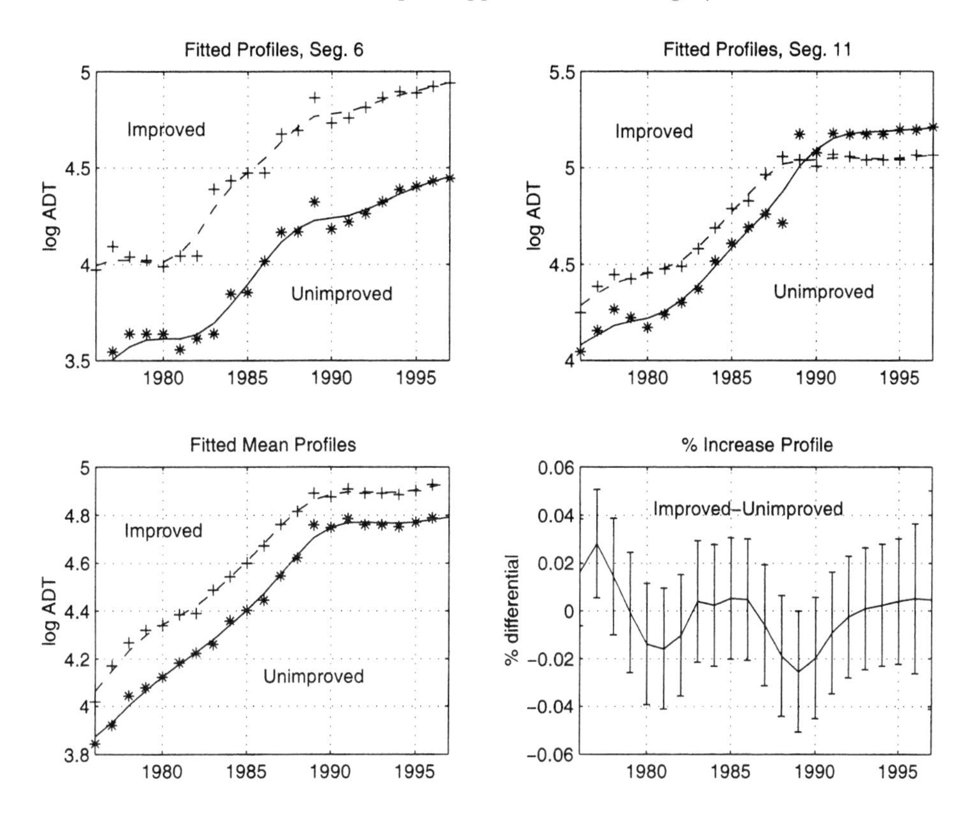

Figure 7.4 Fitted profiles for segments 6 and 11 in the top panel. The bottom left panel contains the fitted mean profiles. The bottom right panel shows the estimated difference between the percentage changes of the improved and unimproved means. The vertical bars are 95% simultaneous intervals using Scheffe's method.

Note that we assume that the replicates over the $\ell = 1, 2, \ldots, N$ segments are independent with common covariance matrices R and Q corresponding to the observation and state equations respectively. Of interest are the smoothed traffic profiles for each segment, $\boldsymbol{x}_{t\ell}^{n}$ and the mean smoothed profiles

$$\bar{d}_t = \sum_{\ell=1}^{n} \boldsymbol{c}'\boldsymbol{x}_{t\ell}^{n}/N \tag{7.10}$$

as an estimator for the mean difference profile in (7.9). The test statistic for the mean difference profiles can be generated from the state-space model by considering the linear contrast. The contrast will have variance

$$\sigma_t^2 = \frac{1}{N^2}\sum_{\ell=1}^{N} \boldsymbol{c}'P_{t\ell}^{n}\boldsymbol{c}. \tag{7.11}$$

Using a Scheffe's method combined with Bonferonni and taking the observed covariance matrices R and Q as known, a $(1 - \alpha)$ simultaneous confidence interval can be constructed for all t as

$$\bar{d}_t \pm \sigma_t z(1 - \alpha/2n). \tag{7.12}$$

Estimation of the two covariance matrices using the EM algorithm requires modifications for replication and for the missing data that may be found in section 7.5.1, in the Technical Appendix (section 5), and in Shumway and Stoffer (2000, sections 4.4 and 4.11).

In order to evaluate formally the parallel tendency of the estimated mean profiles in figure 7.4, we estimated the average mean percentage difference (7.9) by (7.10) and plotted the simultaneous 95% intervals using (7.12). The estimated difference profiles are shown in the bottom right panel of figure 7.4. Note that, after an initial phase between 1976 and 1980 where the improved grew faster, the two grew at roughly the same rates between 1983 and 1987, after which the unimproved grew slightly faster until both leveled off at roughly equal growth. The simultaneous 95% simultaneous confidence intervals exclude zero only for the first year, 1976. The constant multiplier that determines the Scheffe intervals could be reduced by about one-third, but this would still not lead to a statistically larger rate of increase for the improved segments.

Of course, it is possible that our matching of segments has ignored some potentially important covariate that is substantially different for all groups, so that adjusting for this covariate within the model would reduce the variability enough to detect a significant difference. In fact, the addition of fixed covariates to the state-space model introduces no significant additional problems (see Shumway, 2000). The other caveat that might have led to a different result is the fact that many of the improved segments were completed before 1976, the beginning of this particular dataset. What the results do seem to show is that long-term differences in growth rates of unimproved and improved segments are not substantially different so that, at the very least, induced traffic cannot be detected under the above assumptions.

7.6 Conclusions

7.6.1 Final Remarks on the State-Space Approach

We have introduced the basic ideas of the state-space model and have motivated it as a general form that can generate a richly varied collection of standard statistical models such as regression models, autoregressive moving average models, linear mixed effect models for longitudinal data, and dynamic factor analysis models. Furthermore, using the EM algorithm in a maximum-likelihood procedure developed by Shumway and Stoffer (1982) allows estimation for the general state-space model that parallels the application of that procedure to mixed models given by Laird and Ward (1982). Another advantage of the state-space formulation is that it contains within it a simplified method for dealing with missing observations (see next section), allowing unequally spaced data, while retaining options for introducing rich dynamic models for time correlation using the state equations, as discussed in sections 7.1 and 7.2. Furthermore, using the state-space approach, one can avoid inverting large matrices resulting in efficient computations (involving a matrix size of $p \times p$). Mixed model theory would require the inversion of the covariance matrix of the observations from all the subjects together (i.e., a matrix size of $pN \times pN$). This quickly becomes a serious issue in handling intensive longitudinal data if both N and p are large.

The state-space method is based on the structural analysis of the problem, and it is up to the investigators to identify the relevant features (such as monotonic trend, seasonal pattern, shift in the level at some periods) and to build these components into the model. This is not quite the case in the classical Box–Jenkins ARIMA modeling, which is a "black box" data-driven approach. Before the analysis, users need to transform the time series to be stationary, such as by differencing or demeaning. State-space models, on the other hand, allow users to "build in" known or unknown changes in the system in a flexible manner without requiring these preprocessing steps. To end this chapter, we should like to add some remarks on how to modify the state-space method presented above to handle missing data, and nonlinear non-Gaussian data issues which may be encountered in intensive longitudinal data analysis.

7.6.2 Missing Data Modification

It is not uncommon to have missing data in time-series observations or for the data to be observed irregularly over time and space. An attractive feature of the state-space model is its flexibility to handle these situations. At a given time t, we partition the $p \times 1$ observations into two components, $y_t = (y_t^{(1)\prime}, y_t^{(2)\prime})'$, where the first $p_{1t} \times 1$ component is observed and the second $p_{2t} \times 1$ component is unobserved, where $p_{1t} + p_{2t} = p$. The corresponding observation equations in the state-space model can be partitioned in a similar way as

$$\begin{pmatrix} y_t^{(1)} \\ y_t^{(2)} \end{pmatrix} = \begin{pmatrix} A_t^{(1)} \\ A_t^{(2)} \end{pmatrix} x_t + \begin{pmatrix} e_t^{(1)} \\ e_t^{(2)} \end{pmatrix},$$

where $A_t^{(1)}$ and $A_t^{(2)}$ are, respectively, the $p_{1t} \times p$ and $p_{2t} \times p$ partitioned observation matrices, and

$$cov \begin{pmatrix} v_t^{(1)} \\ v_t^{(1)} \end{pmatrix} = \begin{pmatrix} R_t^{(11)} & R_t^{(12)} \\ R_t^{(21)} & R_t^{(22)} \end{pmatrix}$$

denotes the covariance matrix of the measurement errors between the observed and unobserved components. Shumway and Stoffer (1982) presented the filtering equations for handling missing data. At the updating step (7.A.3 and 7.A.4), we need to replace the unobserved component by zero in y and A at time t, and off-diagonal elements of R to be zeros as

$$y_t = \begin{pmatrix} y_t^{(1)} \\ 0 \end{pmatrix}, \quad A_t = \begin{pmatrix} A_t^{(1)} \\ 0 \end{pmatrix}, \quad \text{and} \quad R_t = \begin{pmatrix} R_t^{(11)} & 0 \\ 0 & R_t^{(22)} \end{pmatrix}.$$

The smoother values can then be obtained based on the missing-data filtered values. The maximum likelihood estimators, computed via the EM procedure, also need to be modified to accommodate the missing data. Because of space limitations, interested readers should refer to Shumway and Stoffer (2000, section 4.4) for more details.

7.6.3 Nonlinear and Non-Gaussian State-Space Model

In this chapter, we assumed the state-space model to be linear (i.e., the unobserved states and observed time series are related in a linear fashion) and Gaussian (i.e., in reference to the distribution of the measurement errors and process errors). These assumptions are chosen for simplicity. However, there are situations where the linear Gaussian model fails to provide suitable representation of the data. For example, the number of incidents of emergency visits over time can be modeled more appropriately by Poisson distribution rather than by normal. For such a situation, it is necessary to consider a general version of the state-space model, which may be written as $y_t = H_t(x_t, e_t)$ and $x_t = F_t(x_{t-1}, w_t)$, where F_t and H_t are known functions that may depend on parameters Θ, and e_t and w_t are white noise processes. The likelihood of this model is given by $L_{X,Y}(\Theta) = p_\Theta(x_0)\prod_{t=1}^{n}(x_t|x_{t-1})p_\Theta(y_t|x_t)$. A general method to obtain the estimates for the parameters, Θ, and the unobserved states is through the Markov chain Monte Carlo method. This topic is still a very active research area. Durbin and Koopman (2001, chapters 10–14) present a state-of-the-art discussion on simulation techniques for estimating nonlinear and non-Gaussian state-space models.

ACKNOWLEDGMENTS

We thank Mike Milham for permission to use data and for his help in preparation of the fMRI data used in this chapter. The authors would also like to thank the editors and the reviewer for their helpful comments on an earlier draft of this chapter. This research was funded in part by the National Sciences and Engineering Council of Canada Grant 298244 to Moon-Ho Ringo Ho and NSF Division of Mathematical Sciences Grant 0405243 to Hernando Ombao.

NOTE

1. The analysis in this application was performed by the second author and was based on his collaboration with the Institute of Transportation Studies at the University of California at Davis.

References

Anderson, B.D.O., & Moore, J.B. (1979). *Optimal Filtering.* Englewood Cliffs, NJ: Prentice-Hall.

Aptech Systems Inc. (2003). *GAUSS: User Guide.* Maple Valley, WA: Aptech Systems Inc.

Banich, M.T., Milham, M.P., Atchley, R., Cohen, N.J., Webb, A., Wszalek, T., Kramer, A.F., Liang, Z.P., Wright, A., Shenker, J., & Magin, R. (2000). fMRI studies of stroop tasks reveal unique roles of anterior and posterior brain systems in attentional selection. *Journal of Cognitive Neuroscience, 12,* 988–1000.

Brillinger, D.R. (1975). *Time Series: Data Analysis and Theory.* New York: Holt.

Cavanaugh, J.E., & Shumway, R.H. (1986). On computing the expected Fisher information matrix for state-space model parameters. *Statistics and Probability Letters, 26,* 347–355.

Dempster, A., Laird, N., & Rubin, D. (1977). Maximum likelihood from incomplete data via the EM algorithm. *Journal of the Royal Statistical Society, Series B, 39,* 1–38.

Doornik, J.A. (2001). *Object-Oriented Matrix Programming Using Ox* (4th ed.). London: Timberlake Consultants Press.

Durbin, J., & Koopman, S.J. (2001). *Time Series Analysis by State-Space Models.* New York: Oxford University Press.

Fahrmeir, L., & Tutz, G. (1994). *Multivariate Statistical Modeling Based on Generalized Linear Models.* Berlin: Springer.

Friston, K. (2004). Beyond phrenology: What can neuroimaging tell us about distributed circuity? *Annual Review of Neuroscience, 25,* 221–250.

Geweke, J. (1977). The dynamic factor analysis of economic time series models. In D.J. Aigner & A.S. Goldberger (Eds.), *Latent Variables in Socio-economic Models.* New York: North Holland.

Green, P., & Silverman, B.W. (1994). *Nonparametric Regression and Generalized Linear Models: A Roughness Penalty Approach.* London: Chapman & Hall.

Harvey, A.C. (1989). *Forecasting, Structural Time Series Models and the Kalman Filter.* Cambridge: Cambridge University Press.

Icaza, G., & Jones, R.H. (1999). A state-space EM algorithm for longitudinal data. *Journal of Time Series Analysis, 20,* 537–550.

Insightful Corporation (2002). *S+Finmetrics Reference Manual.* Seattle, WA: Insightful Corporation.

Jazwinski, A.H. (1970). *Stochastic Processes and Filtering Theory.* New York: Academic Press.

Jones, R.H. (1993). *Longitudinal Data with Serial Correlation: A State-Space Approach.* London: Chapman & Hall.

Jöreskog, K.G., & Sörbom, D. (1996). *LISREL 8 User's Reference Guide.* Chicago: Scientific Software International.

Jørgensen, B., Lundbye-Christensen, S., Song, P.X., & Sun, L. (1996). A longitudinal study of emergency room visits and air pollution for Prince George, British Columbia. *Statistics in Medicine, 15,* 823–836.

Koopman, S.J., Shepard, N., & Doornik, J.A. (1999). Statistical algorithms for models in state-space using SsfPack 2.2. *Econometrics Journal, 2*, 113–166.

Koopman, S.J., Shepard, N., & Doornik, J.A. (2001). SsfPack 3.0 beta: statistical algorithms for models in state-space. Unpublished manuscript.

Laird, N., & Ware, J. (1982). Random-effects models for longitudinal data. *Biometrics, 38*, 963–974.

Ljung, G.M., & Box, G.E.P. (1978). On a measure of lack of fit in time series models. *Biometrika, 66*, 67–72.

MacCallum, R., & Ashby, R.G. (1986). Relationships between linear systems theory and covariance structure modeling. *Journal of Mathematical Psychology, 30*, 1–27.

MathWorks (2004). *Getting Started with MATLAB (Version 7)*. Natick, MA: MathWorks, Inc.

Meng, X.L., & Rubin, D.B. (1991). Using EM to obtain asymptotic variance–covariance matrices: The SEM algorithm. *Journal of the American Statistical Association, 86*, 899–909.

Mokhtarian, P.L., Samaniego, F.G., Shumway, R.H., & Willits, N.H. (2002). Revisiting the notion of induced traffic through a matched-pairs study. *Transportation, 29*, 193–220.

Nesselroade, J.R., McArdle, J.J., Aggen, S.H., & Meyers, J.M. (2002). Dynamic factor analysis models for representing process in multivariate time-series. In D.S. Moskowitz & S.L. Hershberger (Eds.), *Modeling Intraindividual Variability with Repeated Measures Data: Methods and Applications* (pp. 235–265). Mahwah, NJ: Lawrence Erlbaum.

Nesselroade, J.R., & Molenaar, P.C.M. (1999). Pooling lagged covariance structure based on short, multivariate time series for dynamic factor analysis. In R.H. Hoyle (Ed.), *Statistical Strategies for Small Sample Research* (pp. 288–306). Thousand Oaks, CA: Sage.

Oakes, D. (1999). Direct calculation of the information matrix via the EM algorithm. *Journal of the Royal Statistical Society, Series B, 61*, 479–482.

Otter, P.W. (1985). *Dynamic Feature Space Modeling, Filtering and Self-Tuning Control of Stochastic Systems*. New York: Springer.

Otter, P.W. (1986). Dynamic structural systems under indirect observation: Identifiability and estimation aspects from a system theoretic perspective. *Psychometrika, 51*, 415–428.

SAS Institute, Inc. (1999). *SAS/ETS Users' Guide*. Cary, NC: SAS Institute, Inc.

Schweppe, F.C. (1965). Evaluation of likelihood functions for Gaussian signals. *IEEE Transactions on Information Theory, IT-4*, 294–305.

Shumway, R.H. (2000). Dynamic mixed models for irregularly observed time series. *Resenhas–Reviews of the Institute of Mathematics and Statistics, 4*, 433–456, University of São Paulo, Brazil: USP Press.

Shumway, R.H., & Stoffer, D.S. (1982). An approach to time series smoothing and forecasting using the EM algorithm. *Journal of Time Series Analysis, 3*, 253–264.

Shumway, R.H., & Stoffer, D.S. (2000). *Time Series Analysis and its Applications*. New York: Springer.

Stock, J.H., & Watson, M.W. (1988). Testing for common trends. *Journal of the American Statistical Association, 83*, 1097–1107.

Stoffer, D.S., & Wall, K. (1991). Bootstrapping state-space models: Gaussian maximum likelihood estimation and the Kalman filter. *Journal of the American Statistical Association, 86*, 1024–1033.

Wecker, W.E., & Ansley, C.F. (1983). The signal extraction approach to non-linear regression and spline smoothing. *Journal of the American Statistical Association, 78*, 81–89.

West, M., & Harrison, J. (1997). *Bayesian Forecasting and Dynamic Models* (2nd ed.). New York: Springer.

Wu, C.F. (1983). On the convergence properties of the EM algorithm. *Annals of Statistics, 11*, 95–103.

APPENDIX

For the properties below, we use the notations $\theta_t^s = E\{\theta_t|Y_s\}$, where $Y_s = \{y_1, \ldots, y_s\}$ denotes the vectors up to time s. For $s = t - 1$, the expectation is a forecast, whereas for $s = t$, the expectation is the Kalman filtered value. For $s = n$, the expectation is conditional on the entire data and is the Kalman smoother. The conditional covariances $P_t^s = E\{(\theta_t - \theta_t^s)(\theta_t - \theta_t^s)'|Y_s\}$ and $P_{tu}^s = E\{(\theta_t - \theta_t^s)(\theta_u - \theta_u^s)'|Y_s\}$ are interpreted in the same way. We summarize the equations for the Kalman filters, smoothers, and their covariances in the three properties below.

Property A1: The Kalman Filter

For the state-space model specified in (7.1) and (7.2) with initial conditions $\theta_0^0 = \mu_0$ with $P_0^0 = \Sigma_0$, for $t = 1, \ldots, n$,

$$\theta_t^{t-1} = \Phi\theta_{t-1}^{t-1}, \tag{7.A.1}$$

$$P_t^{t-1} = \Phi P_{t-1}^{t-1}\Phi' + Q, \text{ with} \tag{7.A.2}$$

$$\theta_t^t = \theta_t^{t-1} + K_t(y_t - \Gamma z_t - A\theta_{t-1}^{t-1}) \text{ and} \tag{7.A.3}$$

$$P_t^t = (I - K_t A)P_t^{t-1}, \tag{7.A.4}$$

where the Kalman gain is

$$K_t = P_t^{t-1}A'[AP_t^{t-1}A' + R]^{-1}. \tag{7.A.5}$$

Property A2: The Kalman Smoother

For the state-space model specified in (7.1) and (7.2) with initial conditions $\theta_n^n = \mu_0$ with P_n^n via Property A1, for $t = n, n-1, \ldots, 1$,

$$\theta_{t-1}^n = \theta_{t-1}^{t-1} + J_{t-1}(\theta_t^n - \theta_t^{t-1}), \tag{7.A.6}$$

$$P_{t-1}^n = P_{t-1}^{t-1} + J_{t-1}(P_t^n - P_t^{t-1})J_{t-1}', \text{ where} \tag{7.A.7}$$

$$J_{t-1} = P_{t-1}^{t-1}\Phi'[P_t^{t-1}]^{-1}. \tag{7.A.8}$$

Property A3: The Lag-One Covariance Smoother

For the state-space model specified in (7.1) and (7.2) with K_t, J_t, $t = 1, \ldots, n$, obtained from Properties A1 and A2 and p_n^n obtained from Properties A1 and A2, and with initial condition

$$P_{n,n-1}^n = (I - K_n A_n)\Phi P_{n-1}^{n-1}, \tag{7.A.9}$$

for $t = n, n-1, \cdots, 2$,

$$P_{t-1,t-2}^n = P_{t-1}^{t-1}J_{t-2}' + J_{t-1}(P_{t,t-1} - \Phi P_{t-1}^{t-1})J_{t-2}'. \tag{7.A.10}$$

8

The Control of Behavioral Input/Output Systems

James O. Ramsay

Managing life is, after all, not so different from running a nuclear power plant; the day-to-day operation should be efficient, productive, and harmonious, but the possibility of a meltdown is always there. No parent, educator, psychologist, or other dabbler in human affairs needs to be convinced that modifying situational variables on the basis of an ongoing evaluation of performance is what we do much of the time. A golfer changes some aspect of how he addresses the ball after noting a persistent slice. A lecturer returns to a body of material with a more elaborate exposition after some poor exam results.

We are, at least in some situations, rather good at these methods for improving performance by changing contextual inputs, but the occasional catastrophe, or even just wanting to do better, leads us to want to reflect on the nature of what is called *control* in engineering and science. In these environments, control is often seen as a response to the cost of variability in the performance of a system. One appreciates that it is cheaper to vary inputs in a way that minimizes output variation than it is to live with the consequences of the variation in the uncontrolled output. In golf, for example, accuracy ultimately counts for more than power, and games are more often won on the green rather than on the fairway. Thus, control can be a matter of transferring variation from the outputs, where it is expensive, to the inputs, where the costs are less.

Control is also required when constraints on the outputs are changed, and it is important to adapt the system both rapidly and smoothly. This can, again, be seen as buying less variation in both time and output level by manipulating input variables because changes in an input variable's level can be faster as well as less expensive. A few seconds invested in considering how to respond to a critical remark can save us many minutes of apology later.

What can we learn from those who use the machinery of mathematics, computer science, and engineering to set up control procedures in situations where the information available is quantitative? Does the rigor and precision of the tools with which control engineers work have anything to tell us about managing behavior? These are hardly new questions, but the context of this volume is an increase in the amount and hopefully precision of the quantitative information available to the behavioral scientist, and there are, as well, recent developments in statistics that may help us to exploit this new information.

There is already a substantial literature on the notion of feedback as a regulator of human and animal behavior. A good deal of this material is inspired by the dramatic progress made in the second half of the last century in what was for a time called the science of *cybernetics*, and Richardson (1991) is a good source on this history. This author especially profited from the readings contained in Levine and Fitzgerald (1992), where a number of illustrations of how we use feedback to modify behavior were considered. The paper by Levine et al. (1992) in those volumes is particularly relevant to this discussion.

This chapter offers a short and informal summary of how control technology works, recast into a behavioral setting. It begins with the essential element that control requires: a model for the dynamics or rate of change in process to be controlled. The natural model for rate of change is a differential equation and, since readers of this book may not have a working knowledge of this branch of mathematics, a small amount of introduction to the topic cannot be avoided. Some conclusions about effective control strategies are drawn, and the chapter concludes with a brief account of some statistical methodology drawn from functional data analysis (Ramsay & Silverman, 2005) that may prove helpful.

8.1 A Typical Input/Output System

We begin with what an engineer would call a process or a *system*, meaning something like a chemical reactor or an electrical circuit. In general, we mean by "system" a relatively encapsulated or autonomous unit that processes over time various identifiable inputs and also produces one or more outputs that are seen to respond in some way to the inputs. People surely qualify; and a musician in an orchestra reading notes with an eye on the conductor while producing music as an output is an example. We shall also mean by *dynamics* the rate of change in an output in response to some manipulation of the input, that is, some order of derivative in the case of a smooth quantifiable output.

I should like to propose as a process a small child playing with its toys or its parent's pots and pans. This example is not intended to be taken seriously, but merely as a means of evoking images within a familiar setting that some readers might find helpful. Other readers will find this exercise intolerably simplistic, and will prefer to substitute a more plausible illustration of how we develop feedback loops to control behavior.

The child takes in one or more inputs, transforms them in some way, and produces one or more outputs as a result. For example, it takes in stimulation and food and outputs a physical impact on its environment and emotionality. We assume here that both the inputs and outputs can be quantified, so that a child's emotionality can be rated on a scale with negative and positive poles.

The process at this primitive level is operating in what is called *open-loop* mode, in which the outputs are mostly functions of the input, and there is consideration of how the output might feed information back that can be used to modify the inputs.

Of course, noise or disturbances also play a role; some of these can be classified as additional exogenous inputs that pass through the process, some are internal to the processor and might be called turbulence, and others act directly on the output downstream of the process. Some disturbances are short–lived, and others are long-lasting or even permanent. The disturbance caused by a telephone call from a grandparent would be transitory, but the impact of an illness would not. The essential thing about noise or disturbances is that they are not subject to any form of control.

Figure 8.1 shows the flow of inputs, disturbances, and outputs in open-loop mode as solid lines. The central box, called the Process, is the child in our example. The upper and lower solid arrows leading to the box indicate stimulation and food inputs, respectively, and it is the upper stimulation arrow that is the target of feedback control. The upper and lower exiting errors show emotionality and impact on the home, respectively, and it is emotionality that is used to produce the feedback that will modify stimulation. Note that although some disturbances may affect inputs upstream of the process, they are lumped together and are shown as coming from a separate source called "Noise" in our diagram.

Now we introduce a control loop that monitors one or more of the output variables via sensors, takes decisions on the basis of the information received

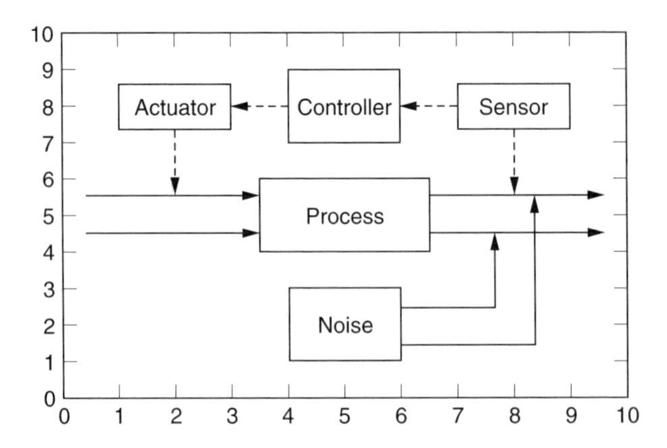

Figure 8.1 A schematic for a process under feedback control. Solid lines indicate relationships that hold in open-loop mode, and dashed lines indicate closed-loop flows of information.

from the sensors, and sends messages to an activator which in turn modifies at least some of the inputs.

Suppose, for example, the child's controlled response is emotionality, and the parent acts as the sensor by assessing its hedonic tone from time to time from a sample of the child's behavior and verbalization. The parent then goes into controller mode, and reflects on possible changes to the child's input. Having decided, for example, that the child has become overstimulated, the parent activates a reduction in stimulation by, perhaps, putting it down for a nap. When the process is connected to the control loop, as it is in the top half of figure 8.1, we say that the process is operating in *closed-loop* mode.

Now a closed-loop system is more complex than an open-loop system. Three new processes have been introduced, each with their own dynamic characteristics and levels of turbulence. With this complexity comes both benefits and possible costs. The benefit, it is hoped, is an emotional life for the child that is better for it and for those with whom it shares its life. But the cost is, in addition to effort on the parent's part, the possibility that the entire system will behave in unexpected ways, and perhaps with consequences that are not at all desirable. The fundamental problem facing the controller is to learn how to achieve satisfactory performance while avoiding potential harmful consequences.

Here goals have to be set. For example, the parent may aim at a happy engaged child but want to avoid the wild excitement that may lead to intolerable impacts on the household. The setting of a level target is a *set point* goal. But variability in emotional tone may be as important as its average; a child that swings between negative and positive moods rapidly will quickly become exhausted. Moreover, the parent will know from much experience that her or his effectiveness as a controller can vary, that a child can be overwhelmed by large disturbances, and that, pushed too far, the child can become unmanageable. The parent aims, therefore, at security or stability in order to avoid tantrums and other unsupportable outcomes. This may cause the parent to "tune down" the control strategy in order to keep the child safely away from emotional explosions.

Of course, all this is too simple, and we ignore the fact that the child will adapt to the parent and is, in other ways as well, an active participant in the process. That is, a more careful analysis will probably require that the child and parent be linked together as a single composite system.

8.2 Modeling System Dynamics

Control works by monitoring change and responding to it; without change, there would be no point to control. A dynamic model for an input/output relationship contains an explicit account of how the system responds to change. If we have such a model, designing an effective control strategy is much less of a hit-or-miss proposition. Moreover, we can try out various control strategies

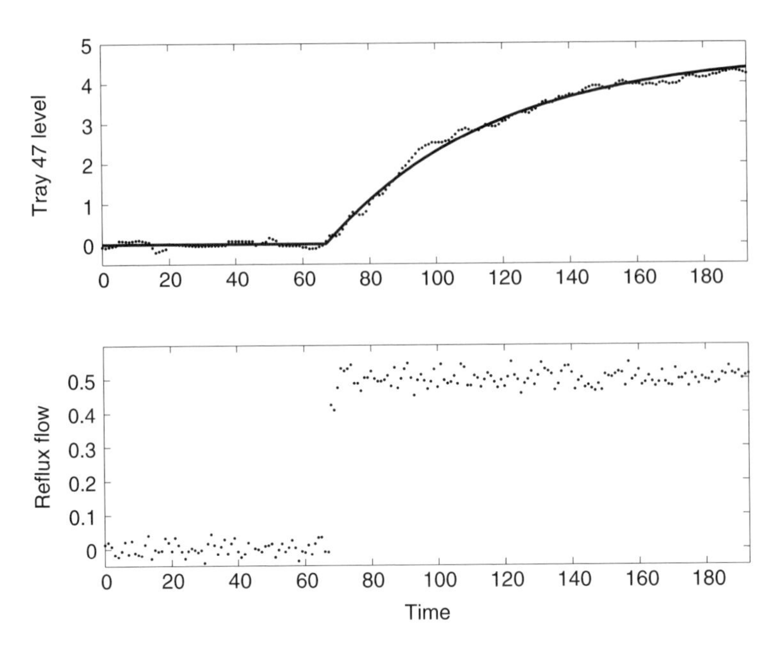

Figure 8.2 The upper panel shows the level of material in a tray of a distillation column in an oil refinery, and the lower level shows the flow of material being distilled into the tray. The points are measured values, and the solid line in the upper panel is the fit to the data defined by differential equation (8.1).

on models that may be quasi-realistic so as to at least understand their qualitative features.

Consider, for example, figure 8.2, which displays how the fluid level in a portion of a distillation column in an oil refinery responds to an input called reflux flow. But why not think of this as how a child's excitement builds up when the parent offers it a new toy? We see in these data two features: first, the *gain* or final change in level, and second, the rate at which the level changes. Both features are critical, but it is the rate of change that will determine whether a control strategy is effective or not, for example, whether the tray will overflow within a given time frame.

The derivative of a function x at a point t, which I shall denote by $Dx(t)$, is its rate of change, and a higher-order derivative such as $D^2x(t)$ is the rate of change of the next lower derivative. It is natural, therefore, that a mathematical model for change will include one or more derivatives. For example, let $u(t)$ be the flow into the tray and let $x(t)$ be the tray's level. The data shown in figure 8.2 can be neatly modeled with the first-order differential equation

$$\tau Dx(t) = -x(t) + Ku(t - \delta), \tag{8.1}$$

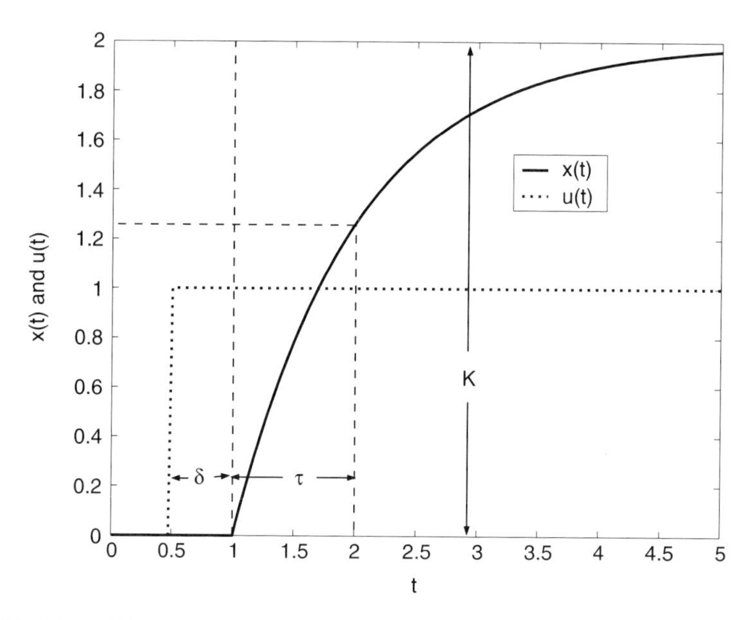

Figure 8.3 The solid line is defined by differential equation (8.1) for initial condition $x(0) = 0$. The dotted line is the input function $u(t)$. The time constant τ, gain K, and dead time δ are as indicated.

where

- τ is the time the fluid takes to reach about 2/3 of its final level, or how long it takes the child to become seriously engaged by the new toy.
- K is ratio of the final change in output x to the change in input u; for example, if the toy is measured by its cost, the ratio of the change in the child's level of delight to its cost.
- δ is called the *dead time*, the time taken before the output begins to respond to the input, that is, how long it takes the child to disengage from what it is doing and begin to attend to the toy.

Figure 8.3 shows the role of these three parameters. The smooth curve in figure 8.2 is defined by $\tau = 0.02 \pm 0.0004$ time units, $K = 0.095 \pm 0.006$ level units per flow unit, and $\delta = 0$ time units.

The simple three-parameter model (8.1) will be sufficient to illustrate most of the important aspects of control. For example, it can offer a first glimpse of instability. Suppose that either u or K is zero. Then the solution of the homogeneous part of the equation, $\tau Dx(t) = -x(t)$, is

$$x(t) = Ce^{-t/\tau}.$$

This is a nice stable system as long as $\tau > 0$, since then $x(t)$ will never grow without limit. But let τ go negative, and exponential decay turns into exponential growth and we know that even our planet is not large enough to support this forever.

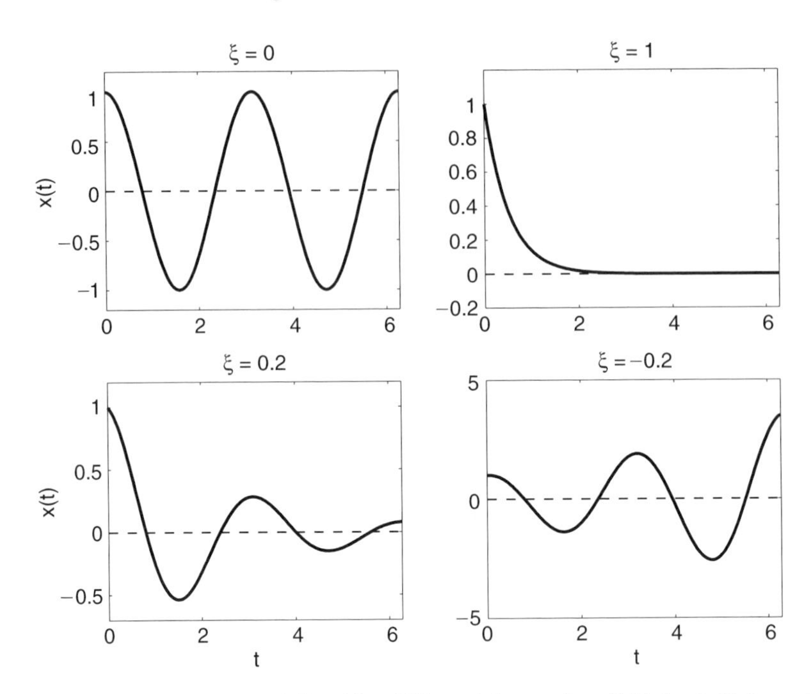

Figure 8.4 The solid lines are defined by differential equation (8.2) for initial condition $x(0) = 1$, time constant $\tau = 0.5$, and various values of the damping factor ξ.

Most real-life processes have some tendency to oscillate under certain circumstances, and to become unstable in others. The following second-order differential equation is a prototype for such systems:

$$\tau^2 D^2 x(t) = -x(t) - 2\xi\tau Dx(t). \tag{8.2}$$

Figure 8.4 shows how the behavior of x is determined by the *damping factor* ξ. The upper left panel displays pure oscillation corresponding to $\xi = 0$. The upper right panel shows another extreme case, called *critically damped*, when $\xi = 1$. The bottom panels show what happens when ξ is close to zero and with both positive and negative signs. When $\xi = 0.2$, the system has an exponentially decaying oscillation, and when $\xi = -0.2$ the oscillation is exponentially increasing in amplitude. The system in the lower right panel is called *unstable*, the system in the upper left panel is said to be at the threshold of stability, and the other two panels show *stable* behavior.

We observed already that closed loops are more complex than open loops. This is because they contain a series of processes, each of which may be as simple as our distillation unit. The sensor, the controller, and the actuator all have their own time constants, gains, and dead times, and by now you will be able to supply your analogs for our kitchen. In effect, each new process increases the order of the differential equation by one since each new process must react to both the level of the previous process and its rate of change, and therefore must compound its own rate of change with that of the previous process. At the end of the day, the other parent greets a differential equation of at least the fourth order, and perhaps can be forgiven for not entirely grasping what is going on.

Dynamic system models like these play both a normative and a descriptive role in applications of the mathematical theory of control. They are a source of textbook situations of varying degrees of realism where the exact implications of various controller design approaches can be worked out, or at least approximated numerically to a high level of accuracy. These models are invaluable test-beds for understanding how control works and training new control designers. Marlin (2000) is a fine example of a text on the topic, and is accompanied by an extensive set of interactive web-based training and software in MATLAB.

However, as in all fields, mathematical models can only take you so far, and a great deal of control design and tuning takes place in contexts where the systems are too complex to model. We shall sketch an approach to modeling actual systems using observed data in the Technical Appendix. We next discuss some general aspects of control systems and how they perform.

8.3 Controller Strategies to Meet an Output Target

I indicated above that the role of the controller is to achieve one or more performance goals. These may be meeting a target, controlling output variability, avoiding instability, and a number of possible goals as well. The sensor process provides the information that the controller works with, and the goal determines what the controller will do with it.

Let us consider for simplicity the goal of meeting a performance target or set point, such as maintaining the child at a contented level, perhaps quantified at modestly above zero. Let this level by denoted by μ. It is possible that the value of μ will be changed at some point, and then we shall want our control strategy to react by raising or lowering the child's hedonic level promptly and without wild swings to the new level.

In this case, the controller will obviously work with the discrepancy between performance and target, $x(t) - \mu$. Strictly speaking, we should replace $x(t)$ here with something like $s[x(t)]$ because the sensor may report to the controller in different units than are used for the performance, but let us keep things simple.

A little reflection will convince you that the controller has three ways of using this information, corresponding to the present, the past, and the future, and designated in the control literature by the letters P, I, and D, respectively. Here they are:

P: the immediate value of $x(t) - \mu$. In this case the controller sends an activation level message $A(t)$ to the actuator that is *proportional* to $-(x(t) - \mu)$. That is, if $x(t)$ is under target, the input will be increased by some fraction of the discrepancy,

$$A(t) = -K_c[x(t) - \mu].$$

The constant K_c is called the *controller gain*.

I: an accumulation or *integral* of past discrepancies,

$$A(t) = -\frac{K_c}{T_I}\int_0^t [x(s) - \mu]\, ds.$$

The larger the accumulated discrepancy between performance and set point, the more the input will be changed. Constant T_I is called the *integral time* and, like K_c for the proportional strategy, it controls the size of the change in the input.

D: this strategy sends an activation message that is proportional to the derivative of the $x(t)$,

$$A(t) = -K_c T_D Dx(t).$$

The logic is that, if the derivative is positive, the response will soon be above target even if the current response is close to target. This is, in other words, a strategy based on a forecast of the future. The *derivative gain* T_D defines how strongly the controller reacts to this predicted discrepancy.

What are the merits and deficiencies of these three strategies? We begin with the simplest of these, the P or proportional strategy.

8.3.1 How Proportional or P-Control Works

Imagine what happens if a new and engaging form of stimulation enters the child's environment, like an exuberant puppy. The child's emotionality rapidly rises to the occasion, and the proportional control strategy decreases the other sources of stimulation accordingly. The discrepancy moves toward zero, which is desired, and so does the decrease in stimulation. This will work fine if the puppy goes away and things return to normal. But what happens if the puppy stays around? Then this strategy will reach an equilibrium state that is above target and below the initial uncontrolled response.

Another problem arises if the controller gain K_c is too high. The controller produces an overreaction and depresses the input so much that the discrepancy switches from positive to negative. The child becomes bored, for example, if the parent puts the dog in a neighboring room and takes away most of the toys. Then the proportional controller swings into enhancement mode, and brings an increase in input. And so, oscillating between negative and positive changes in input until either (1) these oscillations die out and an equilibrium is achieved, or (2) and much worse, the swings become wilder and wilder until the system goes out of control.

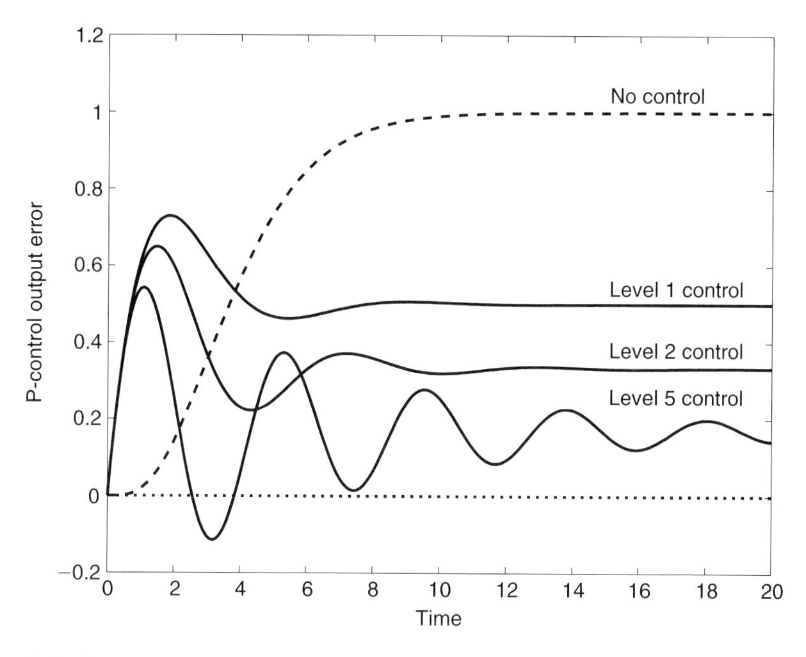

Figure 8.5 The dashed line shows the error in the output of an uncontrolled artificial system after a long-lasting disturbance arrives at time zero. The solid lines indicate error levels for three levels of proportional control.

The proportional controller will react quickly, but will only return the response to the target level if the disturbance is transitory. Moreover, it has the potential for oscillating between positive and negative control phases if the gain is too high and, ultimately, can even produce catastrophic failure.

Figure 8.5 offers a graphical perspective on proportional control. The dashed line shows the error in the response of an artificial system without any control to a long-lasting disturbance that arrives at time zero. The error in output ramps up to its highest value within about 10 time units. The Appendix to this chapter describes the system in terms of transfer functions, but the mathematical details there are not needed to appreciate the plot.

The solid lines show the effects of applying three levels of proportional control to the system. We see that none of these levels is sufficient to bring the output back on target. Moreover, although the higher the level of control, the more the target output is achieved, the cost becomes high. The high level of oscillation in level 5 control would not be acceptable in many industrial control settings, and neither would the time that it takes to reach its final error level, which is 0.2 response units. Finally, if we continue to increase the control level, the response error would actually increase over time as the controlled system became unstable. It looks like a proportional control of level 2 might be a reasonable compromise, and bring the system under its achievable level of control in about 10 time units.

8.3.2 How Integrative or I-Control Works

Integrative control overcomes the main disadvantage of proportional control; it ensures that the long-term change in the target level will eventually be satisfied. This happens because the control message keeps increasing in size as long as the discrepancy remains on one side of zero. Basically, the parent just keeps removing a toy until the child's equilibrium returns to the desired level and stays there.

However, it shares with the proportional controller the possibility that, if T_I is too small, it will overcontrol and produce oscillation about the target, or even spiral out of control.

The integrative strategy is more conservative in the sense that it uses information across time. This necessarily makes it slow to react to a disturbance, and if the disturbance is transitory, it may produce little discernible control. Consequently, it is a poor strategy if short but sharp shocks arrive and the process has a time constant τ sufficiently small to permit a strong reaction to them.

Again, let us see how this looks in a graph. Figure 8.6 plots the response error curves for three levels of integrative control, along with the uncontrolled response. We see that this type of control is able to restore the output to

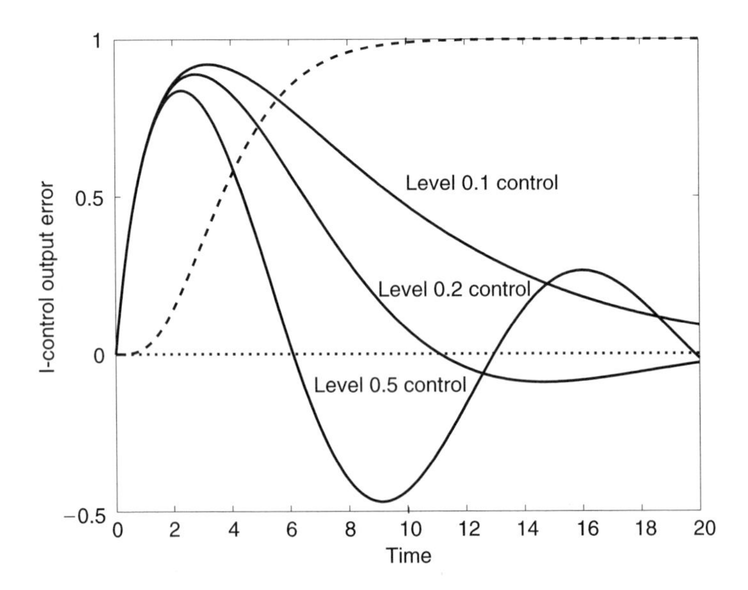

Figure 8.6 The solid lines indicate error levels for three levels of integrative control of the system shown in figure 8.5. The dashed line shows the error in the output of the uncontrolled system.

target in something like 20 time units, or a little more for the lightest control level. Control that is too heavy, however, produces unwanted oscillation, and increasing the level of control further would ultimately lead to unstable behavior. Comparing these curves with the proportional controller curves in figure 8.5 shows us that integral controllers take around twice as long to achieve their best performance.

8.3.3 How Derivative or D-Control Works

Derivative control has the virtue of responding to a sudden disturbance more aggressively and faster than either the proportional or integrative strategies since it applies control based on its forecast of what the uncontrolled steady state is. For example, for the system in figures 8.5 and 8.6 the derivative information within a single time unit is in principle sufficient to cause a derivative controller to apply its maximum level of control.

However, it is difficult in practice to construct a derivative estimate that is not greatly affected by turbulence, noise, or other high-frequency components of variation, and consequently derivative controllers will tend to remain constantly in a flutter in such environments. Moreover, if the disturbance is ongoing and reaches a steady state, this control strategy gives up and returns to zero since it is unable to distinguish between an on–target response and one with a constant error. Consequently, it will not produce a return to target for long-lasting disturbances. And it can, naturally, also produce oscillations or even failure. These problems imply that D-control is almost never used by itself in actual control systems, and we shall not bother to graph its performance.

8.3.4 The Best Strategy: PID Controllers

These three control strategies are defined by the three constants K_c, T_I, and T_D, and a good deal of practical control design focuses on working with all three constants at the same time to find the best blend of strategies, rather like finding a good cup of coffee. A PI controller, for example, will try to combine the capacity to return to target of the I-controller with the faster reaction time of the P-controller. Although D-control is used less frequently, a PID controller will incorporate a touch of the D-controller's capacity to deal with transitory shocks in order to smooth out the response.

Figure 8.7 shows what happens when we use both proportional control at level 2 and integral control at level 0.2. We see that we can combine the fast reaction of the proportional controller with the capacity of the integral controller to return to the set point. The result is clearly superior to either proportional control or integral control taken alone. Adding some derivative control would not help much for these long-lasting disturbances, but could lead to even faster reactions to transitory perturbations.

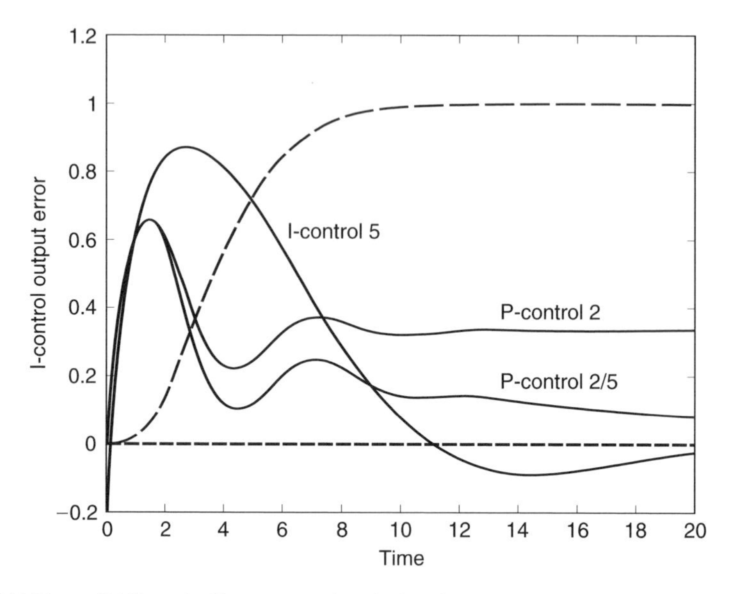

Figure 8.7 The solid lines indicate error levels for the system shown in figure 8.5 of proportional control at level 5, integrative control at level 0.2, and proportional-integrative control by combining these two controllers. The dashed line shows the error in the output of the uncontrolled system.

8.3.5 The All-Important Issue of Stability

Since virtually all systems can go wild under the wrong set of circumstances, and especially when control constants such as K_c, T_I, and T_D are too extreme, a crucial part of control design is ensuring that the final control system will be stable. If a model of the system is available, then the conditions under which the system becomes unstable can be systematically investigated, but otherwise the designer attempts to find a secure and safe range of control settings that will guarantee safety. This usually involves a certain amount of "detuning" of these settings away from optimal values since optimum performance is often only achieved close to the stability boundaries.

8.3.6 When Control Is Difficult

The two conditions that make control difficult in practice are long dead times and a large contrast between the dynamics of the system and the controller. It only takes a few seconds to change the rudder angle on a large ship, but it first just rolls away from the direction of turn for some minutes while carrying on in the same direction. Controlling a freight train is a very different matter from driving an automobile; a long train can take a few kilometers to bring to a stop. Dead times and slow dynamics mean that a substantial amount of time will pass before disturbances can be detected, and in this period the system is essentially running in open-loop mode. If the engineer tries to rely too heavily on derivative control so as to "beat the clock," both estimation error and controller instability can become critical issues. We all know how hard it can be to manage a relationship with someone who resists displaying any emotional reaction for as long as possible, and how chancy it is to rely on tiny little signals that something is going wrong.

8.4 Fitting Dynamic Models to Intensive Longitudinal Data

What are the prospects of applying control design principles and practice to behavioral science data? The obstacles are important. First of all, while people are certainly dynamical systems, their level of complexity in even highly restricted situations goes way beyond what textbook models can describe. Moreover, empirical methods for controller design, involving repeated experimental manipulation of inputs and detailed data on outputs, rarely seems possible where higher organisms are involved. People do not stay in the same state long enough to make repeated experimental manipulation feasible, will seldom tolerate much of this in any case, and may even find explorations of stability limits downright abusive.

On the other hand, there is already an abundance of data on behavioral systems like that shown in figure 8.2. Some of it is very old, as in curves of learning and forgetting, and data from psychophysical experiments. A classic book that still can be a rewarding source on such data is Guilford (1954). More recently, online monitoring of reactions has become commonplace, and, for example, continuous monitoring of perceived tension in music has proven to be an important source of data in music psychology (Krumhansl & Schenck, 1997; Neilsen, 1983; Vines et al., 2004). See Boker and Laurenceau (chapter 9, this volume) for another great example. If we could figure out how to model these data using differential equations, the machinery of controller design would be much more applicable.

Functional data analysis (Ramsay, 2002; Ramsay & Silverman, 2005) is a collection of methods for the analysis of curves, images, and other types of functions as basic observations. Classic techniques such as the analysis of variance, other linear models, and principal components analysis all have their functional analogs. In addition, however, the intrinsic smoothness of functional data makes a number of new methods possible that do not have counterparts elsewhere. Among these is the estimation of differential equations from noisy discrete data.

Ramsay (1996) developed a method for fitting differential equations to data called *principal differential analysis* (PDA) by first converting discrete data to functional form using modern nonparametric curve estimation methods, and then inputting these curves and their derivatives into a functional version of regression analysis. Because the solution to a given differential equation is a linear space rather than a unique function, this analysis also has some of the characteristics of principal components analysis, hence the choice of name.

Principal differential analysis has since evolved greatly in a number of directions, and is now being applied to linear and nonlinear dynamic modeling problems in chemical engineering, where it has been able to yield much more precise parameter estimates than older methods. Some of these new developments are in Ramsay and Silverman (2005), and others are still in the development stages. Applications to the analysis of handwriting can be found in Ramsay (2000) and to motor control in Ramsay et al. (1995). The time constant and gain estimates for the refinery data, for example, were obtained with PDA.

These new methods place us on the edge of a new epoch in the modeling of behavioral dynamical systems, and consequently open up many possibilities for designing control strategies that will enhance performance. Perhaps it will be a long time before we have anything useful to offer the caregivers for small children, but I hope that an improvement in my handwriting is within reach.

TECHNICAL APPENDIX

Here are some notes on some of the more technical aspects of this chapter that are intended to provide readers with the appropriate expertise to find out more.

A.1 How the Figures Were Generated

Control engineers rely heavily on modeling the dynamic response of an industrial process by linear differential equations with constant coefficients. In this situation, it is rather easier to work with *transfer functions*, defined by taking the Laplace transforms of the terms in the differential equation. We use the notation $x^*(s)$ and $u^*(s)$ to indicate the Laplace transforms:

$$x^*(s) = \int_0^\infty e^{-st} x(t)\, dt$$

and

$$u^*(s) = \int_0^\infty e^{-st} u(t)\, dt,$$

respectively. These transforms are applied to inputs and outputs centered on zero by subtracting their steady-state values. For the kinds of inputs and outputs used in engineering models, the Laplace transforms involved are usually ratios of polynomials, or *rational functions*.

For a system with no control, the *system* or *process transfer function* is

$$G_p(s) = \frac{x^*(s)}{u^*(s)},$$

and plays a key role in describing the system. The transfer function that I used in the simple example in the figures was

$$G_p(s) = \frac{1}{(s+1)^3}.$$

We can recover the time course of the output by inverting the Laplace transform, probably with the help of some tables or symbolic computation software such as Mathematica or Maple.

Each of the three types of control procedures is also associated with its own transfer function $G_c(s)$. Proportional control is defined by the constant transfer

$$G_c(s) = K_c,$$

integral transfer by

$$G_c(s) = \frac{K_c}{T_I s},$$

and derivative transfer by

$$G_c(s) = K_c T_D s,$$

and where retaining the common *control gain* K_c facilitates comparing the effects of different types of control, and permits the final PID-control transfer function to be expressed as a sum of the three specific transfer functions. The levels of control in the P-control examples were specified in terms of K_c and in the I-control examples by $1/T_I$.

The transfer function for systems under PID control and subject to a disturbance $d(t)$ has the form

$$\frac{x^*(s)}{d^*(s)} = \frac{1}{1 + G_p(s)G_c(s)}.$$

In the examples, I worked with the disturbance transfer function

$$d^*(s) = \frac{1}{s(s+1)}.$$

A.2 A Sketch of a New Method for Estimating Differential Equations

In order to keep this discussion both concrete and simple, let us confine our attention to differential equation (8.1) with zero dead time ($\delta = 0$). This is the equation used to fit the data in figure 8.2. The method described here can be applied much more widely, however.

Suppose now that we observe function $x(t)$ at a set of n discrete time points $t_j, j = 1, \ldots, n$. At the same time we also observe the values of forcing function $u(t)$ at these sampling points. Assume that the $x(t)$ values are observed with error, so that observation y_j is connected to the true function value $x(t_j)$ by the model

$$y_j = x(t_j) + e_j.$$

To keep things simple, let us assume that the forcing function $u(t)$ is observed without error.

We have two parameters to estimate in this model: time constant τ and gain K. Our task is to go from the n pairs of observations $(y_j, u(t_j))$ to high-quality estimates of these two parameters. As is usual in statistics, we can propose to measure the failure of $x(t)$ to fit the data by the least squares criterion

$$\mathrm{SSE}(\tau, K) = \sum_{j=1}^{n} [y_j - x(t_j|\tau, K)]^2,$$

where the notation $x(t|\tau, K)$ refers to a solution to differential equation (8.1) determined by some candidate values of the parameters. The least squares estimates are the parameter values minimizing $\mathrm{SSE}(\tau, K)$.

What keeps this problem from being a routine nonlinear regression problem is that we do not have a mathematical expression for a solution $x(t_j|\tau, K)$. Of course, we can estimate such a solution given parameter values by numerical methods for solving differential equations, and this is exactly how many current methods proceed. Unfortunately, these methods are both time-consuming and prone to introducing substantial amounts of additional estimation error due to the inaccuracy of the method. Moreover, modern methods for minimizing functions like $\mathrm{SSE}(\tau, K)$ are much faster and more stable if one or more of their derivatives are available, and this is only possible with acceptable speed and accuracy if an expression for the function is available that can be differentiated. Again, this is not possible because we do not have an expression for $x(t)$.

Any specific solution $x(t)$ can be expressed to an arbitrary level of accuracy by a linear combination of M functional building blocks $\phi_m(t)$, called *basis functions*. This linear combination can be expressed as

$$x(t) = \sum_{m=1}^{M} c_m \phi_m(t),$$

where the coefficients c_m define the approximation to $x(t)$. All that is required is a suitable system of basis functions and a sufficiently large number N of these. In most applications involving nonperiodic functions these days, we tend to use the *B-spline* basis system, a description of which can be found in Ramsay and Silverman (2005). The importance of this way of expressing $x(t)$ lies in the fact that we now understand our task as that of finding a suitable set of coefficient values c_m that will minimize $\mathrm{SSE}(\tau, K)$ subject to the constraint that the resulting $x(t)$ is also, at least to some level of approximation, a solution of differential equation (8.1).

We now see that we really have two estimation problems: first, that $x(t)$ will fit the data y_j, and, second, that it will also solve differential equation (8.1). This suggests that we should have a corresponding measure of how well the second objective is achieved.

In order to develop a measure of how well $x(t)$ solves (8.1), let us rearrange the equation to obtain the equivalent equation

$$\tau Dx(t) + x(t) - Ku(t) = 0. \tag{8.3}$$

Now, instead of using a "sum of squared residuals" criterion, we shall replace summation by integration to obtain

$$\mathrm{PEN}(\tau, K) = \int [\tau Dx(t) + x(t) - Ku(t)]^2 \, dt. \tag{8.4}$$

Finally, we need to express a balance between these two objectives. This is because there will probably be no exact solution to the differential equation that fits the noisy data exactly, and some compromise between these two objectives needs to be formulated. We do this by simply taking a weighting of the second objective relative to the first in the following penalized least squares criterion:

$$\begin{aligned} \mathrm{PENSSE}(\tau, K, \mathbf{c}) &= \mathrm{SSE}(\tau, K, \mathbf{c}) + \lambda \mathrm{PEN}(\tau, K, \mathbf{c}) \\ &= \sum_{j=1}^{n} [y_j - x(t_j|\mathbf{c})]^2 + \lambda \int [\tau Dx(t|\mathbf{c}) + x(t|\mathbf{c}) - Ku(t)]^2 \, dt. \end{aligned} \tag{8.5}$$

The new nonnegative *smoothing parameter* λ is what we use to strike the right compromise. If λ is very small, then we are saying that what counts is fitting the data, and we do not care much whether $x(t)$ also solves the differential equation. On the other hand, as λ grows, we are placing more and more emphasis on $x(t)$ solving the equation, and less and less on fitting the data. Naturally, we shall need some approaches that will suggest the right value of λ to us, but we shall pass the reader on to Ramsay and Silverman (2005) for a discussion of this problem.

The notation $x(t|\mathbf{c})$ in (8.5) reminds us that we are now expressing $x(t)$ as a basis function expansion that depends on the coefficient c_m in coefficient vector $\mathbf{c}$ of length M. In effect, we now have $M + 4$ parameters to estimate from the data. This at first seems to have made our problem harder rather than easier.

But, at this point, we ask if it might not be easier to estimate the coefficients in $\mathbf{c}$ if we knew the two equation parameters τ and K. The answer is, at least in a wide range of cases, "yes." That is, if we only need to minimize the composite criterion $\text{PENSSE}(\tau, K, \mathbf{c})$ with respect to $\mathbf{c}$, then there are fairly fast and accurate estimates for achieving this and, in the case of linear differential equations, even a mathematical expression that we can express as $\mathbf{c}(\tau, K)$. This way of expressing a subset of parameters, in this case $\mathbf{c}$, as a function of the remaining parameters is called *profiling* in statistics. The profiling process for linear differential equations is described in detail in Heckman and Ramsay (2000).

With this profiling technology in hand, we can now write the composite criterion as the profiled criterion $\text{PENSSE}_P(\tau, K)$, which has exactly the same expression as $\text{PENSSE}_P(\tau, K, \mathbf{c})$, except that $x(t|\mathbf{c})$ is replaced by $x[t|\mathbf{c}(\tau, K)]$, that is, function $x(t)$ defined by the basis function expansion using the profiled coefficient vector $\mathbf{c}(\tau, K)$.

The rest is fairly routine. We now need only to use a suitable function-minimizing algorithm to minimize $\text{PENSSE}_P(\tau, K, \mathbf{c})$ with respect to the remaining two parameters. Methods that do not need derivatives will work fine when such a small number of parameters is involved, but with more parameters it is also possible in many cases to work out derivative expressions as well. It is also possible to develop confidence limit estimates to indicate how precisely the parameters are estimated, as we did for the data in figure 8.2.

References

Guilford, J.P. (1954). *Psychometric Methods*. New York: McGraw-Hill.

Heckman, N.E., & Ramsay, J.O. (2000). Penalized regression with model based penalties. *Canadian Journal of Statistics, 28*, 241–258.

Krumhansl, C.L., & Schenck, D.L. (1997). Can dance reflect the structual and expressive qualities of music? A perceptual experiment on Balanchine's choreography of Mozart's Divertimento No. 15. *Musicae Scientieae, 1* (Spring), 63–85.

Levine, R.L., & Fitzgerald, H.E. (Eds.) (1992). *Analysis of Dynamic Psychological Systems*, Vols. 1 and 2. New York: Plenum Press.

Levine, R.L., Van Sell, M., & Rubin, B. (1992). System dynamics and the analysis of feedback processes in social and behavioral systems. In R.L. Levine & H.E. Fitzgerald (Eds.), *Analysis of Dynamic Psychological Systems* (Vol. 1, pp. 145–261). New York: Plenum Press.

Marlin, T.E. (2000). *Process Control: Designing Processes and Control Systems for Dynamic Performance* (2nd ed.). Boston: McGraw-Hill.

Neilsen, F.V. (1983). *Oplevelse af Misikalsk Spending* [*The Experience of Musical Tension*]. Copenhagen: Adademisk Forlag.

Ramsay, J.O. (1996). Principal differential analysis: Data reduction by differential operators. *Journal of the Royal Statistical Society, Series B, 58*, 495–508.

Ramsay, J.O. (2000). Functional components of variation in handwriting. *Journal of the American Statistical Association, 95*, 9–15.

Ramsay, J.O. (2002). *Applied Functional Data Analysis*. New York: Springer.

Ramsay, J.O., Flanagan, R., & Wang, X. (1995). A functional data analysis of the pinch force of human fingers. *Applied Statistics, 44*, 17–30.

Ramsay, J.O., & Silverman, B.W. (2005). *Functional Data Analysis* (2nd ed.). New York: Springer.

Richardson, G.P. (1991). *Feedback Thought in the Social Science and Systems Theory*. Philadelphia: University of Pennsylvania Press.

Vines, B., Wanderley, M.M., Nuzzo, R., Levitin, D., & Krumhansl, C. (2004). Performance gestures of musicians: What structural and emotional information do they convey? Unpublished manuscript, McGill University.

9

Dynamical Systems Modeling: An Application to the Regulation of Intimacy and Disclosure in Marriage

Steven M. Boker and
Jean-Philippe Laurenceau

This chapter provides an introduction to coupled differential equations models of self-regulating dynamical systems, describes a method for estimating the parameters of such models, and then works through an application of this method to self-disclosure and feelings of intimacy in a sample of married couples. The methods used include Local Linear Approximation (LLA) of derivatives (Boker & Nesselroade, 2002) and multilevel modeling (i.e., generalized linear mixed modeling or GLMM) (see Walls et al., chapter 1, this volume) to account for and predict individual differences in parameters of differential equations. We have chosen to use LLA and multilevel modeling since this affords a simple and straightforward approach to the estimation of parameters of these models.

Advances in the modeling of longitudinal data have led to the development of theories that are modeled and tested using differential equations. These theories are based on dynamical systems interpretations of social and behavioral phenomena. One type of dynamical systems model attempts to account for self-regulation–the process by which a phenomenon maintains equilibrium by responding to information about change in the phenomenon's state. A more complex dynamical systems model allows regulation in one part of a system to influence the regulation of another part of a system. For instance, one might consider a married couple a system composed of two self-regulating members. The self-regulation of feelings of intimacy of each member of a married couple might influence the self-regulation of feelings of intimacy in the other.

9.1 Self-Regulation and Intrinsic Dynamics

Many psychological constructs show trait-like individual differences, that is, when an individual is measured on many occasions his or her mean score may be distinguishable from other individuals' mean scores. However, there may also be short-term intraindividual variability within each person's score

(Nesselroade, 1991). One may reasonably inquire as to the source of this intraindividual variability. Perhaps this variation is simply some random fluctuation due either to unreliability of a measurement instrument or to the influence of some unmeasured random variable. Or perhaps these fluctuations are due to some sort of process intrinsic to the individual–thus there is some patterning to the intraindividual variability such that the current score for a person is somehow predictive of a future score. In this case, one might consider the patterned variability around the stable mean score to be an instance of an intrinsic dynamical process about an equilibrium value. One may consider such an intrinsic dynamic using the language of self-regulation (Carver & Scheier, 1998), whereby a process variable continuously changes its value so as to remain within some "comfort zone" near its set point, that is, its equilibrium value. Psychological constructs such as well-being (Bisconti et al., 2004) or positive and negative emotions (Chow et al., 2005) may exhibit this type of self-regulating behavior. Physiological variables such as hormone levels in menstrual cycles may couple with behaviorial variables such as eating behavior (Nilsson et al., 2004; Varma et al., 1999) so as to form coupled self-regulating systems.

Dynamical systems theory offers a way to formalize concepts of self-regulation. Let us call the *state* of a system the values of the indicators for a psychological construct at one moment in time and the *trajectory* of a system the continuously evolving state of the system over some interval of time. Suppose a system has a stable *equilibrium state,* in other words a fixed set-point value for the psychological construct that is, given no other information, the expected value for that construct. Now we can define a *linear dynamical system* that has a *basin of attraction* around a *point attractor* by stating that the likelihood that the system's future trajectory turns toward the equilibrium state is proportional to the displacement from the equilibrium state (see, e.g., Kaplan & Glass, 1995). Thus the difference between the equilibrium value and the current value of the psychological construct is *negatively proportional* to the curvature in the construct's trajectory. We next explore this concept in detail because it is fundamental to understanding the correspondence between theories of self-regulation and methods for dynamical systems modeling.

Figure 9.1 plots four trajectories that conform to a model in which the curvature (i.e., the change in the slope) is negatively proportional to the displacement from equilibrium. The slope of the trajectory for some construct x at some selected time t is the tangent to the trajectory of x at time t, that is, the first derivative of x_t with respect to time, and is written either as dx_t/dt or as $\dot{x}_t$. The change in the slope of the construct x at time t (the curvature of the trajectory of x) is the second derivative of x_t and can be written either as d^2x_t/dt^2 or as $\ddot{x}_t$. In this chapter we shall use the notation $\dot{x}_t$ and $\ddot{x}_t$ to represent the first and second derivative of a construct x with respect to time at some selected time t.

We can now formalize a simple linear model of a construct x with a fixed point equilibrium as

$$\ddot{x}_t = \eta x_t, \tag{9.1}$$

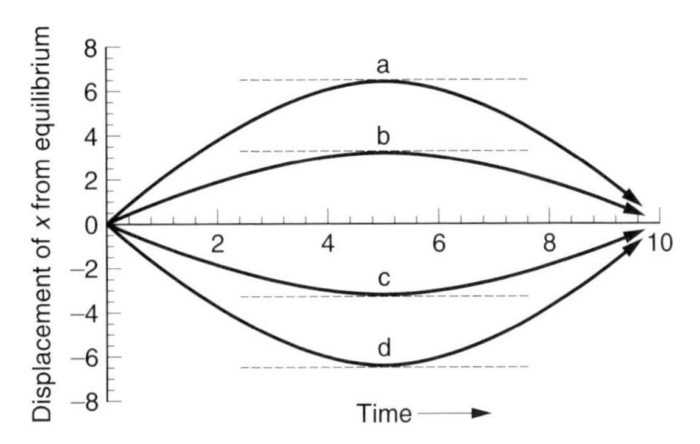

Figure 9.1 Four trajectories in which the curvature of the trajectory of a construct x is negatively proportional to the displacement from an equilibrium state of zero.

where η is some negative valued constant that represents how quickly the trajectory turns back toward its equilibrium state when it is displaced from its equilibrium. In this way, we can see why for this system the equilibrium state is called an attractor–the farther the trajectory is displaced from the equilibrium state, the more it is attracted back toward the equilibrium.

It can be seen why η must be a negative value for this attractor to form by examining figure 9.1, in which a construct x conforming to equation (9.1) has an equilibrium of $x = 0$. Look at figure 9.1 and consider the top trajectory including the point labeled a. This trajectory starts at time $t = 0$ at the equilibrium ($x = 0$) and has some positive slope leading to a positive displacement from equilibrium. Note that at the point labeled a the displacement is just a little greater than 6 and there is a high degree of curvature in the trajectory. Prior to a the slope of the trajectory is positive and after a the slope is negative. Thus, the change in the slope is negative in the neighborhood of a, and therefore the second derivative $\ddot{x}_t$ is negative. If x_t is positive and $\ddot{x}_t$ is negative, then η, the constant coefficient that expresses their proportional relationship, must be negative. One may also verify that when the displacement of x from equilibrium is negative (at points c and d in figure 9.1), then the second derivative is positive.

Although the negative relationship between the displacement of x and the curvature of its trajectory tends to keep the trajectory in the neighborhood of the equilibrium, fluctuations in this simple model do not decrease or increase with time. Figure 9.2 plots the same trajectory that passed through the point a in figure 9.1, but continues for the interval from $t = 0$ to $t = 80$. Note that the points on the trajectory that have a slope of zero (i.e., a, c, and e) have the same absolute value of displacement from equilibrium. After the slope of the trajectory turns from positive to negative in the neighborhood of a, the trajectory continues to have greater and greater negative slope until point b when it crosses through equilibrium. Since the displacement is zero, the change in the slope is also zero at b. Thus, the trajectory continues along the same negative slope and diverges from the equilibrium, having more and more negative displacement and consequently greater positive curvature. By the time the trajectory reaches c, the positive curvature has changed the slope from negative back

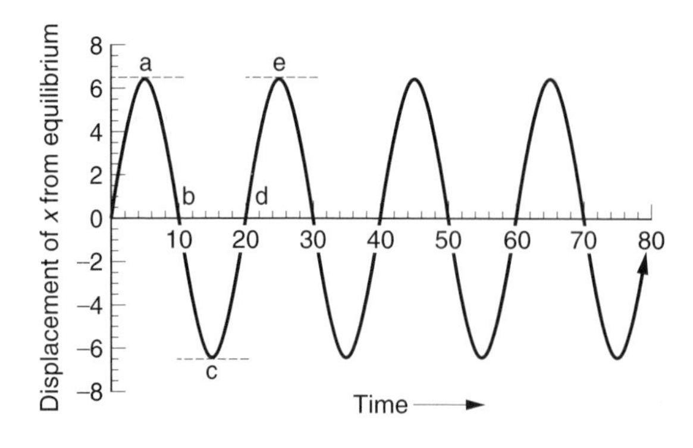

Figure 9.2 A trajectory in which the curvature of the trajectory of a construct x is negatively proportional to the displacement from an equilibrium state of zero, but the fluctuations are not damped.

to positive. But the positive curvature continues until the point d where the trajectory crosses the equilibrium again. As time progresses, the trajectory continues to "overshoot" the equilibrium by the same amount.

When the example trajectory for the construct x crosses its equilibrium, its slope is at a maximum, either as a positive or a negative slope. Equation (9.1) formalized a self-regulation in which displacement from equilibrium induced curvature such that a trajectory moving away from equilibrium would "turn around" and move back toward equilibrium. Another way to say this is that the system responds negatively to displacement from equilibrium. Consider what might happen if the system were to respond negatively to change, that is, a self-regulation mechanism that tended to reduce large absolute values of the slope. In this case, curvature would be negatively proportional to the slope as well as negatively proportional to the displacement from equilibrium. We can formalize this relationship as

$$\ddot{x}_t = \eta x_t + \zeta \dot{x}_t, \tag{9.2}$$

where η and ζ are negative constants (Thompson & Stewart, 1986).

If we plot a trajectory that conforms to equation (9.2), it now damps toward equilibrium as shown in figure 9.3. Although the trajectory continues to overshoot the equilibrium, observe that the points where the slope of the trajectory

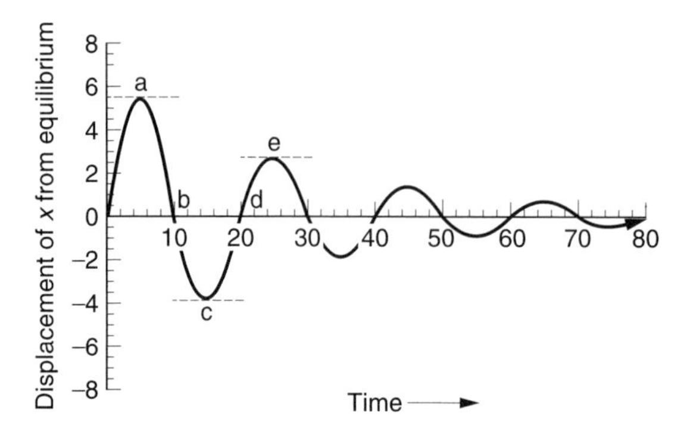

Figure 9.3 A trajectory in which the curvature of the trajectory of a construct x is negatively proportional to the displacement from an equilibrium and also negatively proportional to the slope. Fluctuations from equilibrium are damped and the trajectory settles to the equilibrium.

is zero (i.e., a, c, and e) are closer to zero as time progresses. If some momentary exogenous influence were to displace this self-regulating construct x away from its equilibrium, one might observe a pattern of return to equilibrium similar to that shown in figure 9.3.

Equation (9.2) is an example of a *second-order differential equation*, that is, an equation that expresses relationships between a variable and its first and second derivatives. We have expressed equations (9.1) and (9.2) as being completely deterministic. In other words, these equations do not have any residual term. Of course, this is unrealistic in real-world data. We can add a residual term e_t that conforms to standard regression assumptions (i.e., independent, normally distributed, with a mean of zero) and reexpress equation (9.2) as a regression equation that would allow us to estimate the coefficients η and ζ:

$$\ddot{x}_t = \eta x_t + \zeta \dot{x}_t + e_t. \tag{9.3}$$

Thus, if we were to have estimates of the values of x_t, $\dot{x}_t$, and $\ddot{x}_t$ for a sample of values of time, we can estimate the values of η and ζ using multiple regression (Boker & Nesselroade, 2002). We shall return to this idea in a later section, but first we shall present a short digression on a model that is popular in longitudinal analysis: autoregression.

Many trajectories are possible that still conform to equation (9.3) (see Nesselroade & Boker, 1994, for some examples). One such trajectory is shown in figure 9.4, which can also be modeled more simply as the *first-order differential equation*:

$$\dot{x}_t = \alpha x_t + e_t, \tag{9.4}$$

where α is a negative constant coefficient that expresses the negative proportional relationship between the displacement of a construct x from its equilibrium at time t and the instantaneous slope of x at time t. This results in a negative exponential trajectory that returns toward equilibrium but does not overshoot. If we sample a negative exponential trajectory at discrete intervals of time τ, the value of x_t and $x_{t+\tau}$ will be a simple linear proportion β such that β is between 0 and 1.

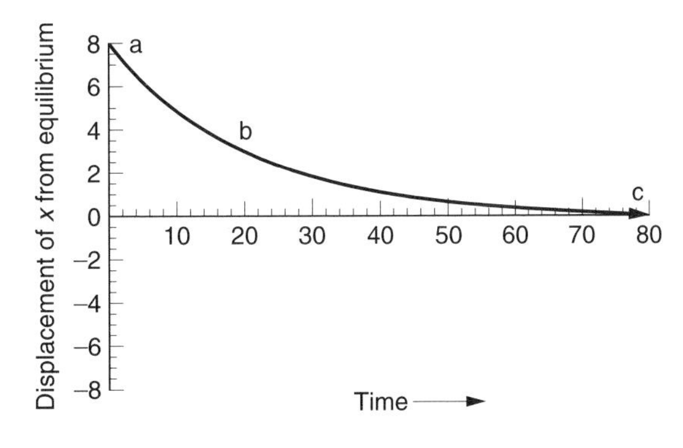

Figure 9.4 A trajectory in which the slope of the trajectory of a construct x is negatively proportional to the displacement from equilibrium. Any displacement from equilibrium leads to a return to equilibrium.

Autoregressive models are often conceptualized as a variable having an influence on itself over time. But one may also consider these models within the framework of self-regulating dynamical systems as a process that regulates back to equilibrium without an overshoot. Equation (9.3) can be used to fit data generated by a model conforming to a first-order autoregressive process. Thus the model in equation (9.3) is more general than autoregressive modeling, although at the expense of being less parsimonious. However, if one suspects that a self-regulating construct might overshoot its equilibrium (i.e., oscillate), then a first-order model (whether autoregressive or differential equation) is inappropriate since its self-regulation does not behave in this way.

In this chapter, we are interested in data from married couples that may show oscillation and that also may show mutual influence between two individuals. In order to build a model for these data we must first consider the possibility that two systems are coupled together. The next sections will explore some possibilities for theoretical models of coupled dynamics and how we might take into account individual differences in equilibria, self-regulation, and coupling between individuals.

9.2 Coupled Regulation and Coupled Dynamics

Suppose two self-regulating systems of the form shown in equation (9.3) not only regulated themselves, but also regulated each other. In this case, the definition of the system we are studying must include both of the self-regulating subsystems as well as their mutual influence on each other. How might we think of such a system?

One commonly used metaphor in dynamical systems is that of a pendulum with friction, a system that can be approximated by equation (9.3). Consider a system composed of two pendulums X and Y as illustrated in figure 9.5. Each pendulum may have its own length rod, thereby determining the frequency at which it would swing if no outside influence were in effect. Each pendulum might also have its own friction at the pivot. The acceleration (i.e., second derivative) due to gravity G is constant, but the acceleration in the direction in which each pendulum can move is proportional to its displacement from equilibrium. Thus, for instance, $\ddot{x}$ is proportional to x as is formalized in equation (9.3).

Now suppose we add a coupling between the two pendulums in the form of a linear spring. In this way, there might be an additional contribution to the second derivative of each pendulum from the displacement and velocity (first derivative) of the other pendulum. We might formalize this relationship as

$$\ddot{x}_t = \eta_x x_t + \zeta_x \dot{x}_t + \gamma_x(\eta_y y_t + \zeta_y \dot{y}_t) + e_{xt}, \tag{9.5}$$

$$\ddot{y}_t = \eta_y y_t + \zeta_y \dot{y}_t + \gamma_y(\eta_x x_t + \zeta_x \dot{x}_t) + e_{yt}, \tag{9.6}$$

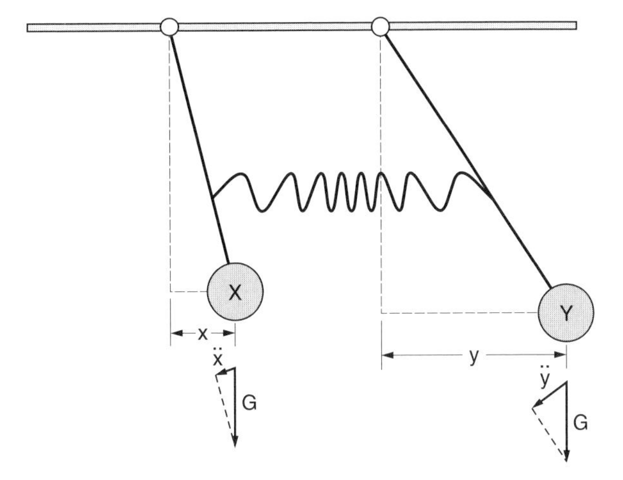

Figure 9.5 A system composed of two pendulums coupled together with a linear spring. Each pendulum may have its own length rod and its own friction at the pivot. The acceleration due to gravity G is constant, but its effect on each pendulum is proportional to the displacement from equilibrium.

where η_x and ζ_x are the frequency and damping coefficients for the x variable, η_y and ζ_y are the frequency and damping coefficients for the x variable, and γ_x and γ_y are the coupling strengths for x and y respectively. Note that in this model the same coefficients η_y and ζ_y are used in the equation for regulating x as well as y. This suggests that the same mechanism for self-regulation is used within a variable as well as in coupling of the variables together.

There is one major difference between the system modeled in equations (9.5) and (9.6) and the pendulums example in figure 9.5. In the pendulums example, the strength of the spring's pull on x is exactly matched by the strength of the spring's pull on y. Thus, the coupling is symmetric in figure 9.5. But in a system such as a married couple, one need not assume that $\gamma_x = \gamma_y$, that is, the influence of the husband on the wife may not be the same as the influence of the wife on the husband. In fact, one may not even wish to assume that these effects have the same sign! Thus, the system in equations (9.5) and (9.6) may potentially exhibit *asymmetric coupling*—a difficult system to build from pendulums and springs.

The coupled dynamical system defined by equations (9.5) and (9.6) can be simulated numerically in order to gain some understanding of its behavior. Figure 9.6 plots the results of a simulation calculated using the numerical integration function NDSolve in the Mathematica (Wolfram Research, 2003) software. The two variables x and y are set to have different intrinsic dynamics: the ratio $\eta_x/\eta_y = 1/1.28$. In addition, ζ_x is negative while ζ_y is positive. Thus one could think of the dynamics of x as being inhibitory whereas the dynamics of y are excitatory. The x and y variables are coupled together with equal and positive coupling coefficients $\gamma_x = \gamma_y = 0.3$, leading to moderately strong coupling. The only difference between the trajectories in figure 9.6(a) and 9.6(b) is in their initial conditions.

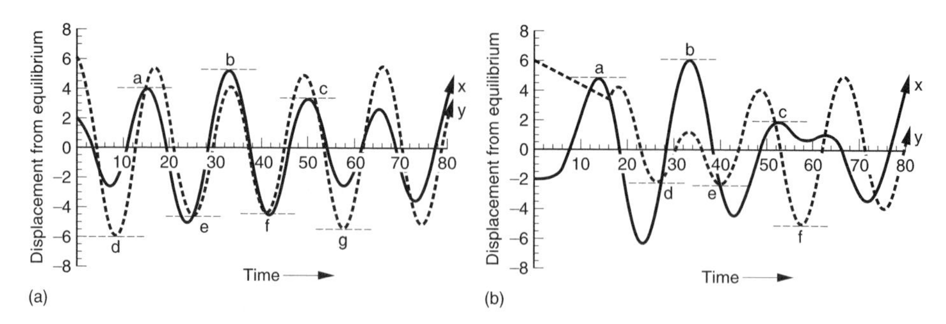

Figure 9.6 Two instances of coupled trajectories, x and y, that exhibit self-regulation as well as mutual influence. Although each figure has parameters that conform to the linear system from equation (9.2), differences in initial conditions can result in apparently irregular trajectories.

Notice how, in figure 9.6(a), the trajectory from the variable x passes through maxima at points a, b, and c. But at first these maxima are increasing from a to b and then the maxima are decreasing from b to c. Similarly, the minima for y, points d, e, f, and g, first appear to be damping to zero from d to e to f and then increase again from f to g. There is not a clear monotonic pattern of increasing or decreasing fluctuations over time. This results from the opposite sign of the two damping parameters where $\zeta_x < 0$ and $\zeta_y > 0$. Also, note that even though the intrinsic frequency of y is faster than x ($\eta_x > \eta_y$), y does not appear to oscillate faster than x. Observe that, at the point a, the maximum of x occurs prior to the maximum of y. Later, at point c, the maximum of y occurs prior to the maximum of x. Then on the next maximum x precedes y again. The two systems are coupled together into a mutually dependent, self-regulated and other-regulated frequency.

Now consider the two simulated trajectories x and y in figure 9.6(b). These two trajectories are generated by the system of equations with the same coefficient values as the two trajectories in figure 9.6(a), and yet the trajectories appear remarkably different. The only difference between figure 9.6(a) and 9.6(b) is that the x trajectory is started at time $t = 0$ at a value $x_0 = 2$ in figure 9.6(a) and $x_0 = -2$ in figure 9.6(b). In this case, the first three maxima at a, b, and c for the trajectory of x vary more widely than they did previously. After the maximum at c the trajectory does not cross the equilibrium before it begins to diverge from equilibrium again. The trajectory for y also differs markedly from the previous figure, including an initial period where the trajectory appears to be following a negative exponential before it changes to an oscillation.

In dynamical systems terminology, figure 9.6(b) exhibits the effect of *transient dynamics*, that is, behavior that is due to some exogenous influence (in this case our arbitrarily setting the values of x and y at time $t = 0$) as well as being due to the dynamics of the system itself. In the real world of psychological systems we must expect that exogenous influences will occur frequently, creating transient perturbations in the observed trajectories of measured psychological constructs. For this reason, in self-regulating systems it is essential to consider the possibility that apparently differing trajectories might not belong to different classes as could be concluded if we had used a latent growth-curve model on the trajectories in figures 9.6(a) and 9.6(b). Instead, it might be that similar intrinsic dynamics (i.e., similar self-regulatory mechanisms) might produce very different trajectories given different exogenous influences. One method for testing this hypothesis lies in the use of *state-space embedding* for modeling the covariances between derivatives in order to estimate coefficients of differential equation models of self-regulation.

9.3 Time-Delay Embedding

In order to estimate coefficients in a differential equation model, for instance equation (9.3), one must find a way to either reparameterize the model so that it does not contain derivatives or find a way to use the data to estimate the effects of the derivatives on one another. In each of these cases one must have multiple occasions of measurement on any one individual; in other words, a time series. Suppose we have measured the individual i on the variable x at p occasions separated by equal intervals of time s. A time series for individual i is a vector of observations $\mathbf{x} = \{x_1, x_2, x_3, \ldots, x_{p-1}, x_p\}$. The intrinsic dynamics of x are evidenced by the ordered sequence of observations; if we were to randomize the sequence, the resulting vector would have the same distribution of scores, but any evolving self-regulation of these observations would be lost. It is the way the x_1 leads to x_2 and x_2 to x_3 that captures the self-regulation of the system. This simple observation led to some formal theorems (Takens, 1985; Whitney, 1936) that show that it is, in theory, possible to recover the dynamics of a system from short, ordered sequences of observations as long as a time-delay constant and number of embedding dimensions is properly chosen. We shall demonstrate this idea using a simple example.

Time-delay embedding, also known as *state-space embedding*, is a method for creating a dataset that will allow the estimation of coefficients of differential equation models (see Ho et al., chapter 7, this volume, for an introduction to state-space methods). The essential idea behind time-delay embedding is easier than it might sound. One begins with a time-series vector of observations $\mathbf{x} = \{x_1, x_2, x_3, \ldots, x_{p-1}, x_p\}$, chooses a time-delay constant, τ, and a number of embedding dimensions d, and then produces a *state-space matrix*. The embedding dimension d is the number of columns in the state-space matrix, in other words, how many observations are in a "short sequence of observations." The time-delay constant τ represents how many observations to skip forward to obtain the next observation in a "short sequence of observations."

As an example, let us suppose we have 40 observations in a vector **x** and we choose a time delay of $\tau = 2$ and embedding dimension $d = 3$. From our time-series vector **x** we then create the embedded state-space matrix **X** as

$$\mathbf{X} = \begin{bmatrix} x_1 & x_3 & x_5 \\ x_2 & x_4 & x_6 \\ x_3 & x_5 & x_7 \\ x_4 & x_6 & x_8 \\ \vdots & \vdots & \vdots \\ x_{34} & x_{36} & x_{38} \\ x_{35} & x_{37} & x_{39} \\ x_{36} & x_{38} & x_{40} \end{bmatrix}. \tag{9.7}$$

Each row of **X** contains three observations, a "short sequence of observations," and could be considered to be a point in a three-dimensional space, a *state-space*. The ordering of the columns is important, since it preserves the time-ordered nature of each short sequence of observations. However, the ordering of the rows of this matrix does not matter. If we have chosen the time delay and embedding dimension correctly (see Sauer et al., 1991), all of the sequential information will be contained in the relationship between the columns. Although there still is sequential dependence between the rows as they are ordered above, that sequential dependence contributes nothing additional to the estimation of parameters for a differential equations model. This means that we might, for instance, select a random sample of rows and reorder it and still be able to obtain estimates of the coefficients of a differential equation that had generated these data (see Boker & Nesselroade, 2002, for a more lengthy discussion).

There are several methods that can be used to estimate differential equation models from state-space embedded data. Stochastic Differential Equations methods (see, e.g., Arminger, 1986; Oud & Jansen, 2000; Singer, 1998) transform the model into an integral form so that coefficients can be estimated directly from the time-delayed observations. While this has some advantages, estimation of such models involve complications arising from integration of the error term. Another approach is to transform the embedded state-space matrix so that explicit estimates of derivatives are obtained and then use standard statistical techniques to obtain parameter estimates (see, e.g., Boker, 2001; Boker & Graham, 1998). This is the approach that will be used in the current chapter. A third method, called Latent Differential Equations (LDE), generates latent variable estimates of the derivatives and then estimates coefficients from the covariances between these latent variables (see Boker et al., 2004, for an introduction). Each of these methods has its advantages and disadvantages. We shall focus on using the second approach, Local Linear Approximation (LLA), to transform our state-space embedded data matrix into explicit estimates of derivatives and discuss how we chose the time delay and embedding dimension for this problem.

We shall then use multilevel modeling to estimate the coefficients of a differential equation model of coupling between husbands and wives. We chose LLA for the estimation due to the simplicity of the approach and its ability to use standard multilevel estimation routines. We chose to use a multilevel model for two reasons: we expect that there may be considerable differences between marriages in the dynamics of self- and other-regulation, and we wish to see if we can account for these individual differences using marital satisfaction as a predictor.

9.4 Accounting for Individual Differences in Dynamics

Dynamical systems modeling focuses on how the current state of a system of variables leads to the future state of those variables. In this way, a dynamical systems analysis is concerned with how change in a system evolves—and this is why differential equations are frequently used to specify dynamical models. Individual differences in a dynamical system might be manifested in several ways. First, there might be differences in equilibria. One might think of this as individual differences in central tendency of a variable over time, and these could be estimated for instance in a mean and slope multilevel model where each individual's central tendency is changing slowly over time. Second, there may be individual differences in the mechanism that regulates how a variable fluctuates about the equilibrium in the absence of other influences. These differences could be estimated as individual differences in the coefficients of a differential equation model of the intrinsic dynamics of a variable. Finally, there might also be individual differences in how variables are coupled together. Some participants might be more reactive or responsive to outside influences than others. Or, in a coupled set of differential equations, there might be individual differences in the strength of the coupling parameter.

Prior to specifying a multilevel differential equations model, we shall need to account for individual differences in equilibrium values by centering each person's time series around their respective equilibrium values. One way to accomplish this is to fit a slope and intercept model to each person's time series and save the residuals from the predicted slope and intercept as input to the differential equations analysis. These slopes and intercepts will not be used in the current analysis, since here we are focusing on the short-term dynamics rather than on individual differences in equilibria. But reliable differences in equilibrium values might also be informative, and if so, one might use methods such as those described elsewhere in this volume (Walls et al., chapter 1) to model them.

We now create a differential equations model of the residuals calculated above by adapting equations (9.5) and (9.6) to account for individual differences in coefficients using a multilevel modeling framework. Suppose we were to be interested in the dynamics of husbands' feelings of intimacy and how they were influenced by the wives' feelings of intimacy. A second-order linear differential equation for the self-regulation and spousal regulation of Husband Intimacy could be written such that

$$
\begin{aligned}
\ddot{x}_{ij} &= \eta_{ix}x_{ij} + \zeta_{ix}\dot{x}_{ij} + \eta_{iy}y_{ij} + \zeta_{iy}\dot{y}_{ij} + e_{ij}, \\
\eta_{ix} &= c_{00} + u_{0i}, \\
\zeta_{ix} &= c_{10} + u_{1i}, \\
\eta_{iy} &= c_{20} + u_{2i}, \\
\zeta_{iy} &= c_{30} + u_{3i},
\end{aligned} \tag{9.8}
$$

where x_{ij} is the ith couple's Husband Intimacy score and y_{ij} is the ith couple's Wife Intimacy score at the jth occasion.[1] We continue to use $\dot{x}$ to indicate the first derivative and $\ddot{x}$ to indicate the second derivative of a variable with respect to time. In the second-level model, the constants c_{00}, c_{10}, c_{20}, and c_{30} represent the mean value of the respective random coefficient and u_{0i}, u_{1i}, u_{2i}, and u_{3i} represent the unique contribution to that random coefficient for the ith couple. A second model of the same form would be needed to specify the self-regulation and spousal regulation of Wife Intimacy.

Specifying the dynamics of a coupled variable in this way has the advantage that once derivatives are estimated from the data, coefficients can be easily estimated using standard mixed-effects software such as SAS PROC MIXED or the lme() function in R (Pinheiro & Bates, 2000). The general advantages of multilevel modeling accrue to this specification as well, including accounting for dependence within a couple and accounting for nonbalanced missing at random incomplete data. The major disadvantage of this specification is that the system is not solved simultaneously for all self-regulated variables and so the solution may be suboptimal. This could lead to bias in the parameter estimates and might obscure potentially reliable coupling effects. However, at this point, we do not yet have a viable alternative to the current specification. We suspect that such an alternative may be developed in the near future.

9.5 Application: Daily Intimacy and Disclosure in Married Couples

The experience of intimacy is the outcome of an interpersonal process of self-revealing disclosure to which the partner responds in a supportive, understanding way (Laurenceau et al., 2004; Reis & Shaver, 1988). An implicit aspect of the intimacy process is that each partner in a relationship has a desired level of intimacy and connectedness that can be conceived as an equilibrium range. In addition, amount of self-disclosure to a partner, which is often a way to start the intimacy process, may also be regulated with respect to a disclosure equilibrium level. Day-to-day experience of intimacy fluctuates around this desired level and the amount of self-disclosure to a partner, which is often a way to start the intimacy process, is regulated with respect to an intimacy equilibrium level.

Some relationship theorists have referred to a phenomenon of intimacy regulation in close relationships, a dyadic process reflecting a balance of the intimacy and autonomy needs of each partner (Prager & Roberts, 2004). As noted, an inherent assumption in this balance is the idea that intimacy is not consistently increasing over the course of a marriage but, rather, fluctuates in accordance with a desired level (Laurenceau et al., 2005). Nevertheless, one spouse's level of intimacy may change not only as a function of their own desired intimacy level but also as a function of the other spouse's intimacy level. Close relationships, such as marriage, exhibit a high degree of this type of interdependence–where the thoughts, feelings, and behaviors of one partner influence the thoughts, feelings, and behaviors of the other (Kelley et al., 1983). In a well-adjusted marriage, the regulation of the experience of intimacy toward one's desired equilibrium level should be facilitated so as to prevent each partner from experiencing long-term extremes in levels of intimacy (i.e., too much or too little). This contention has been observed by Prager and Roberts (2004) regarding intimacy regulation, who note that: "Well-functioning couples make continuous dynamic adjustments in their behavior to avoid emphasizing one pole–intimacy or autonomy–at the expense of the other" (p. 55). Is there an empirical way to capture and examine parameters that influence the dynamics of a putative dyadic intimacy process? Considering a marital dyad as the system, we attempt to model a self-regulating process in each individual partner and a coupling between these dynamic processes.

9.5.1 Application Data

Analysis was conducted on a sample of 96 couples who were married for an average of 9.32 years (SD = 9.50, range = 0.17–52.5). Data were collected using a daily diary sampling method whereby each spouse independently completed a structured diary assessing day-to-day variation in variables tapping intimacy in marriage each evening for 42 consecutive days. Diaries of this type allow participants to give more accurate and focused accounts of actual, everyday social activity and capture the dynamic nature of the process of intimacy that appears static with the use of more conventional, cross-sectional designs.

9.5.1.1 Procedure

The married couple participants were recruited from a region of central Pennsylvania to participate in a "study on daily experiences in marital relationships." Advertisements were placed in the local area newspaper and flyers were posted at various public locations.

One research assistant was assigned to each couple and visited their home three times over the course of the study. At the first visit, the research assistant obtained informed consent, collected demographic information, and administered cross-sectional measures. Next, spouses were instructed to complete independently a daily diary questionnaire during the evening on each of 42 consecutive days (6 weeks). The research assistant explained the procedure for completing the diary and defined various terms on the diary form. Each partner was given a written set of diary study instructions and definitions for reference throughout the study.

To help preserve confidentiality and ensure response integrity and honesty, each spouse was given a set of 42 adhesive labels with which to seal closed each completed daily diary form. Spouses were instructed to fold each completed diary form in thirds and use the adhesive label to seal it shut. At the end of the first visit, the members of each couple were given a sufficient number of diaries to take them through the mid-point of the 42-day recording period (i.e., 21 days) and a tentative appointment for the second visit was made. The research assistant phoned couples the following evening and spoke to each spouse individually in order to answer any questions that may have come up about the diary procedures. Couples were also called on a weekly basis to help ensure they were following the study procedure and completing diaries appropriately and to remind couples of the importance of completing the diaries independently.

The second visit was conducted at approximately the mid-point of the 42-day recording period. At this visit, the research assistant collected each spouse's completed diaries for the first half of the recording period and scheduled a tentative final visit. Upon completion of the final week of diary recordings, the research assistant visited each couple a final time at their home to collect the completed diaries for the second half of the recording period, and to provide couples with remuneration for their participation in the study.

9.5.1.2 Measures

A daily diary measure was constructed to assess the variables theorized as per Reis and Shaver's (1988) interpersonal process model of intimacy and was modeled after the diary form used by Laurenceau et al. (1988). Responses to diary items were all rated using 5-point Likert scales (e.g., 1 = very little, 5 = a great deal). Being part of a larger diary form, only the diary variables relevant to the current study are reported here.

Intimacy Spouses rated the amount of closeness that they experienced across the marital interactions with their spouse that day. We chose to use the term "closeness" rather than "intimacy" to ensure that participants were rating the degree of psychological, rather than physical or sexual, proximity. Based on the identified strong link between intimacy and perceptions of a partner's responsiveness (Reis & Shaver, 1988), we also included items assessing perceived partner responsiveness. Spouses rated the degree to which he/she felt understood by their partner (one item), validated by their partner (one item), accepted by their partner (one item), and cared for by their partner (one item) across daily marital interactions. Based on factor analysis, it appeared that these four items had substantial shared variance and therefore were aggregated to create a single daily intimacy score.

Disclosure Spouses rated the amount that they disclosed facts and information (one item), the amount that they disclosed their thoughts (one item), and the amount that they disclosed their feelings (one item) across all the interactions that they had with their spouse during the day. Spouses also rated the amount they perceived that their partner disclosed facts and information (one item), the amount of perceived disclosure of their partner's thoughts (one item), and the amount of perceived disclosure of feelings (one item) across all the interactions that they had with their spouse during the day. A disclosure summary variable was created using the sum of these six items.

Dyadic Adjustment Scale (DAS) The DAS is a commonly administered, 32-item self-report measure used to assess global marital satisfaction (Spanier, 1976). Scores range from 0 to 151 with higher scores indicating greater marital satisfaction, and this measure was completed by a spouse prior to beginning the 42-day diary recording period. The mean DAS score for husbands in this sample was 112.64 (SD = 12.73), while the mean of the wives was 113.92 (SD = 14.34). A matched pairs t-test indicated that husbands and wives did not differ in their levels of global marital satisfaction ($t(95) = -1.06, p = 0.29$). Cronbach's alphas for the husband and wife DAS scales were 0.90 and 0.91, respectively.

9.5.2 Modeling the Application Data

Daily diaries were completed by 96 couples for 42 consecutive days. Two couples were excluded due to low response rates. Of the remaining 94 couples, the overall complete husband and wife response rate was 97% for the disclosure measures and 96% for the intimacy measures. Figure 9.7 plots the husband and wife scores for intimacy and disclosure from four example couples. By inspection, it does not seem unreasonable that each individual may have a preferred equilibrium value or set point for each of these scores. In addition, we note that there appear to be short-term *synchronization events* in which the husbands' and wives' scores seem to be displaced far from equilibrium on the same day. These events appear to be relatively short-term, in that the scores return to near equilibrium within an interval of a few days. Given our inspection of the plotted data, we decided that it was not unreasonable to test a model in which husbands' and wives' scores were coupled together such that they regulated each other as well as themselves.

An intercept-only mixed-effects model grouping by dyad was fitted separately to husbands' and wives' scores for intimacy and disclosure as

$$\begin{aligned} x_{ij} &= b_i + e_{ij}, \\ b_i &= c_0 + u_i, \end{aligned} \tag{9.9}$$

where x_{ij} is the score (e.g., wives' intimacy) for individual i at occasion j, b_i is the intercept for individual i, c_0 is the mean value for the intercepts, and u_i is the unique within-individual contribution to the intercept. Two values of interest in this equation are the standard deviation of u_i, the between-persons variance in intercepts, and the standard deviation of e_{ij}, a measure of the mean

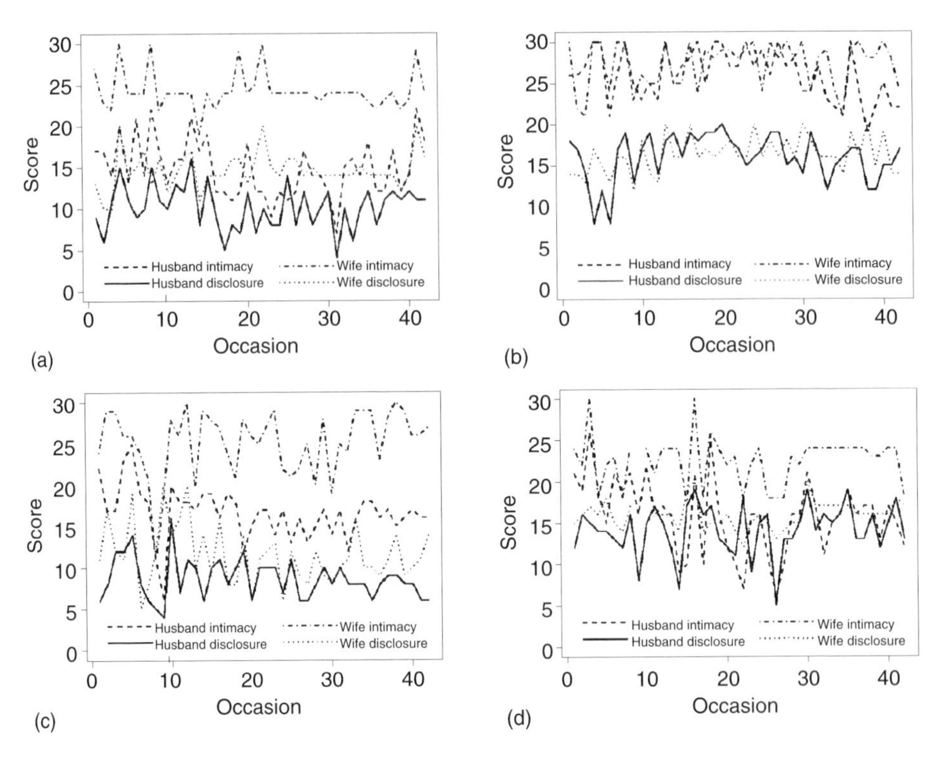

Figure 9.7 Daily husband and wife intimacy and disclosure time series for four couples. Note that there appear to be individual differences in equilibrium values as well as between-couple differences in the variability around the equilibrium.

within-person variability. As shown in table 9.1, the standard deviations of the between-person differences (e.g., wives' intimacy between person SD $= 3.66$) were approximately the same as the within-person variability (e.g., wives' intimacy residual SD $= 3.87$) for each of the variables. This suggests that although there may be individual differences in equilibrium values, the intraindividual variability is of approximately equivalent magnitude in this sample and thus it is warranted to test whether the substantial within-person component of variance is patterned as a self-regulatory system.

In this study, there is no single event that we can use to align these intensively measured variables in time. Thus, there is no way to meaningfully assign $t = 0$, as this may be different for each couple or could even change for a couple during the course of the study. Outside events and daily stressors such as problems at work or sickness of a parent might occur at unpredictable intervals and these stressors might influence the intimacy and disclosure scores for a couple (see Boker & Nesselroade, 2002; Ramsay, 2002, for discussions of *curve registration* or *phase problems*). For this reason, a state-space model is a better choice for these data than would be a growth curve model. A state-space model is relatively insensitive

Table 9.1 Intercept-only mixed-effects models grouped by dyad and each predicting one variable

Variable	Intercept	SE	Between SD	Within SD
Husband Intimacy	21.50	0.40	3.84	3.32
Wife Intimacy	21.62	0.38	3.66	3.87
Husband Disclosure	13.07	0.27	2.56	2.56
Wife Disclosure	12.83	0.28	2.64	2.94

to the timing of influences exogenous to the system whereas a growth curve model will incorporate the unpredictable intervals of the external stressors as part of the individual differences in trajectories.

We elected to use Local Linear Approximation (LLA) to explicitly estimate derivatives and then to use random coefficients (i.e., mixed effects or HLM) modeling to estimate parameters. First and second derivatives for a time series $\mathbf{x} = \{x_1, x_2, x_3, \ldots, x_p\}$ can be estimated using LLA by first removing the linear trend from each individual's data and then creating a three-column time-delay embedded state-space matrix $\mathbf{X}$ of order $(p - 2\tau) \times 3$ from these residuals as discussed above. The derivatives $\dot{x}_k$ and $\ddot{x}_k$ for the kth row of the matrix $\mathbf{X}$ can be calculated as

$$\dot{x}_k = (x_{k3} - x_{k1})/2\tau \tag{9.10}$$

and

$$\ddot{x}_k = (x_{k3} + x_{k1} - 2x_{k2})/\tau^2, \tag{9.11}$$

where τ is the lag offset used to create the embedded state-space matrix $\mathbf{X}$ (Boker & Nesselroade, 2002).

Once these derivatives are calculated, a second-order linear differential equation mixed-effects model can be fitted as

$$\begin{aligned} \ddot{x}_{ij} &= \eta_{ix} x_{ij} + \zeta_{ix} \dot{x}_{ij} + e_{ij}, \\ \eta_{ix} &= c_{00} + u_{0i}, \\ \zeta_{ix} &= c_{10} + u_{1i}, \end{aligned} \tag{9.12}$$

where x is one of the four variables: Husband Disclosure, Husband Intimacy, Wife Disclosure, or Wife Intimacy. In order to choose a value of τ that is appropriate for these data, we fitted the model in equation (9.12) to each of the four variables using values for $\tau = \{1, 2, \ldots, 8\}$. The resulting mean explained variance (r^2) over 93 individuals' data and the lower 95% confidence interval for the mean explained variance is plotted for each value of τ in figure 9.8. The horizontal line at $r^2 = 0.656$ is the expected value of r^2 for uncorrelated measurement error. Thus, we reject the hypothesis that the residual intraindividual variability from a linear trend is measurement error when $\tau \geq 4$.

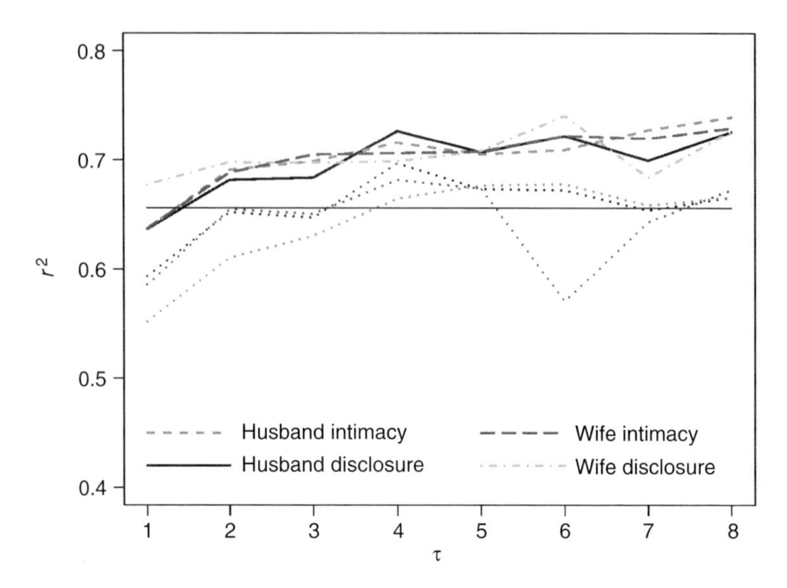

Figure 9.8 Mean explained within-individual variance (r^2) for univariate damped linear oscillator models for husbands' and wives' disclosure and intimacy. Dotted lines are lower 95% confidence interval for r^2 for each variable. Horizontal line is expected value for uncorrelated measurement error.

However, note that the largest gain in mean r^2 occurs between $\tau = 1$ and $\tau = 2$. The mean r^2 for all measures are near their peak value by $\tau = 2$. Previous simulations have suggested that minimum bias estimates for frequency (η) parameters are achieved at the minimum τ when the r^2 first nears its maximum value. Combining this observation with the previous observation that high-displacement co-occurring "events" happened over a period of just 3 to 5 days suggests that $\tau = 2$ may be the best choice. Larger values of τ would tend to obscure these short-term episodes.

Thus, when modeled with an uncoupled linear oscillator, these data are unlikely given a null hypothesis that the data consist solely of measurement error. However, we might suspect that an uncoupled model is incomplete since it ignores the possibility of husbands and wives regulating each other's behavior. A better model would include coupling parameters. We next present the results of a coupled model for each of the four variables.

Husbands' intimacy regulation was modeled using a mixed-effects model. After time-delay embedding using a delay of $\tau = 2$, we calculated derivatives using LLA and predicted the second derivative of husbands' intimacy, $\ddot{x}$, as

$$\ddot{x}_{ij} = \eta_{ix}x_{ij} + \zeta_{ix}\dot{x}_{ij} + \eta_{iy}y_{ij} + \zeta_{iy}\dot{y}_{ij} + e_{ij},$$

$$\eta_{ix} = c_{00} + c_{01}h_i + u_{0i},$$

$$\zeta_{ix} = c_{10} + c_{11}h_i + u_{1i},$$

$$\eta_{iy} = c_{20} + c_{21}w_i + u_{2i},$$
$$\zeta_{iy} = c_{30} + c_{31}w_i + u_{3i}, \tag{9.13}$$

where x_{ij} is the intimacy score for husband i on day j, y_{ij} is the intimacy score for wife i on day j, and $\dot{x}_{ij}$ and $\dot{y}_{ij}$ are their respective first derivatives. The variable h_i is the marital satisfaction score for husband i and the variable w_i is the marital satisfaction score for wife i. The constant coefficients c_{00}, c_{10}, c_{20}, and c_{30} are the intercept values (fixed effects) for the random coefficients η_{ix}, ζ_{ix}, η_{iy}, and ζ_{iy}, respectively. The constant coefficients c_{01}, c_{11}, c_{21}, and c_{31} represent the effects of husbands' and wives' marital satisfaction on their respective random coefficients. And finally, u_{0i}, u_{1i}, u_{2i}, and u_{3i} are the unique contributions of each husband or wife on his or her random coefficient.

Once the data have been time-delay embedded into a three-dimensional state space, the rows are reduced by three for every missing observation. Thus, in our example, the total observations across the 93 individuals is 3,006 rather than the 3,722 complete observations in the dataset. This reduction of the state-space matrix by incompleteness in the data is a problem that must be considered prior to using a dynamical systems method: one missing observation creates two missing first derivatives and three missing second derivatives.

In the hope of clarifying the discussion, we have adopted the following notation for the parameters of subsequent models. Frequency parameters (i.e., η parameters) are denoted Husband Intimacy (HI), Husband Disclosure (HD), Wife Intimacy (WI), and Wife Disclosure (WD). Damping parameters (i.e., ζ parameters) are similarly denoted dHI, dHD, dWI, and dWD for the husbands' and wives' intimacy and disclosure scores, respectively. Finally, interactions with the DAS scale are denoted HI×HDAS, dHI×HDAS, HD×HDAS, dHD×HDAS, WI×WDAS, dWI×WDAS, WD×WDAS, and dWD×WDAS for the husbands' and wives' DAS scores interacting with their respective intimacy and disclosure scores.

The results of fitting the mixed effects model from equation (9.13) (to derivatives calculated with LLA using a lag of $\tau = 2$ on the Husband Intimacy residuals after removing a linear trend) are presented in table 9.2. The Husband η parameter (HI) value is different from zero, suggesting a lawful patterning of oscillation in Husband Intimacy from day to day. The Husband η effect is consistent with the intimacy trajectory showing the greatest curvature when the displacement is farthest from equilibrium. There is no evidence of intrinsic damping as reflected in a nonsignificant Husband ζ parameter (dHI). Examining the coupled effects of Wife Intimacy regulation on Husband Intimacy curvature, higher levels of Wife Intimacy (WI) is associated with less Husband Intimacy curvature, indicating that the Husband Intimacy would turn around and move back toward equilibrium less quickly than usual. Interestingly, this effect is moderated by marital satisfaction, where greater Wife marital satisfaction has a negative effect on Wife η (WI×WDAS). Thus, the effect of greater wife displacement on husband curvature for wives reporting higher satisfaction is diminished, allowing Husband Intimacy to turn around back toward equilibrium as per his intrinsic η parameter. The damping parameters reflected no significant effects on Husband curvature.

Table 9.2 Mixed-effects model predicting second derivative of Husband Intimacy score using intrinsic intimacy self-regulation, wives' intimacy regulation, and marital satisfaction scores (AIC = 9,129, BIC = 9,243, $N = 3{,}006$, Groups = 94, mean $r^2 = 0.693$, $\tau = 2$, $\eta\tau^2 = -1.614$)

	Value	SE	DF	t	p
dHI	−0.0012	0.1440	2905	−0.01	0.993
HI	−0.4034	0.0789	2905	−5.11	<0.001
dWI	0.0490	0.1330	2905	0.37	0.713
WI	0.1216	0.0482	2905	2.52	0.012
HI × HDAS	−0.0008	0.0006	2905	−1.22	0.224
dHI × HDAS	0.0000	0.0012	2905	0.02	0.984
WI × WDAS	−0.0011	0.0004	2905	−2.64	0.008
dWI × WDAS	−0.0007	0.0011	2905	−0.63	0.526

Examining the model for Husband Disclosure contained in table 9.3, Husband η (HD) reflects a significant oscillation parameter. However, unlike the results for Husband Intimacy, Wife Disclosure regulation parameters are not coupled to Husband Disclosure curvature. Moreover, marital satisfaction is not a moderator of η or ζ effects.

The model for Wife Intimacy contained in table 9.4 reflects the same pattern of results as for Husband Intimacy. This model demonstrated a significant η parameter, indicating patterned oscillations around an equilibrium. Husband Intimacy showed a coupled effect on Wife curvature, where higher levels of Husband Intimacy (HI) is associated with less Wife Intimacy curvature, keeping Wife Intimacy higher than she would want it to be. The strength of this coupled effect was moderated by Husband satisfaction, with greater levels of Husband satisfaction

Table 9.3 Mixed-effects model predicting second derivative of Husband Disclosure score using intrinsic disclosure self-regulation, wives' disclosure regulation, and marital satisfaction scores (AIC = 7,385, BIC = 7,500, $N = 3{,}006$, Groups = 94, mean $r^2 = 0.683$, $\tau = 2$, $\eta\tau^2 = -1.738$)

	Value	SE	DF	t	p
dHD	0.0982	0.1388	2905	0.707	0.4793
HD	−0.4346	0.0750	2905	−5.789	0.0000
dWD	0.0081	0.1174	2905	0.069	0.9446
WD	0.0352	0.0423	2905	0.831	0.4057
HD × HDAS	−0.0004	0.0006	2905	−0.647	0.5173
dHD × HDAS	−0.0009	0.0012	2905	−0.780	0.4351
WD × WDAS	−0.0004	0.0003	2905	−1.238	0.2156
dWD × WDAS	−0.0002	0.0010	2905	−0.220	0.8254

Table 9.4 Mixed-effects model predicting second derivative of Wife Intimacy score using intrinsic intimacy self-regulation, husbands' intimacy regulation, and marital satisfaction scores (AIC = 9,812, BIC = 9,926, N = 3,006, Groups = 94, mean $r^2 = 0.690$, $\tau = 2$, $\eta\tau^2 = -1.955$)

	Value	SE	DF	t	p
dWI	0.0607	0.1481	2905	0.410	0.6818
WI	−0.4887	0.0769	2905	−6.354	0.0000
dHI	0.1273	0.1620	2905	0.785	0.4321
HI	0.1230	0.0604	2905	2.035	0.0418
WI × WDAS	−0.0001	0.0006	2905	−0.213	0.8312
dWI × WDAS	−0.0007	0.0012	2905	−0.546	0.5851
HI × HDAS	−0.0011	0.0005	2905	−2.191	0.0285
dHI × HDAS	−0.0008	0.0014	2905	−0.600	0.5479

Table 9.5 Mixed-effects model predicting second derivative of Wife Disclosure score using intrinsic disclosure self-regulation, husbands' disclosure regulation, and marital satisfaction scores (AIC = 8,264, BIC = 8,378, N = 3,006, Groups = 94, mean $r^2 = 0.701$, $\tau = 2$, $\eta\tau^2 = -2.55$)

	Value	SE	DF	t	p
dWD	0.0623	0.1343	2905	0.463	0.6431
WD	−0.6381	0.0747	2905	−8.539	0.0000
dHD	−0.0233	0.1627	2905	−0.143	0.8862
HD	0.1320	0.0652	2905	2.022	0.0432
WD × WDAS	0.0011	0.0006	2905	1.672	0.0946
dWD × WDAS	−0.0006	0.0011	2905	−0.543	0.5866
HD × HDAS	−0.0013	0.0005	2905	−2.218	0.0266
dHD × HDAS	0.0002	0.0014	2905	0.150	0.8802

(HI×HDAS) being associated with greater Wife Intimacy curvature. Damping was not a significant parameter in this model.

In contrast to the results predicting Husband Disclosure curvature, table 9.5 shows that Wife Disclosure was coupled to Husband Disclosure (HD) and the strength of this coupling was predicted by Husband satisfaction (HD×HDAS).

9.6 Discussion

An undamped linear oscillator performed well as a dynamic process model of husbands' and wives' intimacy and disclosure trajectories. Some degree of mutual dependence between husband and wife scores might be expected, but may not be equal. We discovered *symmetric coupling* between husband and wife intimacy, whereby the strength of the coupling was moderated by marital satisfaction. Findings also revealed *asymmetric coupling* between spouse disclosure scores, such that Husband Disclosure was not coupled to Wife Disclosure, but Wife Disclosure was coupled to Husband Disclosure. Damping was not a significant parameter in any of these models.

In some ways, it may not be surprising that an undamped linear oscillator would be a reasonable model for the trajectory of intimacy, rather than a model with damping to an equilibrium range. Intimacy is the outcome of an interpersonal process where the experience of intimacy for one partner (A) is dependent upon self-disclosing acts by partner A, supportive responses by the other partner (B), the perception of responsiveness by A from B, the timing of the exchanges between A and B, and influences outside of A and B (e.g., intimacy-facilitating vs. nonfacilitating situations). The number of potential inputs to the process may lead to a situation where intimacy is unlikely to remain at an equilibrium level long, but rather is constantly oscillating around it. If an individual shoots upward past equilibrium until she is feeling too intimate, then she may actively engage in intimacy-distancing tactics (e.g., reductions in self-disclosure, inattention to partner attempts at responsiveness) to allow the regulation process to come down toward equilibrium. Our findings suggest that individuals with spouses who are highly satisfied with their marriages also have spouses who perhaps facilitate intimacy regulation toward equilibrium.

Based on theory in the close relationships literature (Reis & Shaver, 1988), intimacy is a construct that shows qualities of both constancy and change. Constancy is reflected in an assumed desired level of intimacy that can be considered an equilibrium range that may be different across individuals. Change is reflected in the day-to-day variability in the experience of intimacy that fluctuates around an individual's equilibrium range. Moreover, as a likely consequence of the inherent interdependence that exists in close relationships, such as marriage, we found mutual influence (symmetric coupling) between the self-regulating dynamics of both spouses' intimacy trajectories. We believe that the current application of coupled differential equations models of dynamic systems to intimacy in married couples is a way to parameterize the argument that intimacy is best conceptualized as a process reflecting variability, change, and fluctuation over time.

A bottom-line conclusion from this work is that in couples reporting greater marital adjustment, intimacy regulation is facilitated. If the goal of an intimacy regulation system is to stay within an equilibrium range, then our findings suggest that this regulation is more apparent in satisfied couples. This type of dynamic mutual influence may be exemplified in no better context than that of close dyadic relationships, such as marriage.

ACKNOWLEDGMENT

Funding for this work was provided in part by NIH grants 1R29 AG14983 and K01 MH64779.

NOTE

1. Although the notation used here for the multilevel models differs from the notation presented elsewhere in this volume, this notation is popular in many areas of psychology, especially educational psychology. We feel that it simplifies the presentation of the ideas while maintaining an accurate account of the modeling. The equivalence between the notation used here and the generalized linear mixed models notation is explained in detail by Walls et al. in chapter 1 of this volume.

References

Arminger, G. (1986). Linear stochastic differential equation models for panel data with unobserved variables. In N. Tuma (Ed.), *Sociological Methodology 1986* (pp. 187–212). San Francisco: Jossey Bass.

Bisconti, T.L., Bergeman, C.S., & Boker, S.M. (2004). Emotion regulation in recently bereaved widows: A dynamical systems approach. *Journal of Gerontology: Psychological Sciences*, *59*, 158–167.

Boker, S.M. (2001). Differential structural modeling of intraindividual variability. In L. Collins & A. Sayer (Eds.), *New Methods for the Analysis of Change* (pp. 3–28). Washington, DC: American Psychological Association.

Boker, S.M., & Graham, J. (1998). A dynamical systems analysis of adolescent substance abuse. *Multivariate Behavioral Research*, *33*, 479–507.

Boker, S.M., Neale, M.C., & Rausch, J. (2004). Latent differential equation modeling with multivariate multi-occasion indicators. In K. van Montfort, H. Oud, & A. Satorra (Eds.), *Recent Developments on Structural Equation Models: Theory and Applications* (pp. 151–174). Dordrecht, Netherlands: Kluwer.

Boker, S.M., & Nesselroade, J.R. (2002). A method for modeling the intrinsic dynamics of intraindividual variability: Recovering the parameters of simulated oscillators in multi–wave panel data. *Multivariate Behavioral Research*, *37*, 127–160.

Carver, C.S., & Scheier, M.F. (1998). *On the Self-Regulation of Behavior.* New York: Springer.

Chow, S.M., Ram, N., Boker, S.M., Fujita, F., Clore, G., & Nesselroade, J.R. (2005). Capturing weekly fluctuation in emotion using a latent differential structural approach. *Emotion*, *5*, 208–225.

Kaplan, D., & Glass, L. (1995). *Understanding Nonlinear Dynamics.* New York: Springer.

Kelley, H.H., Berscheid, E., Christensen, A., Harvey, J.H., Huston, T.L., Levenger, G., et al. (1983). *Close Relationships.* New York: Freeman.

Laurenceau, J.-P., Barrett, L.F., & Pietromonaco, P.R. (1998). Intimacy as an interpersonal process: The importance of self-disclosure, and perceived partner responsiveness in interpersonal exchanges. *Journal of Personality and Social Psychology*, *74*, 1238–1251.

Laurenceau, J.-P., Feldman Barrett, L., & Rovine, M.J. (2005). The interpersonal process model of intimacy in marriage: A daily-diary and multilevel modeling approach. *Journal of Family Psychology*, *19*, 314–323.

Laurenceau, J.-P., Rivera, L.M., Schaffer, A.R., & Pietromonaco, P.R. (2004). Intimacy as an interpersonal process: Current status and future directions. In D.J. Mashek & A. Aron (Eds.), *Handbook of Closeness and Intimacy* (pp. 61–78). Mahwah, NJ: Lawrence Erlbaum.

Nesselroade, J.R. (1991). Interindividual differences in intraindividual changes. In J.L. Horn & L. Collins (Eds.), *Best Methods for the Analysis of Change: Recent Advances, Unanswered Questions, Future Directions* (pp. 92–105). Washington, DC: American Psychological Association.

Nesselroade, J.R., & Boker, S.M. (1994). Assessing constancy and change. In T.F. Heatherton & J.L. Weinberger (Eds.), *Can Personality Change?* (pp. 121–147). Washington, DC: American Psychological Association.

Nilsson, M., Naessen, S., Dahlman, I., Hirschberg, A., Gustafsson, J.A., & Dahlman-Wright, K. (2004). Association of estrogen receptor beta gene polymorphisms with bulimic disease in women. *Molecular Psychiatry*, *9*, 28–34.

Oud, J.H.L., & Jansen, R.A.R.G. (2000). Continuous time state space modeling of panel data by means of SEM. *Psychometrica*, *65*, 199–215.

Pinheiro, J.C., & Bates, D.M. (2000). *Mixed-Effects Models in S and S-Plus.* New York: Springer.

Prager, K.J., & Roberts, L.J. (2004). Deep intimate connection: Self and intimacy in couple relationships. In D. Mashek & A. Aron (Eds.), *Handbook of Closeness and Intimacy* (pp. 43–60). Mahwah, NJ: Lawrence Erlbaum.

Ramsay, J.O. (2002). Multilevel modeling of longitudinal and functional data. In D. Moskowitz & S. Hershberger (Eds.), *Modeling Intraindividual Variability with Repeated Measures Data: Methods and Applications* (pp. 87–107). Mahwah, N.J.: Lawrence Erlbaum.

Reis, H.T., & Shaver, P. (1988). Intimacy as an interpersonal process. In S. Duck (Ed.), *Handbook of Personal Relationships* (pp. 367–389). Chichester: John Wiley.

Sauer, T., Yorke, J., & Casdagli, M. (1991). Embedology. *Journal of Statistical Physics*, *65*, 95–116.

Singer, H. (1998). Continuous panel models with time dependent parameters. *Journal of Mathematical Sociology*, *23*, 77–98.

Spanier, G. (1976). Measuring dyadic adjustment: New scales for assessing the quality of marriage and similar dyads. *Journal of Marriage and the Family*, *38*, 15–28.

Takens, F. (1985). Detecting strange attractors in turbulence. In A. Dold & B. Eckman (Eds.), *Lecture Notes in Mathematics 1125: Dynamical Systems and Bifurcations* (pp. 99–106). Berlin: Springer.

Thompson, J.M.T., & Stewart, H.B. (1986). *Nonlinear Dynamics and Chaos.* New York: John Wiley.

Varma, M., Chai, J.K., Meguid, M.M., Laviano, A., Gleason, J.R., Yang, Z.J., & Blaha, V. (1999). Effect of estradiol and progesterone on daily rhythm in food intake and feeding patterns in Fischer rats. *Physiology and Behavior*, *68*, 99–107.

Whitney, H. (1936). Differentiable manifolds. *Annals of Mathematics*, *37*, 645–680.

Wolfram Research (2003). *Mathematica 5.0.* Champaign–Urbana, IL: Wolfram Research.

10

Point Process Models for Event History Data: Applications in Behavioral Science

Stephen L. Rathbun, Saul Shiffman, and Chad J. Gwaltney

Some human behavior is characterized by repeated events that occur at discrete points in time. For example, figure 10.1 shows the times, represented by vertical lines, at which cigarettes were lit by an anonymous smoker over a seven-day period as recorded on an electronic diary. These data were obtained from a study of smoking patterns by Shiffman et al. (2002). The pattern of smoking events reflects the diurnal sleep/wake cycle of this smoker, with most cigarettes being smoked during the afternoon and evening. The bold lines are actually composed of two or more vertical lines, clustered so closely in time that they cannot be separated at the temporal scale at which this picture was drawn. Thus, this smoker exhibits occasional smoking binges, but otherwise the cigarettes appear to be well spaced throughout each day.

The pattern of times over which discrete events occur over a specified time interval is called a *temporal point pattern*. The points at which the behavior of interest is observed (here, the lighting of a cigarette) are referred to as the *events* of the pattern to distinguish these from arbitrary points in time (Diggle, 1983, p. 1). In the social science literature, temporal point patterns fall under the domain of event history data (Blossfeld & Rohwer, 2002). Event history modeling often focuses on the durations of time between successive events, or counts of the numbers of events that occur in discrete time intervals. The former involves the application of survival analysis (e.g., Allison, 1984; Blossfeld & Rohwer, 2002; Vermunt, 1997), whereas the latter may involve Poisson regression (e.g., King, 1988, 1989; Minkoff, 1997; Olzak, 1992). Survival analysis tends to ignore the actual times on a clock or calendar at which the events have occurred, reducing the data to the durations of time between successive events. Daily, weekly, and seasonal patterns are often ignored. On the other hand, the reduction of an event history to counts over arbitrary intervals results in a loss of information. The results may depend on the intervals selected, and may produce biased estimates of model parameters.

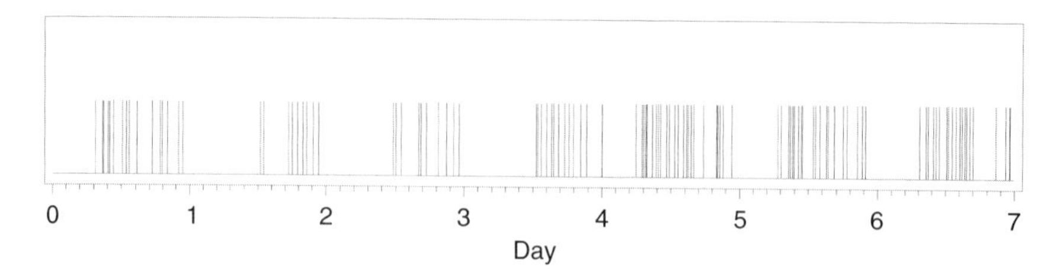

Figure 10.1 Times at which cigarettes were lit by an anonymous smoker over a one-week period (Thursday through Wednesday).

This chapter considers point process models for event history data. A point process is a stochastic mechanism for generating the times of the events of a point pattern. Point process models are closely related to survival models (Lawless, 2003, pp. 568–570). Indeed, the application of survival models to temporal point patterns is legitimate. Connections between point process and survival models will be presented in section 10.2. Point processes are concerned with the times of events as they may appear on a calendar, whereas survival models focus on the durations of time between successive events. Point process modeling focuses on the estimation of the rate of event occurrences, expressed as numbers of events per unit time. Investigation of event rates may lend insight into the mechanisms regarding the behavior of interest. For example, daily activities may induce diurnal patterns of variation in smoking rates, and the impact of temporally varying covariates on event rates may be quantified. Point process models also allow for the consideration of the impact of all events preceding the event of interest, not just the duration since the most recent event.

Examples of point pattern data may be found throughout the behavioral and social sciences. Investigations of substance abuse have considered the effects of mood (positive and negative affect) on patterns of smoking (Shiffman et al., 2002) and alcohol consumption (Collins et al., 1998). Investigations of children have monitored the timing of aggressive behavior (i.e., hitting, pushing, kicking, and spitting) toward their siblings (Jones et al., 1992), and disruptive behavior (e.g., calling out, disputing with the teacher) in a classroom setting (Hall et al., 1971; Pfiffner & O'Leary, 1987). Interest may focus on the timing of significant events occurring within the lifetimes of individuals, including marriages, divorces, and job changes. For example, Willet and Singer (1995) consider successive entries of teachers into teaching. Processes involving larger social groupings may also be of interest. For example, Olzak (1992) considers the timing of riots and protests in her investigation of racial and ethnic confrontations. Organizational ecology may focus on the times at which firms are founded (Lomi, 1995), or go bankrupt (Carroll & Delacroix, 1982). Along those same lines, Minkoff (1997) considers the founding of social organizations. Additional examples are cited by Blossfeld and Rohwer (2002).

Ideally, point pattern data are comprised of a complete list of the times of all events of interest. In practice, however, it is often difficult to obtain such information. Retrospective surveys may involve questionnaires, asking participants to recall the dates at which events occurred. Such data are likely to contain errors, resulting from imperfect memory regarding event dates. In addition, not all events may be recalled when the survey is conducted. In their investigations of ethnic conflicts, Olzak et al. (1996) identify candidate events from the *New York Times* index. Snyder and Kelly (1977) discuss potential sources of errors in such data. For example, selection bias in newspaper accounts can result in errors of omission, and content bias may result in errors in the classification of ethnic conflicts. Data on organizational foundings and firm bankruptcies are often in the form of aggregate annual counts (Lomi, 1995); information on seasonal variation in such foundings is lost. Investigations of human behavior may be assisted by tape recorders (Hall et al., 1971) or videotape (Hsu & Fogel, 2003). Pfiffner and O'Leary (1987) observed the interactions between children and their teachers from behind a one-way mirror in their investigation of disruptive behavior. Tape recorders, videotape, and one-way mirrors are often used in laboratory settings, where behavior may or may not reflect what would have occurred in the natural environment of the subjects.

Recent advances in electronics have made it possible to record the events of interest in the natural environment of the participants. Stone and Shiffman (1994) have defined a collection of methods they collectively call Ecological Momentary Assessment (EMA). These methods involve collecting human behavioral data in real time and in real-world environments. While not logically requiring electronic diary methods, practical applications of EMA have been facilitated by the availability of palmtop computers that research participants can carry with them for data recording. Some EMA studies use signaling devices (palmtop computers, beepers) to implement experimenter-designed sampling schemes, for example, beeping participants at random to sample their condition. Another use of EMA methods and devices is to record the occurrence of discrete events (Wheeler & Reis, 1991: event-contingent recording; McFall, 1977: self-monitoring), such as episodes of eating or social interaction. These event records are of particular interest for point process models.

Section 10.3 will illustrate point process modeling using EMA smoking data, described in section 10.1. Our intent is not to build a single model explaining variation in smoking rates, but to illustrate, in some detail, the variety of point process models available for event history data. Emphasis is placed on the kinds of questions that may be asked by such models. How do the diurnal and weekly patterns of smokers' lives affect smoking rates? What is the quantitative effect of emotional distress on smoking rate? In addition, we shall use point process analyses to explore priming (i.e., "chain smoking"), and nicotine regulation models of cigarette smoking patterns.

Before considering specific point process models, section 10.2 will introduce the intensity function of a point process. Heuristically, the intensity function measures event rates, expressed in units of numbers of events per unit time, just as velocity measures the distance traveled per unit time. The inhomogeneous Poisson process is defined, a model yielding independent counts in successive time intervals. This model is used to model weekly and diurnal patterns and the effects of time-varying covariates in section 10.3. The conditional intensity function is also defined in section 10.2, allowing for the modeling of dependence among events of a single type. Section 10.4 summarizes what we have learned by fitting point process models to the smoking data. Finally, section 10.5 describes the extension of point processes to multivariate data, where either multiple types of events are observed on individual subjects, or the same event type may be observed simultaneously on multiple subjects.

10.1 Ecological Momentary Assessment of Smoking

10.1.1 Relevance

One area in which EMA methods have been pioneered has been the study of cigarette smoking. Cigarette smoking is responsible for 430,000 deaths annually in the United States alone (CDC, 2002) and approximately 5 million deaths annually worldwide (WHO, 2004). Although behavioral accounts of smoking are complex, the fundamental structure of smoking–that smokers periodically consume tobacco by smoking cigarettes on multiple discrete occasions each day–is simple and lends itself to EMA data collection and to point process analysis.

Most analyses of cigarette smoking highlight the role of both pharmacological and behavioral factors (see U.S. Surgeon General, 1988). Smoking is, in part, motivated by attempts to maintain some minimum level of nicotine in the body, in order to avoid withdrawal. In theory, this should result in a fairly regular schedule of smoking. However, smoking is observed to be more unevenly distributed over time. Behavioral accounts highlight the role of situational factors in prompting smoking, and many prominent theories emphasize the role of emotional states in smoking. Most smokers report that they smoke more when they are upset and to reduce or manage their distress. Inferences about the association between mood and smoking have generally been drawn from global summaries such as are elicited on questionnaires. However, there is reason to doubt that smokers can accurately recall and reliably summarize their smoking patterns; autobiographical memory processes introduce both error and systematic bias into such reports (Shiffman, 1993). Real-time EMA data require neither recall, summary, nor inference from participants, and therefore have the potential to overcome these difficulties and shed light on the variables that influence smoking.

10.1.2 Methods

Participants were 304 smokers who enrolled in a smoking cessation research study. To qualify, participants had to smoke at least 10 cigarettes per day, to have been smoking for at least 2 years, and to report high motivation and overall efficacy to quit during a screening interview. Details regarding the demographic profile of the participants and their smoking habits may be found in Shiffman et al. (2002).

Upon enrollment, participants were trained to use an electronic diary (ED; for a description of the ED system, see Shiffman et al., 2002) designed to allow data collection in near-real time. Participants monitored ad libitum smoking for 16 days prior to a designated quit date; they were instructed not to change their smoking during this time.

During the monitoring period, participants were instructed to record each cigarette on the ED, immediately before smoking. On about 4–5 randomly selected smoking occasions per day (M = 4.1, SD = 2.5), the ED administered an assessment (see section 10.3 for the sampling algorithm). Participants were also prompted audibly by the ED 4–5 times per day (M = 4.5, SD = 2.3) to complete a similar assessment while they were not smoking (nonsmoking assessments). The timing of the prompts was random, with the constraint that no prompts were issued for 10 minutes after a cigarette entry.

Participants could use an ED option to suspend random prompting, for example, when driving or to avoid intruding on important meetings. This option was used infrequently: an average of once every 2.5 days, for an average of 24.2 minutes per day. Subjects used a similar feature for naps once every 5 days, for a daily average duration of 18.4 minutes. Thus, although subjects had access to features that allowed for some time-out from observation, the features were sparingly used and likely do not substantially bias the data. Continuous monitoring could be interrupted if the participant failed to respond to the audible prompts. Such noncompliance was rare: participants responded to 91% of all prompts within the 2 minutes allowed. Each evening, participants had the opportunity to note if they had failed to record some cigarettes in real time. On average, they reported failing to enter 0.7 cigarettes per day (SD = 1.4). As the time these cigarettes were smoked is not known, they do not enter into the analysis. Additional information on compliance can be found in Shiffman et al. (2002).

10.1.3 Assessments

Cigarette and nonsmoking assessments incorporated identical assessments of situation, activity, and mood, completed in approximately 1–3 minutes. Each assessment was time-stamped in order to unambiguously identify when the assessment was completed.

Although participants completed items assessing multiple features of the situation (e.g., setting, activity; see Shiffman et al., 2002), only the affect items are described here and used in the following analyses. Participants rated mood adjectives derived from the circumplex model of affect (e.g., J. Russell, 1980), which specifies that affect consists of two bipolar dimensions: positive–negative affect and arousal. We also included bipolar items on affect and arousal to directly tap these key circumplex dimensions, as well as affect items drawn from the DSM-IV (American Psychiatric Association, 1994) criteria for tobacco withdrawal.

Factor analyses of the mood data (based on 66,230 assessments from the pre- and post-quit period) yielded three orthogonal factors (see Shiffman et al., 1996). The first two replicated those expected under the circumplex model: negative affect ($\alpha = 0.87$) and arousal ($\alpha = 0.79$). A third factor, labeled attention disturbance ($\alpha = 0.64$), captured reports of difficulty concentrating. Nicotine withdrawal items did not factor separately, but loaded cleanly on the negative affect factor. However, a single item regarding "restlessness" did not load heavily on any of the three factors, but seemed to tap a unique variance, related to nicotine withdrawal syndrome, and was independently associated with relapse (Shiffman et al., 1996); we therefore retained this as a separate item for analysis. In section 10.3, factor scores are used as time-varying covariates to assess their impacts on smoking rates. Measurement errors in estimating the corresponding latent variables may lead to biased estimates of model parameters, but a detailed assessment of the magnitude of potential biases falls beyond the scope of the current chapter.

10.2 Point Process Models

10.2.1 Counting Measure, Poisson Process, and Intensity Function

The data from a temporal point pattern can be presented as a listing of the times at which the events of the pattern have occurred. However, for modeling purposes, another useful description can be obtained using the counting measure $N(A)$, defined here as the number of events (e.g., cigarettes lit) in an arbitrary set of times A. Here, A may be an arbitrary time interval $[t_1, t_2]$, so that $N[t_1, t_2]$ is the number of cigarettes smoked between times t_1 and t_2. For example, if A is the week of January 7–13, 2004, then $N(A)$ is the number of cigarettes smoked during that week.

The homogeneous Poisson process is the standard null model to which point patterns are often compared. Under the *homogeneous Poisson process* with intensity λ, the numbers of events observed in any finite collection of nonoverlapping intervals are independently sampled from Poisson distributions with means proportional to the lengths of their respective intervals. Thus, the probability that n events will be observed in a given interval $[a, b]$ is

$$\Pr\{N(A) = n\} = \frac{e^{-\lambda|b-a|}}{n!}(\lambda\,|b-a|)^n .$$

The parameter λ denotes the mean number of events per unit time, for example, the mean number of cigarettes smoked per hour.

Figure 10.2 shows a realization of a homogeneous Poisson process with intensity $\lambda = 16.1$ cigarettes per day, corresponding to the average number of cigarettes smoked by the anonymous smoker depicted in figure 10.1. Despite their completely random locations, the events appear to show moderate clustering.

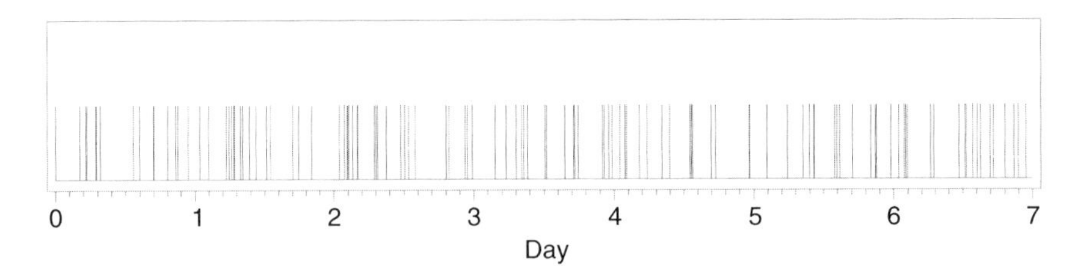

Figure 10.2 Realization of a Poisson process with intensity $\lambda = 16.1$ cigarettes per day.

This apparent clustering is due to the property of the homogeneous Poisson process which states that the time intervals between successive events is exponentially distributed with mean $1/\lambda$. That is, if T is the time since the most recent smoking event, then the probability that T exceeds τ is given by the survival function

$$S(\tau) = \Pr(T \geq \tau) = e^{-\lambda\tau},$$

with a constant hazard rate of λ.

The homogenous Poisson process is analogous to the constant mean model in univariate statistics where the data are assumed to be independently sampled from a normal distribution with mean μ and variance σ^2. As such, the homogeneous Poisson process is of limited interest. What is desired are models that allow for temporal dependence in event rates, dependence on covariates, and event rates that depend on the past history of events.

Before considering such models, define the intensity function

$$\lambda(t) = \lim_{\delta \to 0} \frac{E\{N[t, t+\delta]\}}{\delta}.$$

The numerator gives the mean number of events in a small interval of time $[t, t+\delta]$. Dividing by δ, we obtain the mean rate of events per unit time. Taking the limit as $\delta \to 0$ yields a momentary rate that can be expressed in units of events per unit time. The intensity function plays a role analogous to that of the mean of a random variable, here allowing that mean to depend on time. Times with high intensities will tend to contain large numbers of events, while times with low intensities will tend to contain few events.

10.2.2 Models

The inhomogeneous Poisson process is the simplest alternative to the homogeneous Poisson process. A point process is an *inhomogeneous Poisson process* with intensity $\lambda(t)$ if:

1. The numbers of events observed in any finite collection of nonoverlapping intervals are independently distributed.

2. The number of events $N[a,b]$ in any interval $[a,b]$ is Poisson distributed with mean

$$E\{N[a,b]\} = \Lambda[a,b] = \int_a^b \lambda(t)\,dt.$$

That is, the probability that there are n events in the interval is

$$\Pr\{N[a,b] = n\} = \frac{e^{-\Lambda[a,b]}}{n!}\{\Lambda[a,b]\}^n.$$

The first of these properties is the so-called independent increments property of the inhomogeneous Poisson process. It is analogous to the assumption of independent random variables in univariate statistics. Thus, if smoking events were realized from an inhomogeneous Poisson process, the numbers of cigarettes smoked on different hours will be independently distributed. The same could be said for the numbers of cigarettes smoked on different days, weeks, or years. Note that if the intensity function $\lambda(\tau)$ is expressed as a function of the time τ since the most recent event, and T is the time since the most recent event, then the inhomogeneous Poisson process is equivalent to a survival model with survivor function

$$S(\tau) = \Pr\{T \geq \tau\} = \exp\left\{-\int_0^\tau \lambda(u)\,du\right\}$$

and a time-varying hazard rate equal to the intensity function $\lambda(\tau)$.

Typically, the intensity function $\lambda(t;\boldsymbol{\theta})$ is assumed to be a function of a vector of parameters $\boldsymbol{\theta}$. If smoking rate depends on day of the week, for example, we may take $\lambda(t;\boldsymbol{\theta}) = \theta_i$ if t belongs to the ith day of the week. Under a modulated Poisson process (Cox, 1972a), the intensity function

$$\lambda(t;\boldsymbol{\beta}) = \exp\{\beta_0 + \beta_1 x_1(t) + \beta_2 x_2(t) + \cdots + \beta_p x_p(t)\} \tag{10.1}$$

depends on the time-varying covariates $x_1(t), x_2(t), \ldots, x_p(t)$. If the data are reduced to daily counts (or alternatively, weekly, monthly, or annual counts), and the values of the covariates are aggregated into daily means, then Poisson regression (McCullagh & Nelder, 1989) may be used to estimate model parameters. This would yield consistent estimates of the model parameters (in the sense that the estimates will converge to the true values of their respective parameters as sample size increases) provided that the covariates are constant within each day. In the present application, however, the covariates are not piecewise constant, but vary with time of day. The application of Poisson regression in this case introduces bias into the parameter estimates, bias that does not diminish with increasing sample size. Section 10.3.4 developes a method for fitting the modulated Poisson process that retains the actual event times and produces a consistent estimator for model parameters.

The analysis of longitudinal data in the social sciences often involves the fitting of a hierarchical linear model, including random subject effects for intercepts and slopes. The inclusion of such random effects in (10.1) yields a random intensity function. Poisson processes with random intensity functions are called Cox processes (Cox, 1955; Lundberg, 1940). Point processes defined through conditional intensity functions form an important class of Cox processes.

Under the independent increments property of the Poisson process, the number of events in any time interval is independent of the past history of events. However, there may be theoretical reasons to expect dependence among the events of the process. For example, a model for nicotine addiction suggests that there may be negative dependence among smoking events, with smoking occurring at regular intervals as blood nicotine levels fall over time following the previous cigarette. Conversely, contagion can result in positive dependence among riots (Myers, 2000), resulting in their clustering over time. Likewise, in a classroom setting, disruptive behavior by one student may trigger further disruptive acts by his or her colleagues.

Dependence among the events of a point process may be modeled using the *conditional intensity function*

$$\lambda(t; H_t) = \lim_{\delta \to 0} \frac{E\{N[t, t+\delta] \mid H_t\}}{\delta}.$$

Here, the numerator gives the expected number of events in the time interval $[t, t+\delta]$, conditional on the past history of events $H_t = \{t_i : t_i < t\}$, the times of all events before time t. Dividing by δ obtains the conditional mean number of events per unit time. Then, taking the limit as $\delta \to 0$ yields an instantaneous rate of event occurrence, now depending on the past history of events. Note that the definition above for the conditional intensity function is similar to the definition of the hazard function in survival analysis. While the hazard is expressed as a function of duration since the most recent event, the conditional intensity is expressed as a function of time on the clock or calendar. Moreover, the conditional intensity function may depend on events prior to the most recent event, and thus offers a more flexible class of models, including the self-exciting and stress-release models discussed in sections 10.3.5 and 10.3.6, respectively.

10.2.3 Parameter Estimation

Suppose that temporal point patterns are observed for m subjects. Let $[0, T_i]$ denote the time interval over which the pattern is observed, and let

$$0 < t_{i1} < t_{i2} < \cdots < t_{in_i} < T_i$$

denote the times at which the events were observed for subject i. For each subject, assume that the times of the events are realized from independent point processes with respective (conditional) intensity functions $\lambda_i(t; \boldsymbol{\theta}, H_{ti})$, where $\boldsymbol{\theta}$ is a vector of unknown parameters to be estimated, and H_{ti} is the history of past events at time t for subject i. If the data are realized from a Poisson process, then the conditional intensity function does not depend on the history H_{ti} of past events, and so is equal to the unconditional intensity.

The log-likelihood for a point process with (conditional) intensity $\lambda_i(t; \boldsymbol{\theta})$ is

$$L(\boldsymbol{\theta}) = \sum_{i=1}^{m} \left(\sum_{j=1}^{n_i} \log \lambda_i(t_{ij}; \boldsymbol{\theta}) - \int_0^{T_i} \lambda_i(t; \boldsymbol{\theta})\, dt \right) \tag{10.2}$$

(Cox & Lewis, 1966). The maximum likelihood estimator is then obtained by finding $\widehat{\boldsymbol{\theta}}$ that maximizes $L(\boldsymbol{\theta})$. With a few exceptions (i.e., section 10.3.1), there is no closed-form expression for the maximum likelihood estimator. When a closed-form expression does not exist, numerical methods (e.g., the Newton–Raphson algorithm) must be applied to obtain the maximum likelihood estimate. Moreover, there is no exact finite-sample expression for the standard errors of the elements of $\widehat{\boldsymbol{\theta}}$. The large-sample properties of $\widehat{\boldsymbol{\theta}}$ are outlined in Appendix A. Berman and Turner (1992) discuss methods for obtaining approximate maximum likelihood estimates that may be implemented using software written for generalized linear models.

10.3 Application: An EMA Study of Smoking Data

10.3.1 Models

The following considers a sequence of increasingly complex point process models, illustrating the variety of questions that may be addressed by such models. The simplest model investigates the effect of day of the week on the numbers of cigarettes smoked. This model yields a closed-form expression for the intensity function and its standard error. Next, the effect of time of day is explored using a nonparametric estimator of the intensity function. A modulated Poisson point process is used to assess the effects of time-varying covariates on smoking rates. As special cases of the inhomogeneous Poisson process, these models all share the independent increments property.

Two models will be borrowed from seismology to illustrate the modeling of dependent events. Positive dependence during chain smoking may be modeled using Hawkes' (1971) self-exciting point process. Conversely, the stress-release model is used to illustrate negatively dependent regularly spaced smoking events, as would result if smokers are attempting to maintain their blood nicotine levels.

The following analyses focus on the routine behavior of smokers, and not on what patterns may emerge after they attempt to quit at the end of day 17. Therefore, all of the following analyses were carried using only data collected from days 4–16 of the investigation. The first three days were designed to allow participants to become acclimated with the electronic diary.

10.3.2 Day of the Week

Smoking may be influenced by weekly rhythms. For example, smoking may be suppressed by smoking restrictions that most smokers face at work. Conversely, smoking may be promoted by the weekend environment, which is more likely to include stimuli that promote smoking, such as alcohol consumption and social contact with other smokers.

Table 10.1 summarizes the overall amount of smoking by day of the week. For a given day of the week, the number of person-days D_i is defined to be the number of days data were available for that day of the week, summed over the 304 subjects in the study. The corresponding number N_i is the total number of cigarettes smoked on that day of the week.

The simplest inhomogeneous Poisson process has an intensity function that depends on day of the week:

$$\lambda(t) = \theta_i \quad \text{if } t \text{ belongs to the } i\text{th day of the week.} \tag{10.3}$$

Here, θ_i gives the smoking rate as a function of day of the week i. Here, the log-likelihood simplifies to

$$L(\boldsymbol{\theta}) = \sum_{i=1}^{7} \left(N_i \log \theta_i - D_i \theta_i\right).$$

Taking the derivative of $L(\boldsymbol{\theta})$ with respect to θ_i, setting the resulting expression equal to zero, and then solving for $\widehat{\theta}_i$, we can obtain a closed-form expression for the maximum likelihood estimator. In this case:

$$\widehat{\theta}_i = \frac{N_i}{D_i}.$$

This estimator is equal to the mean number of cigarettes smoked by the subjects on the ith day of the week. Thus, smoking rate is expressed in units of cigarettes smoked per day. To express this rate in number of cigarettes smoked per hour,

Table 10.1 Number of person-days and total number of cigarettes smoked for each day of the week

Day (i)	Person-days (D_i)	Total cigarettes (N_i)
Monday	600	12,617
Tuesday	596	12,302
Wednesday	302	6,330
Thursday	603	12,932
Friday	603	12,744
Saturday	599	12,174
Sunday	599	12,032
Total	3,902	81,131

simply divide this figure by 24 hours per day. Under the model, the total number of cigarettes N_i smoked on the ith day of the week is Poisson distributed with mean $D_i\theta_i$. For Poisson random variables, the variance is equal to the mean, so the variance of $\widehat{\theta}_i$ is

$$\text{var}\left(\widehat{\theta}_i\right) = \frac{\theta_i}{D_i}.$$

Substituting $\widehat{\theta}_i$ into the expression above and taking the square root, we obtain the standard error

$$\text{SE}\left(\widehat{\theta}_i\right) = \frac{\sqrt{N_i}}{D_i}.$$

To determine if smoking rate depends on day of the week, compute the likelihood ratio test statistic

$$G = 2\left\{\sum_{i=1}^{7} N_i \log\frac{N_i}{D_i} - N\log\frac{N}{D}\right\},$$

where $N = \sum_{i=1}^{7} N_i$ is the total number of cigarettes smoked, and $D = \sum_{i=1}^{7} D_i$ is the total number of person-days over which data are available. Under the null hypothesis that the smoking rate is independent of day of the week, this test statistic is approximately chi-square distributed with 6 degrees of freedom.

Smoking rates varied significantly across the days of the week ($G = 39.19$; DF $= 6$; $p < 0.0001$). Figure 10.3 plots the estimated smoking rate as a function of day of the week for the EMA smoking data. The vertical bars give 95% confidence intervals for the rates. The dashed line gives the overall average smoking rate of 20.79 cigarettes per day. Although there is statistically significant variation

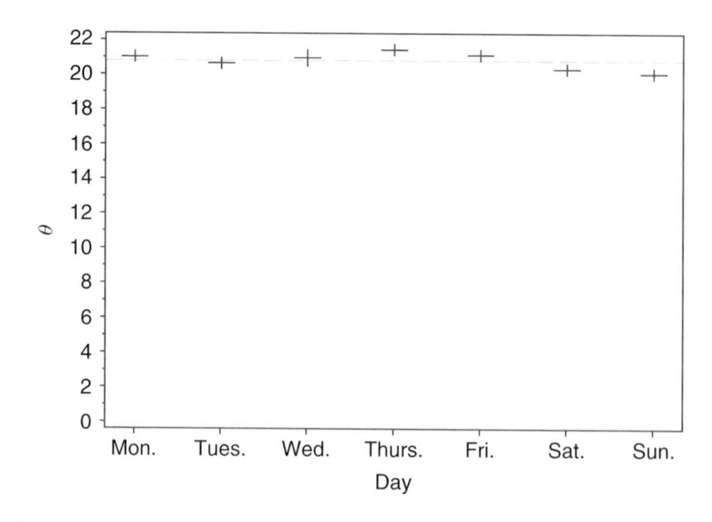

Figure 10.3 Relationship between smoking rate and day of the week.

Table 10.2 Sample mean and sample variance for the number of cigarettes smoked on each day of the week

Day (i)	Mean ($\widehat{\theta}_i$)	Variance
Monday	21.0	87.7
Tuesday	20.6	92.4
Wednesday	21.0	92.0
Thursday	21.4	93.7
Friday	21.1	94.5
Saturday	20.3	104.6
Sunday	20.1	94.0

in smoking rates with day of the week, the magnitude of that variation is not great. The peak smoking rate of 21.4 cigarettes per day occurs on Thursdays, while minimum smoking rates of 20.1–20.3 cigarettes per day are found on the weekends.

The goodness of fit of an inhomogeneous Poisson process with intensity (10.3) may be assessed by comparing the sample variance to the sample mean number of cigarettes smoked on each day the week (table 10.2). Under the assumed model, the sample variances should be equal to the sample means. For the EMA smoking data, however, the variances are approximately 4–5 times the magnitudes of the means, indicating a significant lack of fit. This lack of fit may be attributed to exclusion of important covariates explaining variation in the numbers of cigarettes smoked each day, variation in baseline smoking rates among subjects, and/or the clustering of smoking events. Such clustering would occur under chain-smoking behavior, where the lighting of a cigarette enhances the likelihood that additional cigarettes will be lit shortly thereafter.

10.3.3 Time of Day

Human behavior is likely to show periodic patterns of variation. Our daily activities may induce diurnal patterns of variation, the work-week may induce patterns that depend on day of the week, and climate may induce annual patterns of variation across the four seasons. To investigate such patterns, Poisson processes with periodic intensity functions may be considered. An intensity function is *periodic* with period τ if it satisfies

$$\lambda(t) = \lambda(t + k\tau)$$

for all integers k. Thus, the intensity function is identical for all times separated by the period τ. An inhomogeneous Poisson process is said to be *cyclic* if it has a periodic intensity function. The model presented in section 10.3.1 is a special case, where $\tau = 1$ week, so that the intensity of smoking depends on day of the week. To model the effect of time of day on smoking rate, take $\tau = 24$ hours.

We shall consider a nonparametric estimator for the periodic intensity of a cyclic Poisson process. Using a nonparametric estimator avoids any parametric assumptions regarding the shape of the periodic intensity function, so the resulting estimator will be completely informed by the data. Our estimator is a small modification of the nonparametric estimator recently proposed by Helmers, Mangku, and Zitikis (2003), who considered data from only a single subject. Our estimator averages the Helmers–Mangku–Zitikis HMZ estimator across subjects to allow for longitudinal data across multiple subjects, implicitely assuming that there is no subject-to-subject variation in smoking patterns. The original HMZ estimator may be applied to each of the 304 subjects, but this would not yield a useful summary of the data. Let t_{ij} denote the time of day for the jth smoking event of subject i, let N_i denote the total number of smoking events recorded for subject i, let T_i denote the number of days on which subject i is observed, and let m denote the total number of subjects. Compute $T = \sum_{i=1}^{m} T_i$, the total number of subject-days available in the data. Then the modified HMZ estimator for the intensity at time t during the day is

$$\widehat{\lambda}(t) = \frac{\tau}{T} \sum_{i=1}^{m} \sum_{j=1}^{N_i} \kappa_h \left(\left\| t_{ij} - t \right\| \right); \; 0 \leq t \leq \tau, \tag{10.4}$$

where

$$\left\| t_{ij} - t \right\| = \begin{cases} \left| t_{ij} - t \right|; & \text{if } \left| t_{ij} - t \right| \leq \frac{\tau}{2} \\ \tau - \left| t_{ij} - t \right| & \text{if } \left| t_{ij} - t \right| > \frac{\tau}{2}. \end{cases}$$

Taking $\tau = 24$ hours, this yields an estimate of the smoking rate in units of cigarettes per hour. The metric $\left\| t_{ij} - t \right\|$ measures the difference in time between the observation t_{ij} and the time of day t, expressed on a 24-hour clock. The second choice in the expression above is written so that, for example, an observation at 10 P.M. is 4 hours away from 2 A.M. rather than 20 hours.

The selection of an appropriate bandwidth is more important than the choice of kernel. The Epanichnikov (1969) kernel

$$\kappa_h(r) = \begin{cases} \frac{3}{4h} \left\{ 1 - \left(\frac{r}{h} \right)^2 \right\}; & \text{if } |r| \leq h \\ 0; & \text{if } |r| > h \end{cases}$$

is optimal for density function estimation. In practice, a wide variety of bandwidths h should be considered, plotting the estimated intensity $\widehat{\lambda}(t)$ against time of day t for each bandwidth. For example, figure 10.4 shows such plots for three different bandwidths, ranging from 15 minutes to 2 hours. Note that the smoothness of the plotted curves decreases with increasing bandwidth. Bandwidth is selected so as to yield the most meaningful interpretation of the data. Selecting too small a bandwidth may result in the smoothing out of important features in the periodic intensity function. Conversely, selecting too large a bandwidth may produce a lot of uninterpretable noise.

Using a wide bandwidth of 2 hours yields a very smooth curve, revealing a broad diurnal pattern in smoking behavior, with very few cigarettes smoked

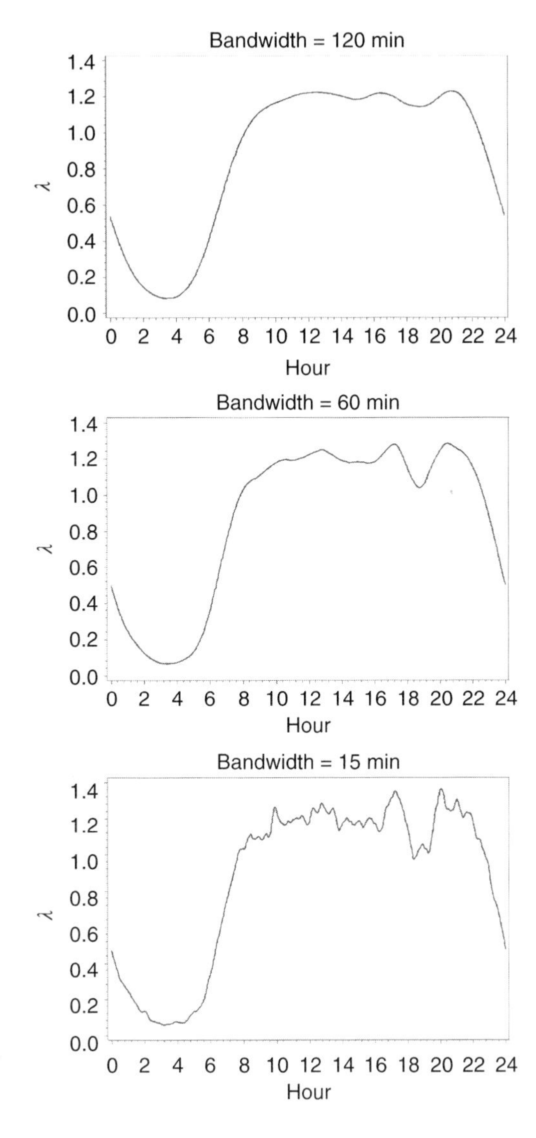

Figure 10.4 Nonparametric estimate of the relationship between smoking rate (cigarettes per hour) and time of day for three different bandwidths.

during the early morning hours. Smoking rate increases rapidly after 4 A.M., reaching a broad plateau that extends over the workday. There is a slight dip in smoking rate around 3 P.M., before it reaches a peak of 1.22 cigarettes per hour at around 4:25 P.M. Smoking rate then drops to a local minimum of 1.14 cigarettes per hour at 6:50 P.M., before rising to a local maximum of 1.23 cigarettes per hour at 8:45 P.M.

Reducing the bandwidth to 60 minutes reveals more subtle patterns in smoking behavior. Here local maxima were found at 10:37 A.M., 12:49 P.M., 3:16 P.M.,

5:12 P.M., and 8:27 P.M. These local maxima appear to correspond to morning coffee, lunch, afternoon coffee, after work, and after dinner. The peak smoking rate of 1.28 cigarettes per hour at 5:12 P.M. is particularly dramatic. A marked decrease in smoking rate is observed thereafter, reaching a local minimum of 1.04 cigarettes per hour at 6:45, dinnertime. The smoking rate then rises to a local maximum of 1.28 cigarettes per hour at 8:27 P.M., after which it drops as smokers retire for the night.

A narrow bandwidth of 15 minutes produces a very rough curve with more local minima and maxima that can be readily interpreted. In contrast, the wide bandwidth of 120 minutes only revealed the broadest diurnal patterns in smoking behavior, and so may be deemed too wide. The bandwidth of 60 minutes revealed a pattern that appears to reflect the pattern of work breaks and mealtimes, yielding the best interpretation of the data.

The relationship between smoking rate and time of day differs between weekdays and weekends (figure 10.5). On weekdays, the diurnal pattern of smoking resembles the general pattern found in figure 10.4. By breaking out the weekdays, an additional local maximum is observed at 8:20 A.M. However, the most striking difference between weekdays and weekends is the dramatic variation in smoking in the early evenings of weekdays, a pattern that is not observed on weekends. On weekends, smoking rate increases less rapidly in the early morning hours than on

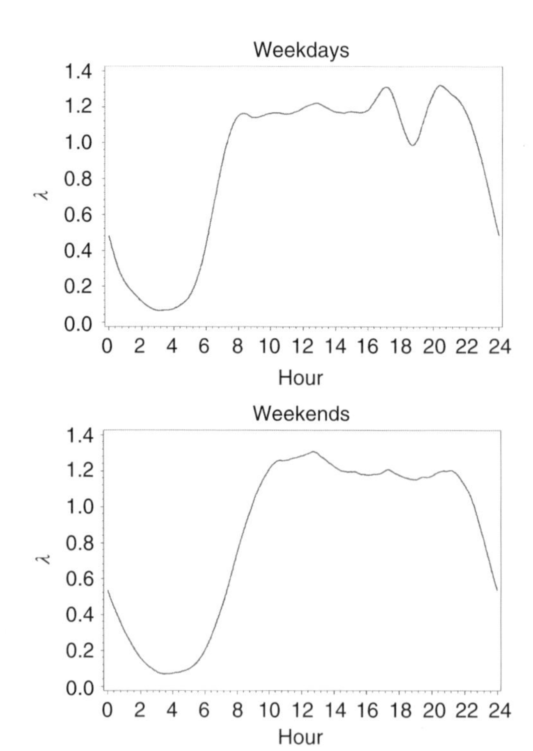

Figure 10.5 Relationship between smoking rate and time of day for weekdays and weekends.

weekdays, a pattern that may be attributed to smokers sleeping in on the weekend. Weekend smoking rates are higher in the late morning hours than in the afternoon and evening.

10.3.4 Covariates

The modulated Poisson process (Cox, 1972a) may be used to model the effects of covariates such as mood (positive and negative affect) on smoking rate. Most smokers report that they smoke more when they are distressed, and that smoking relaxes them (Kassel et al., 2003; Shiffman, 1993). These reports have been subject to various interpretations. Some have suggested that nicotine genuinely relieves psychological distress through its central nervous system effects. Others have suggested that nicotine does not have such general effects, but only relieves the nicotine withdrawal syndrome experienced by dependent smokers when deprived, including depression, tension, irritability, and restlessness, as well as difficulty concentrating. Smokers may experience mild versions of withdrawal even between cigarettes while smoking ad libitum, and these may cue more smoking. Alternatively, it has been proposed that after repeated cycles of withdrawal and withdrawal-relief, smokers may come to respond to mood changes with smoking, even if the mood changes were not due to withdrawal. Thus, whether nicotine actually improves mood or not, motivation to smoke is thought to increase when smokers experience emotional distress. This suggests that smoking rate may vary with mood.

The modulated Poisson process is a special case of the Poisson process with log intensity function

$$\log \lambda (t) = \beta_0 + \beta_1 x_1 (t) + \beta_2 x_2 (t) + \cdots + \beta_p x_p (t),$$

where $x_1 (t), x_2 (t), \ldots, x_p (t)$ are temporally varying covariates, and $\beta_0, \beta_1, \beta_2, \ldots, \beta_p$ are regression coefficients. The log-linear form guarantees that this intensity function is nonnegative as required. Maximum likelihood estimation of the parameters of the modulated Poisson process model requires that the values of the covariate be known not only for each of the smoking events, but also for all times during which the subjects were monitored. In the present application, it was not practical to obtain the values of the covariates (mood variables) for all smoking events, as this would have placed an undue burden on the subjects. Moreover, it is generally not possible to obtain the values of covariates for all times during any interval. Appendix B describes a new approach to parameter estimation that takes these data limitations into account.

Table 10.3 shows the results of fitting four modulated Poisson process models, each predicting smoking rate from time-varying covariates. Model 1 considers negative affect, arousal, and attention, the three factors identified by the factor analysis described in section 10.1.3. The results suggest that smoking rate is a decreasing function of arousal and an increasing function of attention disturbance. The estimated intercept corresponds to an average smoking rate of $\exp\{-0.05898\} = 0.94$ cigarettes per hour when all covariates take null values.

Table 10.3 Fitted modulated Poisson process models, predicting smoking rates as functions of time-varying covariates; $\hat{L}^*(\boldsymbol{\beta})$ is the approximate log-likelihood

	$\hat{L}^*(\boldsymbol{\beta})$	Parameter	Estimate	SE	z	p-value
Model 1	−15758.48	Intercept	−0.05898	0.00836		
		Negative affect	0.02059	0.01075	1.92	0.0553
		Arousal	−0.02569	0.01075	−2.39	0.0169
		Attention	0.03900	0.01137	3.43	0.0006
Model 2	−15766.96	Intercept	−0.06234	0.00831		
		Negative affect	0.01760	0.01074	1.64	0.1013
		Arousal	−0.02648	0.01086	−2.44	0.0148
		Interaction	−0.00707	0.01118	−0.63	0.5271
Model 3	−15770.78	Intercept	−0.07296	0.01146		
		Negative affect	0.00748	0.01296	0.58	0.5638
		Negative affect squared	0.00889	0.00797	1.12	0.2647
Model 4	−15598.62	Intercept	−0.05924	0.00839		
		Negative affect	0.01950	0.01077	1.81	0.0702
		Arousal	−0.01594	0.01078	−1.48	0.1392
		Attention	−0.01787	0.01198	1.49	0.1358
		Restlessness	0.21017	0.01577	13.33	<0.0001

The estimated regression coefficient for arousal corresponds to a decrease in smoking rate by a multiplicative factor of $\exp\{-0.02569\} = 0.97$ for every unit increase in arousal. That even such a small effect was found to be statistically significant suggests that our data are of sufficient size, and the proposed approach has sufficient power to detect even subtle covariate effects.

Research into the structure of emotions suggests that emotional experience is organized around two bipolar dimensions (J. Russell, 1980). Valence captures the range from positive to negative affect, represented in our data by negative affect. Arousal captures a range from somnolent to energetic states. Under this "circumplex" model of affect, specific emotional states (e.g., anxiety, sadness, elation) fall on particular coordinates on the two-dimensional space defined by valence and arousal. For example, anxiety and sadness are both negative affect states, but differ in that anxiety is associated with high arousal, whereas sadness is marked by low arousal. Similarly, anxiety and elation are both high-arousal states, but the first have negative valence while the second has positive valence. This two-dimensional structure, capturing the full range of emotional state on smoking rates, was captured by including the interaction between negative affect and arousal in model 2. The results continue to show an inverse relationship between arousal and smoking rate, but no significant interaction between negative affect and arousal was detected. Thus, the impact of arousal on smoking rate does not appear to depend on the level of negative affect.

Smoking rate may be expected to be highest among smokers who are either feeling particularly bad or good. To test for this, the log-smoking rate was modeled as a quadratic function of negative affect. The results (model 3) do not support the hypothesis that smoking rates are a function of negative affect. Neither the coefficients for negative affect nor for negative affect squared are statistically significant.

The single item, restlessness, did not load heavily on any of the three factors, and thus can be treated separately. Moreover, the results of Shiffman et al. (1996) suggest that this variable was associated with relapse among smokers who are attempting to quit. The results of adding this variable to the model indicate that restlessness has a strong impact on smoking rate (model 4). Each unit increase in restlessness is estimated to increase smoking rate by a multiplicative factor of $\exp(0.21017) = 1.24$. Moreover, when restlessness is included in the model, arousal and attention disturbance are no longer found to have a significant impact on smoking rate.

To take into account heterogeneity among subjects in baseline smoking rates, fixed effects for subjects may be readily included in the model by taking the intensity function for the ith subject to be

$$\lambda_i(t) = \mu_i \exp\left\{\beta_1 x_{i1}(t) + \beta_2 x_{i2}(t) + \cdots + \beta_p x_{ip}(t)\right\},$$

where μ_i denotes that smoker's baseline smoking rate. With 304 subjects, however, this would result in a heavily parameterized model. Alternatively, a mixed-effects model may be considered under which $\{\mu_i\}$ are independently sampled from the gamma distribution, yielding a random intensity function for a Cox process. Since the gamma distribution is conjugate for the Poisson process, a closed-form expression for the log-likelihood may be obtained. As for the modulated Poisson process, maximum likelihood estimation requires that the values of the covariates be known for each subject at all points in time during the study period.

An alternative approach to the above analysis would have been to fit a Cox proportional hazards model with time-varying covariates (e.g., Lawless, 2003, pp. 355–358). Under this model, the semiparametric intensity function takes the form

$$\lambda(t) = \lambda_0(t) \exp\left\{\beta_1 x_1(t) + \beta_2 x_2(t) + \cdots + \beta_p x_p(t)\right\},$$

where the baseline intensity $\lambda_0(t)$ is to be estimated nonparametrically. When expressed as a function of the lifetime since the most recent event, this is the so-called proportional hazards function. Let τ_i denote the lifetime of the ith smoking event, and let $\mathbf{x}_i(\tau)$ denote the corresponding vector of covariates at time τ. Then the parameter vector $\boldsymbol{\beta}$ may be estimated by maximizing the partial likelihood

$$L(\boldsymbol{\beta}) = \prod_{i=1}^{n} \frac{\exp\left\{\boldsymbol{\beta}'\mathbf{x}_i(\tau_i)\right\}}{\sum_{j \in R_i} \exp\left\{\boldsymbol{\beta}'\mathbf{x}_j(\tau_i)\right\}}$$

(Cox, 1972b). Here, the product is over all n smoking events, and the sum in the denominator is over all events in the risk set R_i of all smoking events with lifetimes exceeding that of smoking event i. Note that this requires observations for the covariates $\mathbf{x}_i(\tau_i)$ not only at the times of the smoking events τ_i, but also observations $\mathbf{x}_i(\tau_j)$ at all times in the risk sets R_j of the remaining events. Such observations are not available for the current data, nor are they likely to be available for large event history datasets. Although it may be feasible to impute the missing information, such imputations would require a joint model for the covariates and all the prerequisite assumptions. The point process modeling approach proposed in this section does not require such imputations. Moreover, the proportional hazards model takes the baseline hazard to be a function of the duration of time since the most recent event, whereas it may be more appropriate to consider a baseline intensity that is a function of time of day as in section 10.3.3.

10.3.5 Self-Exciting Point Process

Hawkes' (1971) self-exciting point process may be used to model clustered events, such as may occur during chain smoking. The act of smoking itself may cue further smoking, through a process known as "priming" (Stewart et al., 1984): it has been demonstrated that delivering a small dose of drug to an animal actually cues further drug-seeking (e.g., de Wit & Stewart, 1981). This results in patterns of clustered drug administration or "binging" or, in the case of smoking, "chain smoking." This model for dependent events has the conditional intensity function

$$\lambda(t) = \rho + \mu \sum_{t_i < t} g(t - t_i),$$

where the sum is over all events occurring prior to time t, and $g(\cdot)$ is a probability density function. Under this model, "parent" events are first generated according to a Poisson process at a rate defined by intensity ρ. Each event produces a Poisson number of offspring with mean $\mu < 1$ to ensure stationarity. The offspring are independently distributed about their parent events according to the probability density function $g(\cdot)$. Each generation of offspring in turn generates a successive generation of offspring in the same manner. The final process is composed of all generations of events, where the times of each successive generation serve as cluster centers for the next.

Figure 10.6 illustrates a single realization of a self-exciting point process with intensity function

$$\lambda(t) = \rho + \mu \sum_{t_i < t} \gamma \exp\{-\gamma(t - t_i)\}. \tag{10.5}$$

Here, parent events are generated at a background rate of $\rho = 0.6$ events per hour, each parent event produces an average of $\mu = 0.5$ offspring, and offspring are generated about the parent events according to an exponential distribution with mean $1/\gamma = 40$ minutes. (Note that for chain smoking, we

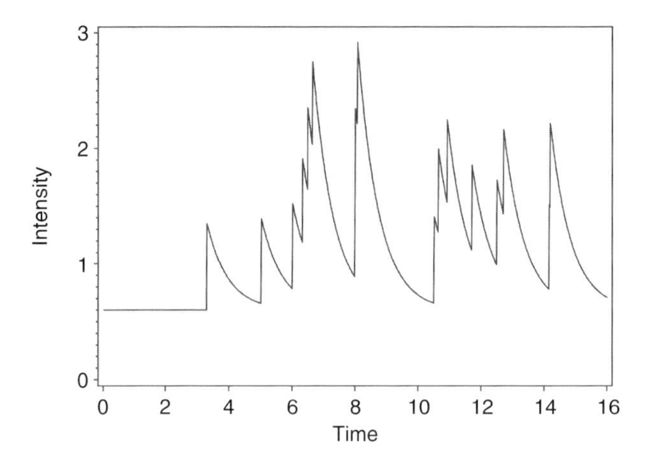

Figure 10.6 Conditional intensity for a self-exciting point process.

would expect a mean interval between parents and offspring to be on the order of 5–10 minutes; a longer mean is used here to produce a more easily visualized graph.) The lighting of each cigarette results in a vertical increase in smoking rate, triggering the smoking of additional cigarettes, resulting in a clustered pattern of smoking.

Maximum likelihood estimation of the parameters of the above self-exciting point process model requires complete records on all cigarettes smoked. Such records are not available for days on which the smoker put the electronic diary into suspend mode. Methods for handling such incomplete or censored data go beyond the scope of this chapter. Therefore, within each subject, data from days on which the diary was suspended were eliminated from the following analysis. The elimination of such data left 2,291 person-days over 298 subjects for the analysis.

Table 10.4 gives the maximum likelihood estimates of the parameters of the self-exciting point process model (10.5). The baseline smoking rate is estimated to be $\widehat{\rho} = 1$ cigarette per hour. Each cigarette produces a Poisson number of offspring with mean of $\widehat{\mu} = 0.346$. The mean time between a cigarette and its offspring is estimated to be $1/\widehat{\gamma} = 5.29$ hours, which far exceeds what would be expected under chain smoking. This suggests that the self-exciting point process

Table 10.4 Fitted Hawkes' self-exciting point process model

Parameter	Estimate	SE
ρ	1.00003	0.01043
μ	0.34604	0.01062
γ	0.18917	0.00238

does not adequately describe the observed pattern of smoking. This inadequacy may be attributed to attempting to fit a model for clustered events to data that are regularly spaced over time. The next section will consider a model appropriate for regularly spaced smoking events.

10.3.6 Stress-Release Point Process

It has been shown that smokers regulate their blood nicotine levels, presumably to avoid nicotine withdrawal (M.A. Russell, 1980). This suggests that the craving for a cigarette may be linked to the amount of nicotine in the bloodstream. Smoking cigarettes introduces nicotine into the blood, reducing the craving for further cigarettes. As time passes following the last cigarette smoked, nicotine levels decline increasing the craving for cigarettes, and hence the likelihood that another cigarette will be lit. To put this into the framework of a point process model, assume that the nicotine concentration in the blood of a smoker at time t takes the form

$$Y(t) = Y(0) + \delta \sum_{t_i < t} \exp\{-\gamma (t - t_i)\}, \tag{10.6}$$

(Moran, 1967), where the sum is over all cigarettes smoked prior to time t, and t_i is the time of the ith cigarette. As illustrated in figure 10.7, nicotine level jumps by a fixed amount δ when each cigarette is smoked. Then nicotine levels decline at an exponential rate, governed by the parameter γ. Physiological evidence suggests that nicotine has a half-life of about 2 hours (Benowitz et al., 1982), corresponding to $\gamma = -0.5 \log 0.5 = 0.347$. The model (10.6) takes the same form as a storage model used to describe water level behind dams (e.g., Prabhu, 1998), where water is input into a reservoir following rainstorms. In contrast to storage models where the arrival of rainstorms is considered to be independent of the current store of water, and the amount of water added by each storm is treated as a random variable, smoking rate is likely to be a function of the current

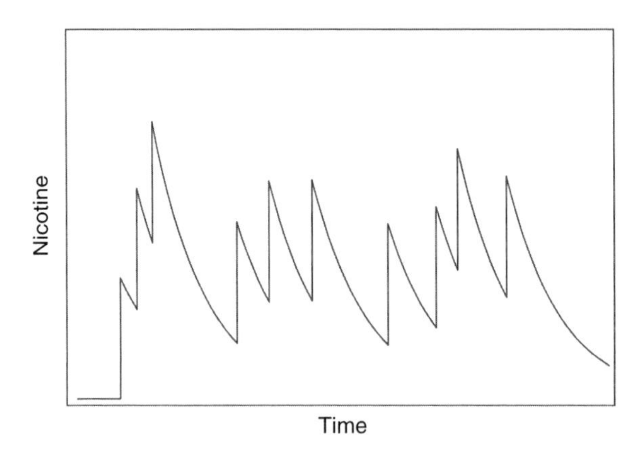

Figure 10.7 Storage model for blood nicotine.

store of nicotine in the blood, and each cigarette is considered to add a fixed amount δ of nicotine to the blood.

Given the above model for the temporal dynamics of blood nicotine levels, smoking rate may be modeled through the conditional intensity function

$$\lambda(t) = g\{Y(t)\},$$

where $g(\cdot)$ is a nonnegative function, describing the relationship between smoking rate and blood nicotine level. Note that this intensity is similar to that of the stress-release model in seismology (e.g., Zheng & Vere-Jones, 1994), which describes the rate of earthquake occurrences as a function of tectonic stress in place of the blood nicotine levels considered here. Very little is known about the relationship between cigarette craving and blood nicotine levels. To ensure a nonegative intensity, we shall follow Zheng and Vere-Jones' suggestion, and take

$$\lambda(t) = \exp\{\alpha_0 - \alpha_1 Y(t)\}. \tag{10.7}$$

Substituting the model (10.6) for blood nicotine into the expression above, and assuming zero blood nicotine at time zero (waking), we obtain the following model for smoking rate:

$$\lambda(t) = \exp\left\{\alpha_0 - \alpha_1 \delta \sum_{t_i < t} \exp\{-\gamma(t - t_i)\}\right\}.$$

Note that the parameters α_1 and δ are not identifiable in this model; if $\widehat{\alpha}_1$ and $\widehat{\delta}$ are estimates of these parameters, then the estimates $\widetilde{\alpha}_1 = c\widehat{\alpha}_1$ and $\widetilde{\delta} = \widehat{\delta}/c$ will fit just as well for any constant $c \neq 0$. Therefore, we shall take $\rho = \alpha_1\delta$, yielding the reparameterized model

$$\lambda(t) = \exp\left\{\alpha_0 - \rho \sum_{t_i < t} \exp\{-\gamma(t - t_i)\}\right\}$$

for smoking rate.

Maximum likelihood estimation of the parameters of the above stress-release model requires complete records on all cigarettes smoked. Therefore, the following considers only those days on which the electronic diary was never suspended. The log-likelihood of −29,377.72 for the stress-release point process model was considerably larger than the log-likelihood of −33,458.53 for the self-exciting point process model, suggesting that the stress-release model provides a better description of the data than the self-exciting model.

Maximum likelihood estimates of the parameters of the stress-release model are presented in table 10.5, and figure 10.8 shows the fitted model for an anonymous smoker. The baseline smoking rate is estimated to be $\exp\{\widehat{\alpha}_0\} = 1.55$ cigarettes per hour when there is no nicotine in the blood. Each cigarette results in a precipitous drop in smoking rate to a value equal to $\exp\{-\widehat{\rho}\} \times 100 = 2.36\%$

Table 10.5 Fitted stress-release model

Parameter	Estimate	SE
α_0	0.43611	0.00548
ρ	3.74466	0.10934
γ	15.56989	0.37451

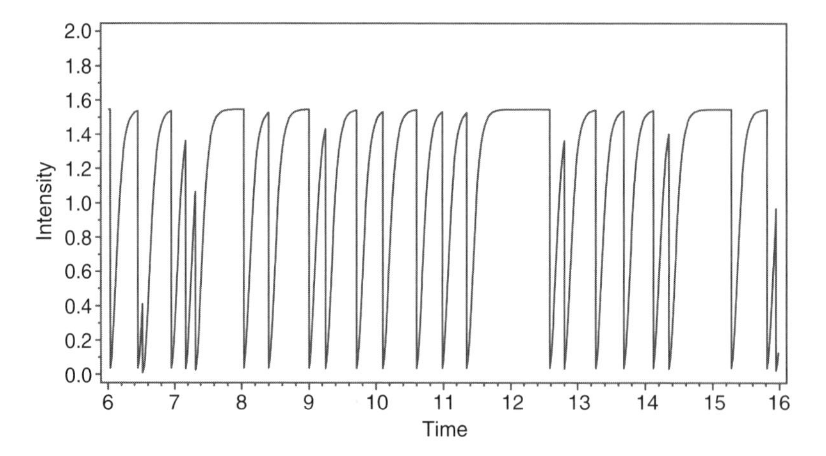

Figure 10.8 Conditional intensity of the fitted stress-release model.

of the rate immediately preceding the cigarette. This is followed by a rapid recovery governed by the high estimate of γ. Most cigarettes were not lit until that recovery was nearly complete, resulting in a more regular spacing between successive cigarettes than would be expected under complete temporal randomness. This pattern suggests that, at least for this smoker, the pattern of cigarettes largely matches the stress-release model, where smoking occurs in response to falling blood nicotine levels and concomitant emerging withdrawal symptoms. The half-life of the recovery toward the background smoking rate is estimated to be $-\widehat{\gamma}^{-1} \log 0.5 = 0.0445$ hours or approximately 2.67 minutes. Moreover, about 95% of that recovery occurs within $3/\widehat{\gamma} = 0.193$ hours, or approximately 11.6 minutes.

10.4 Discussion of Results

Understanding patterns of events in the natural environment is a significant goal of the behavioral and social sciences. In this chapter, we have demonstrated the utility of point process models in predicting patterns of cigarette smoking over time. The point process models differ from our previous analyses of smoking patterns (Shiffman et al., 2002, 2004), in that the current analyses model smoking rate over the course of time (in this case approximately 2 weeks), rather than contrasting smoking and nonsmoking occasions. Therefore, the point process models can (1) quantify the influence of covariates on smoking rate, whereas previous analyses (using generalized estimating equations) were only capable of estimating the odds of an assessment being a cigarette assessment, and (2) more effectively account for the influence of time. Indeed, our previous analyses only attempted to control for time, statistically equating all observations in time, rather than exploring potential influences of time on smoking patterns.

Initial analyses focused on the influence of day of the week on smoking rate. Results suggested that smoking rate is largely constant over the days of the week; there is no evidence that smoking is decreased during the work-week in response to smoking restrictions, nor increased on weekends due to relaxed smoking restrictions or social cues (e.g., being with other smokers at a bar).

The point process models uncovered interesting within-day variability in smoking rate. Of course, the greatest within-day influence on smoking rate is whether or not an individual is asleep. However, significant variability in smoking rate was also observed during the waking day. This variability may be explained by smoking restrictions (either formal regulations or incompatibility of work and smoking) during the work-week. On weekdays, smoking rate peaked at times that coincide with breaks in the workday and was particularly high around 5 P.M.—exactly the time when individuals would leave work. While smoking rate seems to peak at the end of the workday during the work-week, it accelerates more quickly in the morning and then levels off on weekends. Such a pattern could be explained by the lack of smoking restrictions on weekends: in other words, our data do not suggest that smokers smoke more on the weekends, but that their pattern of smoking becomes more "regular," demonstrated by the smooth curve seen in figure 10.5. These data complement and extend our previous analyses demonstrating the influence of smoking restrictions on the odds of smoking (Shiffman et al., 2002). They also are interesting in light of previous findings suggesting that smoking lapses (episodes of smoking occurring during an attempt to quit smoking) in this sample are more likely to occur later in the day, primarily after 8 P.M. (Shiffman et al., 1996). Although speculative, this lapse pattern may occur because smoking regulations are relaxed later in the day, therefore increasing the probability that smoking will occur, and/or because situations encountered later in the day (bar/restaurant, home) have become associated with smoking via conditioning processes and elicit emotional states that may motivate smoking (e.g., craving, negative affect).

The point process analyses also allow for an examination of how changes in affect state during the course of a day influence smoking rate. Given the close connection between affect and smoking in theoretical models of smoking, examining and understanding such relationships are important goals of smoking research. The results of the point process analyses largely mirrored our previous results (Shiffman et al., 2002, 2004), in that restlessness emerged as the most potent predictor of smoking rate. In contrast to our previous analyses, however, the point process models allow for an estimation of smoking rate per hour conditional on restlessness level. These data suggest that baseline smoking rate is less than one cigarette per hour (0.94). Increases in restlessness substantially increase this rate; when restlessness is at its maximum, the hourly smoking rate approximately doubled to 1.91 cigarettes per hour. This finding further supports the role of restlessness, and challenges the role of other negative affect states, in motivating smoking.

Conditional intensity functions were used to take an exploratory look at two theoretical models of smoking patterns: a "priming" model, where smoking engenders further smoking in an escalating manner (self-exciting model), and a nicotine regulation model, where nicotine intake during smoking decreases smoking rate initially and then the probability of smoking gradually recovers over time as nicotine levels fall (stress-release model). Neither model yielded parameter estimates that were consistent with theoretical models for smoking. The fitted self-exciting model estimates a mean time of 5.29 hours between each cigarette and the additional cigarettes it triggered, a duration of time far exceeding what would be expected in a priming model. The stress-release model fitted the data better than the self-exciting model. However, it yielded an estimated recovery time of 11.6 minutes, a figure that much more closely reflects the time it takes to smoke a cigarette than the time it takes for nicotine to be purged from the blood. In his review of investigations of both laboratory and field measurements of smoking times, Moody (1980) reported that smoking times ranged from 8.25 to 10.58 minutes among field observations. The time it takes to smoke each cigarette may be masking the effects of nicotine regulation that the model was originally intended to detect. It is also possible that the sensory aspects of smoking a cigarette place limits on continuous nicotine self-administration, acting as a sort of "stop" signal, but one that decays fairly rapidly. A stress-release model including terms for both the variable times it takes to smoke a cigarette, as well as for nicotine regulation, is currently under investigation.

The current versions of both conditional intensity models also do not take into account diurnal patterns of variation, or variation in important explanatory variables like restlessness. It is also possible that these models may fit some smokers better than others. For example, the stress-release model may fit highly addicted smokers (who require regular nicotine intake in order to prevent withdrawal effects) well, but not smokers who are not nicotine addicted. Smoking prohibitions may adversely affect the fit of both the self-exciting and stress-release models. Priming effects may not be seen, because smoking is unable to "escalate" in situations where it is not allowed. Additionally, regularly spaced smoking in response to falling nicotine blood levels may not be seen, because smoking occurs at times when it is possible to smoke (during breaks), rather than at times dictated by nicotine blood levels. In future analyses, it will be interesting to address the effects of smoking regulations in the context of the self-exciting and stress-release models.

10.5 Multivariate Point Patterns

Many applications involve more than a single type of event. For example, an investigation of substance abuse may consider the times at which cigarettes are lit together with the times at which alcohol is consumed or drugs are partaken. The pattern of times over which multiple types of events occur is called a *multivariate point pattern.* Other examples of multivariate point patterns include the following: Investigations of aggressive behavior among children may distinguish between type of aggressive behavior, such as hitting, pushing, or name-calling. Likewise different types of disruptive behavior, such as calling out or disputing with the teacher, may be distinguished in a classroom setting. A life-course analysis may consider dependences among women's relational, reproductive, and employment events (Budig, 2003). On a larger scale, Olzak (1992, p. 100) considers the effect of immigration on the rates of ethnic conflicts and labor strikes. In organizational ecology, different types of firms may be distinguished in investigations of firm foundings or bankruptcies.

In some applications, it may be useful to treat repeated events of a single type on multiple subjects as forming a multivariate point pattern. Such an approach can allow for dependences in the timing of events among clustered observations (Vermunt, 1997, p. 171). For example, the lighting of a cigarette by one smoker may increase the likelihood that other smokers in the same room will light up. An aggressive act by one child in a classroom may invoke similar acts of misbehavior among their classmates. At a larger sociological scale, the pattern of riots in one city may impact the pattern of riots in other cities in a region.

Just as in the univariate case, survival analysis is often used to analyze multivariate event history data; see Vermunt (1997) for a review of such methods. Competing risk and multivariate hazard models may be fitted to the data, focusing on the durations between successive events. An alternative approach would be to fit a multivariate point process model, focusing on the times of the events on a clock or calendar. The simplest such model would treat the events of each type as being realized from independent univariate point processes. In this case, each event type may be analyzed separately using the methods described in the previous sections of this chapter. Another approach would be to generate the events, regardless of type, from a point process model. Then the types may be independently sampled from a discrete distribution function.

The most natural approach to modeling dependence among events of different types or for different members of the same cluster is through the conditional intensity functions. Let $N_i\left[t, t+\delta\right]$ denote the number of events of type i (or subject i in a cluster) in the time interval $[t, t+\delta]$. Then the conditional intensity for events of type i is

$$\lambda_i\left(t; H_t\right) = \lim_{\delta \to 0} \frac{E\left\{N_i\left[t, t+\delta\right] | H_t\right\}}{\delta}.$$

Here, the numerator gives the expected number of events of type i occurring in the interval $[t, t+\delta]$, conditional on the history H_t of events of all types occurring before time t. Dividing by δ obtains the conditional mean number of events by unit time in the interval. Then taking the limit as $\delta \rightarrow 0$ yields an instantaneous rate of event occurrence, depending on the past history of events of all types. Note that the above definition for the conditional intensity function is similar to that for the multiple-risk hazard function. The multiple-risk hazard is expressed as a function of duration since the most recent event (of any type), the conditional intensity is expressed as a function of time on the clock or calendar. Moreover, the conditional intensity function may depend on times of events prior to the most recent event, and again offers a more flexible class of models.

Ogata et al. (1982) extend Hawkes' self-exciting point process model to the analysis of the bivariate point pattern of deep and shallow earthquakes. To extend this model further to the multivariate point process setting, let t_{ij} denote the timing of the jth event of type i. Then a multivariate version of the Hawkes' self-exciting point process model has conditional intensity

$$\lambda_i\left(t; H_t\right) = \rho_i + \sum_{k=1}^{p} \mu_{ik} \sum_{t_{ij}<t} g_{ik}\left(t - t_{ij}\right),$$

where $\mu_{ik} < 1$ and $g_{ik}(\cdot)$ is a probability density function, defined for each pair of event types i and k. Under this model, events of each type i are independently generated according to a homogeneous Poisson process with respective intensities ρ_i. Each event of type k generates a Poisson number of offspring of type i with mean μ_{ik}. Offspring of type i are then distributed about their parents of type k according to the probability density function $g_{ik}(\cdot)$. Just as in the univariate case, each generation of offspring has the potential to generate offspring of their own. Events of each type are thus clustered around the events of other types as well as the events of the same type, yielding positive dependences among the types of events.

APPPENDIX A: STATISTICAL INFERENCE FOR MAXIMUM LIKELIHOOD ESTIMATOR

This appendix reviews the inferential properties of the maximum likelihood estimator for point process models defined by a (conditional) intensity function. As a function of the random data, the maximum likelihood estimator is a random vector, and therefore has a joint probability density function, a population mean, and population variance–covariance matrix. In the most interesting applications, the maximum likelihood estimator has no closed-form expression, so inference must rely on the large-sample approximations. See the model for the effect of day of week for an exception.

Suppose that the intensity function takes a parametric form $\lambda_i(t; \boldsymbol{\theta})$, and let $\widehat{\boldsymbol{\theta}}$ denote the maximum likelihood estimator, obtained by maximizing $L(\boldsymbol{\theta})$ in expression (10.2). Kutoyants (1984, 1998) explored the large-sample properties of the maximum likelihood estimator for inhomogeneous Poisson processes, and Ogata (1978) explored these properties for point processes with a conditional intensity function. In either case, the maximum likelihood estimator $\widehat{\boldsymbol{\theta}}$ is approximately normally distributed with mean equal to the true value of the parameter $\boldsymbol{\theta}$, and variance–covariance matrix $\mathbf{V}(\boldsymbol{\theta}) = \left\{\mathbf{J}(\boldsymbol{\theta})\right\}^{-1}$, where the

Fisher information matrix $\mathbf{J}(\boldsymbol{\theta})$ has elements

$$\mathbf{J}_{jk}(\boldsymbol{\theta}) = \sum_{i=1}^{m} \int_{0}^{T_i} \frac{(\partial/\partial\theta_j)\,\lambda_i(t;\boldsymbol{\theta})(\partial/\partial\theta_k)\,\lambda_i(t;\boldsymbol{\theta})}{\lambda_i(t;\boldsymbol{\theta})} dt.$$

What is meant by a large sample depends on whether or not the data are realized from a Poisson process. For data realized from a Poisson process, the sample may be deemed large if the number of subjects m is large, and/or the lengths of time T_i over which the process is observed is large for all subjects. If the intensity depends on the history of past events, it does not suffice to have a large number of subjects to obtain the above results. In such cases, the lack of information regarding events that have occurred before time zero, the start of the study, results in an edge effect introducing bias into the maximum likelihood estimator. As the length of the study increases, the impact of this edge effect diminishes, reducing that bias. So for models of the conditional intensity function, a sample is deemed large if the lengths of time T_i over which the process is observed is large for each subject.

APPENDIX B: STATISTICAL INFERENCE FOR THE MODULATED POISSON PROCESS

This appendix proposes an approach to estimating the parameters of a modulated Poisson process with partially observed covariates. Assuming that the subjects are independent, the log-likelihood for the modulated Poisson process is

$$L(\boldsymbol{\beta}) = \sum_{i=1}^{m} \sum_{j=1}^{d_i} \left\{ \boldsymbol{\beta}' \sum_{k=1}^{N_{ij}} \mathbf{x}_i(t_{ijk}) - \int_{T_{ij}} \exp\left\{\boldsymbol{\beta}'\mathbf{x}_i(t)\right\} dt \right\},$$

where, from the left, the first sum is over the m subjects in the study, and the second sum is over the number of days d_i over which subject i was observed. Inside the brackets, the sum is over the N_{ij} cigarettes smoked by subject i on day j, and the integral is over the collection of times T_{ij} on which that subject was observed on that day. The latter excludes times when the subject was sleeping, napping, or had put the electronic diary in suspension. The vector $\mathbf{x}_i(t) = (1, x_{i1}(t), x_{i2}(t), \ldots, x_{ip}(t))'$ is composed of the time-varying covariates, $\boldsymbol{\beta} = (\beta_0, \beta_1, \beta_2, \ldots, \beta_p)'$ is the vector of model parameters, and t_{ijk} is the time of the kth cigarette smoked by subject i on day j. The maximum likelihood estimator is obtained by finding $\widehat{\boldsymbol{\beta}}$ that maximizes $L(\boldsymbol{\beta})$.

Maximum likelihood estimation requires that the values of the covariates be known for each event in the data, and the computation of integrals of the form

$$\mathbf{I}_{ij}(\boldsymbol{\beta}) = \int_{T_{ij}} \exp\left\{\boldsymbol{\beta}'\mathbf{x}_i(t)\right\} dt, \qquad \text{(A.10.1)}$$

which requires that the values of the covariate be known for all times in the set T_{ij}. In the present application, the values of the covariates were only observed for a random sample of smoking events. Moreover, it is generally not possible to obtain the values of covariates for all times during any interval. The following describes a new approach to parameter estimation that takes these data limitations into account. This approach uses information on the covariates collected at randomly sampled assessment times during the study intervals for each subject. Careful attention to the details regarding the selection of cigarettes to be assessed, and regarding the selection of random assessment times, is required.

Consider the random selection of cigarettes to be assessed. It was desired to obtain an average of five cigarette assessments per day. To achieve this objective, smoking events for subject i on day j were independently selected for assessment with known probability $p_{ij} = 5/N_{i,j-1}$. Thus, the times of the randomly assessed cigarettes are realized from a thinned point process. It is well known that a thinned inhomogeneous Poisson process is also an inhomogeneous Poisson process (Cressie, 1991, p. 689). Conditional on $N_{i,j-1}$, the assessed cigarettes of subject i are realized from an inhomogeneous Poisson process with intensity

$$\lambda^*(t) = p_{ij} \exp\left\{\boldsymbol{\beta}'\mathbf{x}_i(t)\right\}$$

for times t on day j. The log-likelihood of the assessed cigarettes then becomes

$$L^*(\boldsymbol{\beta}) = \sum_{i=1}^{m}\sum_{j=1}^{d_i}\left\{N_{ij}^* \log p_{ij} + \boldsymbol{\beta}'\sum_{k=1}^{N_{ij}^*}\mathbf{x}_i\left(t_{ijk}^*\right) - p_{ij}\int_{T_{ij}} \exp\left\{\boldsymbol{\beta}'\mathbf{x}_i(t)\right\} dt\right\}, \tag{A.10.2}$$

where t_{ijk}^* denotes the kth assessed cigarette for subject i on day j, and N_{ij}^* is the number of cigarettes assessed for subject i on day j.

For each value of the parameter $\boldsymbol{\beta}$, the integral $I_{ij}(\boldsymbol{\beta})$ in expression (A.10.1) may be regarded as the population total for the variable $Y_i(t;\boldsymbol{\beta}) = \exp\left\{\boldsymbol{\beta}'\mathbf{x}_i(t)\right\}$, where the population is composed of all times available for assessment on day j for subject i. Likewise, the data

$$\mathbf{x}_i\left(s_{ij1}\right), \mathbf{x}_i\left(s_{ij2}\right), \ldots, \mathbf{x}_i\left(s_{ijM_{ij}}\right)$$

on the covariates collected at randomly selected assessment times $s_{ij1}, s_{ij2}, \ldots, s_{ijM_{ij}}$ may be regarded as a sample from that population, where M_{ij} denotes the number of random assessments carried out for subject i on day j. Assuming that the covariates take fixed, nonrandom values at each point of time, the only source of variability in the covariates arises from the random selection of assessment times, suggesting a design-based inferential approach to estimating $I_{ij}(\boldsymbol{\beta})$. If a simple random sampling design is used to select the assessment times, then a design-unbiased estimator for $I_{ij}(\boldsymbol{\beta})$ is given by

$$\widehat{I}_{ij}(\boldsymbol{\beta}) = \frac{\left|T_{ij}\right|}{M_{ij}}\sum_{k=1}^{M_{ij}} \exp\left\{\boldsymbol{\beta}'\mathbf{x}_i\left(s_{ijk}\right)\right\}, \tag{A.10.3}$$

where $\left|T_{ij}\right|$ is the duration of time available for assessment of subject i on day j. Substituting (A.10.3) into expression (A.10.2), we obtain the estimated log-likelihood

$$\widehat{L}^*(\boldsymbol{\beta}) = \sum_{i=1}^{m}\sum_{j=1}^{d_i}\left\{N_{ij}\log p_{ij} + \boldsymbol{\beta}'\sum_{k=1}^{N_{ij}^*}\mathbf{x}_i\left(t_{ijk}^*\right) - \frac{p_{ij}\left|T_{ij}\right|}{M_{ij}}\sum_{k=1}^{M_{ij}}\exp\left\{\boldsymbol{\beta}'\mathbf{x}_i\left(s_{ijk}\right)\right\}\right\}. \tag{A.10.4}$$

The proposed estimator is obtained by finding $\widetilde{\boldsymbol{\beta}}$ that maximizes $\widehat{L}^*(\boldsymbol{\beta})$. The estimator $\widetilde{\boldsymbol{\beta}}$ also solves the estimating equation $\widehat{\Psi}(\boldsymbol{\beta}) = \mathbf{0}$, where

$$\begin{aligned}\widehat{\Psi}(\boldsymbol{\beta}) &= \frac{\partial}{\partial\boldsymbol{\beta}}\widehat{L}^*(\boldsymbol{\beta}) \\ &= \sum_{i=1}^{m}\sum_{j=1}^{d_i}\left\{\sum_{k=1}^{N_{ij}^*}\mathbf{x}_i\left(t_{ijk}^*\right) - \frac{p_{ij}\left|T_{ij}\right|}{M_{ij}}\sum_{k=1}^{M_{ij}}\mathbf{x}_i\left(s_{ijk}\right)\exp\left\{\boldsymbol{\beta}'\mathbf{x}_i\left(s_{ijk}\right)\right\}\right\}.\end{aligned}$$

Now consider the inferential properties of $\widetilde{\boldsymbol{\beta}}$ as the number of subject-days $d = \sum_{i=1}^{m} d_i$ increases, and as the random sampling intensity M/d increases, where $M = \min_{i,j} \{M_{ij}\}$. In spatial statistics, such large-sample properties fall under the rubric of mixed infill/increasing domain asymptotics. For fixed numbers of subject-days d, methods similar to those outlined in Rathbun (1996) may be used to demonstrate that $\widetilde{\boldsymbol{\beta}}$ converges to the maximum likelihood estimator $\widehat{\boldsymbol{\beta}}$ as the sampling intensity $M/d \to \infty$. Moreover, the maximum likelihood estimator converges in probability to the true value $\boldsymbol{\beta}$, as the number of subject-days $d \to \infty$. Putting these two results together shows that $\widetilde{\boldsymbol{\beta}}$ converges to the true value $\boldsymbol{\beta}$, as $M/d \to \infty$ and $d \to \infty$.

Using a Taylor series expansion, we may obtain the approximation

$$
\begin{aligned}
\widetilde{\boldsymbol{\beta}} - \boldsymbol{\beta} &= \left\{\widehat{\mathbf{J}}(\widetilde{\boldsymbol{\beta}})\right\}^{-1} \widehat{\Psi}(\boldsymbol{\beta}) \\
&= \left\{\widehat{\mathbf{J}}(\widetilde{\boldsymbol{\beta}})\right\}^{-1} \left\{\Psi(\boldsymbol{\beta}) + \widehat{\Psi}(\boldsymbol{\beta}) - \Psi(\boldsymbol{\beta})\right\}
\end{aligned}
$$

for large m, where

$$
\widehat{\mathbf{J}}(\boldsymbol{\beta}) = -\sum_{i=1}^{m}\sum_{j=1}^{d_i} \frac{p_{ij}\left|T_{ij}\right|}{M_{ij}} \sum_{k=1}^{M_{ij}} \mathbf{x}_i(s_{ijk})\,\mathbf{x}_i'(s_{ijk}) \exp \boldsymbol{\beta}'\mathbf{x}_i(s_{ijk}) \tag{A.10.5}
$$

and

$$
\Psi(\boldsymbol{\beta}) = \sum_{i=1}^{m}\sum_{j=1}^{d_i} \left\{ \sum_{k=1}^{N_{ij}^*} \mathbf{x}_i\left(t_{ijk}^*\right) - p_{ij} \int_{T_{ij}} \mathbf{x}_i(t) \exp\left\{\boldsymbol{\beta}'\mathbf{x}_i(t)\right\} dt \right\}.
$$

By the strong law of large numbers, $\widehat{\mathbf{J}}(\boldsymbol{\beta}) \to \mathbf{J}(\boldsymbol{\beta})$ almost surely as $M/d \to \infty$ for fixed d, where $\mathbf{J}(\boldsymbol{\beta})$ is the Fisher information,

$$
\mathbf{J}(\boldsymbol{\beta}) = -\sum_{i=1}^{m}\sum_{j=1}^{d_i} p_{ij} \int_{T_{ij}} \mathbf{x}_i(t)\,\mathbf{x}_i'(t) \exp\left\{\boldsymbol{\beta}'\mathbf{x}_i(t)\right\} dt.
$$

For large d, $\Psi(\boldsymbol{\beta})$ is approximately multivariate normally distributed with mean zero, and variance–covariance matrix $\mathbf{J}(\boldsymbol{\beta})$. Moreover, for large M/d, $\widehat{\Psi}(\boldsymbol{\beta}) - \Psi(\boldsymbol{\beta})$ is approximately multivariate normally distributed with mean zero, and variance–covariance matrix

$$
\mathbf{S}(\boldsymbol{\beta}) = \sum_{i=1}^{m}\sum_{j=1}^{d_i} \frac{p_{ij}^2\left|T_{ij}\right|^2}{M_{ij}} \mathbf{S}_{ij}(\boldsymbol{\beta}),
$$

where

$$
\mathbf{S}_{ij}(\boldsymbol{\beta}) = \frac{1}{\left|T_{ij}\right|} \int_{T_{ij}} \left(\mathbf{y}_i(t;\boldsymbol{\beta}) - \boldsymbol{\mu}_{ij}(\boldsymbol{\beta})\right)\left(\mathbf{y}_i(t;\boldsymbol{\beta}) - \boldsymbol{\mu}_{ij}(\boldsymbol{\beta})\right)' dt
$$

and

$$
\boldsymbol{\mu}_{ij}(\boldsymbol{\beta}) = \frac{1}{\left|T_{ij}\right|} \int_{T_{ij}} \mathbf{y}_i(t;\boldsymbol{\beta})\, dt
$$

are the population variance–covariance matrix and mean vector of $\mathbf{y}_i(t;\boldsymbol{\beta}) = \mathbf{x}_i(t) \exp\left\{\boldsymbol{\beta}'\mathbf{x}_i(t)\right\}$, $t \in T_{ij}$. Putting all of the above results together, we find that the proposed estimator $\widetilde{\boldsymbol{\beta}}$ is approximately multivariate normally distributed with mean $\boldsymbol{\beta}$ and variance–covariance matrix

$$\mathrm{var}(\widetilde{\boldsymbol{\beta}}) = \mathbf{J}^{-1}(\boldsymbol{\beta}) + \mathbf{J}^{-1}(\boldsymbol{\beta})\,\mathbf{S}(\boldsymbol{\beta})\mathbf{J}^{-1}(\boldsymbol{\beta}). \tag{A.10.6}$$

Thus, the variability in $\widetilde{\boldsymbol{\beta}}$ may be partitioned into two sources of uncertainty, that which is inherent to the modulated Poisson process model as quantified by the inverse of the Fisher information matrix $\mathbf{J}(\boldsymbol{\beta})$, and that due to the random sampling of the covariate.

Expression (A.10.6) is a function of $\mathbf{J}(\boldsymbol{\beta})$ and $\mathbf{S}(\boldsymbol{\beta})$, both of which are unknown and must be estimated from the data. The Fisher information matrix $\mathbf{J}(\boldsymbol{\beta})$ may be unbiasedely estimated by either $\widehat{\mathbf{J}}(\boldsymbol{\beta})$ or by

$$\widetilde{\mathbf{J}} = \sum_{i=1}^{m}\sum_{j=1}^{d_i}\sum_{k=1}^{N_{ij}} \mathbf{x}_i(t_{ijk})\,\mathbf{x}_i'(t_{ijk}).$$

The attractive feature of the latter is that it does depend on an estimate for $\boldsymbol{\beta}$. The variance–covariance matrix $\mathbf{S}(\boldsymbol{\beta})$ may be estimated by

$$\widehat{\mathbf{S}}(\boldsymbol{\beta}) = \sum_{i=1}^{m}\sum_{j=1}^{d_i} \frac{p_{ij}^2\,|T_{ij}|^2}{M_{ij}}\widehat{\mathbf{S}}_{ij}(\boldsymbol{\beta}),$$

where, for each (i,j), $\widehat{\mathbf{S}}_{ij}(\boldsymbol{\beta})$ is the sample variance–covariance matrix of $\mathbf{y}_{ijk}(\boldsymbol{\beta}) = \mathbf{x}_i(s_{ijk})\exp\{\boldsymbol{\beta}'\mathbf{x}_i(s_{ijk})\}$, $k = 1, \ldots, M_{ij}$; that is,

$$\widehat{\mathbf{S}}_{ij}(\boldsymbol{\beta}) = \frac{1}{M_{ij}-1}\sum_{k=1}^{M_{ij}}\left(\mathbf{y}_{ijk}(\boldsymbol{\beta}) - \overline{\mathbf{y}}_{ij\cdot}(\boldsymbol{\beta})\right)\left(\mathbf{y}_{ijk}(\boldsymbol{\beta}) - \overline{\mathbf{y}}_{ij\cdot}(\boldsymbol{\beta})\right)',$$

where

$$\overline{\mathbf{y}}_{ij\cdot}(\boldsymbol{\beta}) = \frac{1}{M_{ij}}\sum_{k=1}^{M_{ij}}\mathbf{y}_{ijk}(\boldsymbol{\beta}).$$

Putting all of this together, and substituting the estimator $\widetilde{\boldsymbol{\beta}}$ for $\boldsymbol{\beta}$, we obtain the following estimator for the variance–covariance matrix of $\widetilde{\boldsymbol{\beta}}$:

$$\widehat{\mathrm{var}}\,(\widetilde{\boldsymbol{\beta}}) = \widetilde{\mathbf{J}}^{-1} + \widetilde{\mathbf{J}}^{-1}\widehat{\mathbf{S}}(\widetilde{\boldsymbol{\beta}})\widetilde{\mathbf{J}}^{-1}. \tag{A.10.7}$$

References

Allison, P.D. (1984). *Event History Analysis: Regression for Longitudinal Data*. London: Sage.

American Psychiatric Association (1994). *Diagnostic and Statistical Manual of Mental Disorders* (4th ed.). Washington, DC: American Psychiatric Association.

Benowitz, N.L., Jacob, P., Jones, R.T., & Rosenberg, J. (1982). Interindividual variability in the metabolism and cardiovascular effects of nicotine in man. *Journal of Pharmacology and Experimental Therapeutics, 221*, 368–372.

Berman, M., & Turner, T.R. (1992). Approximating point process likelihoods with GLIM. *Applied Statistics, 41*, 31–38.

Blossfeld, H.-P., & Rohwer, G. (2002). *Techniques of Event History Modeling*. Mahwah, NJ: Lawrence Erlbaum.

Budig, M.J. (2003). Are women's employment and fertility histories interdependent? An examination of causal order using event history analysis. *Social Science Research, 32*, 376–401.
Carroll, G.R., & Delacroix, J. (1982). Organizational mortality in the newspaper industries of Argentina and Ireland: An ecological approach. *Administrative Science Quarterly, 27*, 169–198.
Centers for Disease Control (CDC) (2002). Annual smoking-attributable mortality, years of potential life lost, and economic costs—United States, 1995–1999. *MMWR, 51(14)*, 300–303.
Collins, R.L., Morsheimer, E.T., Shiffman, S., Paty, J.A., Gnys, M., & Papandonatos, G.D. (1998). Ecological momentary assessment in a behavioral drinking moderation training program. *Experimental and Clinical Psychopharmacology, 6*, 306–315.
Cox, D.R. (1955). Some statistical methods related with series of events. *Journal of the Royal Statistical Society, Series B, 17*, 129–157.
Cox, D.R. (1972a). The statistical analysis of dependencies in point processes. In P.A.W. Lewis (Ed.), *Stochastic Point Processes* (pp. 55–66). New York: John Wiley.
Cox, D.R. (1972b). Regression models with life tables. *Journal of the Royal Statistical Society, Series B, 34*, 187–220.
Cox, D.R., & Lewis, P.A.W. (1966). *The Statistical Analysis of Series of Events*. London: Chapman & Hall.
Cressie, N. (1991). *Statistics for Spatial Data*. New York: John Wiley.
de Wit, H. & Stewart, J. (1981). Reinstatement of cocaine-reinforced responding in the rat. *Psychopharmacology, 75*, 134–143.
Diggle, P.J. (1983). *Statistical Analysis of Spatial Point Patterns*. New York: Academic Press.
Epanechnikov, V.A. (1969). Non-parametric estimation of multivariate probability density. *Theory of Probability and its Applications, 14*, 153–158.
Hall, R.V., Fox, R., Willard, D., Goldsmith, L., Emerson, M., Owen, M., Davis, F., & Procia, E. (1971). The teacher as observer and experimenter in the modification of disputing and talking-out behaviors. *Journal of Applied Behavior Analysis, 4*, 141–149.
Hawkes, A.G. (1971). Spectra of some self-exciting and mutually exciting point processes. *Biometrika, 58*, 83–90.
Helmers, R., Mangku, W., & Zitikis, R. (2003). Consistent estimation of the intensity of a cyclic Poisson process. *Journal of Multivariate Analysis, 84*, 19–39.
Hsu, H.-C., & Fogel, A. (2003). Stability and transitions in mother-infant face-to-face communication during the first 6 months: A microhistorical approach. *Developmental Psychology, 39*, 1061–1082.
Jones, R.N., Sloane, H.N., & Roberts, M.W. (1992). Limitations of "don't" instructional control. *Behavior Therapy, 23*, 131–150.
Kassel, J.D., Stroud, L.R., & Paronis, C.A. (2003). Smoking, stress, and negative affect: Correlation, causation, and context across stages of smoking. *Psychological Bulletin, 129*, 270–304.
King, G. (1988). Statistical models for political science event counts: Bias in conventional procedures and evidence for the exponential Poisson regression model. *American Journal of Political Science, 32*, 838–863.
King, G. (1989). Event count models for International Relations: Generalizations and applications. *International Studies Quarterly, 33*, 123–147.
Kutoyants, Yu.A. (1984). *Parameter Estimation for Stochastic Processes*. Berlin: Helderman.
Kutoyants, Yu.A. (1998). *Statistical Inference for Spatial Poisson Processes*. New York: Springer.
Lawless, J.F. (2003). *Statistical Models and Methods for Lifetime Data* (2nd ed.). New York: John Wiley.

Lomi, A. (1995). The population and community ecology of organizational founding: Italian co-operative banks, 1936–1989. *European Sociological Review, 11*, 75–98.

Lundberg, O. (1940). *On Random Processes and their Application to Sickness and Accident Statistics*. Uppsala: Almqvist & Wiksells.

McCullagh, P., & Nelder, J.A. (1989). *Generalized Linear Models* (2nd ed.). New York: Chapman & Hall.

McFall, R.M. (1977). Parameters of self-monitoring. In R.B. Stuart (Ed.), *Behavioral Self-Management: Strategies, Techniques, and Outcome* (pp. 196–214). New York: Brunner/Mazel.

Minkoff, D.C. (1997). The sequencing of social movements. *American Sociological Review, 62*, 779–799.

Moody, P.M. (1980). The relationships of quantified human smoking behavior and demographic variables. *Social Science and Medicine, 14A*, 49–54.

Moran, P.A.P. (1967). Dams in series with continuous release. *Journal of Applied Probability, 4*, 380–388.

Myers, D.J. (2000). The diffusion of collective violence: Infectiousness, susceptibility, and mass media networks. *American Journal of Sociology, 106*, 173–208.

Ogata, Y. (1978). The asymptotic behavior of maximum likelihood estimators for stationary point processes. *Annals of the Institute for Statistical Mathematics, 30A*, 243–261.

Ogata, Y., Akaike, H., & Katsura, K. (1982). The application of linear intensity models to the investigation of causal relations between a point process and another stochastic process. *Annals of the Institute for Statistical Mathematics, 34B*, 373–387.

Olzak, S. (1992). *The Dynamics of Ethnic Competition and Conflict*. Stanford: Stanford University Press.

Olzak, S., Shanahan, S., & McEneaney, E.H. (1996). Poverty, segregation, and race riots: 1960 to 1993. *American Sociological Review, 61*, 590–613.

Pfiffner, L.J., & O'Leary, S.G. (1987). The efficacy of all-positive management as a function of the prior use of negative consequences. *Journal of Applied Behavior Analysis, 20*, 265–271.

Prabhu, N.U. (1998). *Stochastic Storage Processes: Queues, Insurance Risk, Dams, and Data Communcation*. New York: Springer.

Rathbun, S.L. (1996). Estimation of Poisson intensity using partially observed concomitant variables. *Biometrics, 52*, 226–242.

Russell, J. (1980). A circumplex model of affect. *Journal of Personality and Social Psychology, 37*, 341–356.

Russell, M.A. (1980). Nicotine intake and its regulation. *Journal of Psychosomatic Research, 24*, 253–264.

Shiffman, S. (1993). Assessing smoking patterns and motives. *Journal of Consulting and Clinical Psychology, 61*, 732–742.

Shiffman, S., Gwaltney, C.J., Baladanis, M.H., Liu, K.S., Paty, J.A., Kassel, J.D., Hickcox, M., & Gnys, M. (2002). Immediate antecedents of cigarette smoking: An analysis from ecological momentary assessment. *Journal of Abnormal Psychology, 111*, 531–545.

Shiffman, S., Paty, J.A., Gnys, M., Kassel, J.D., & Hickcox, M. (1996). First lapses to smoking: Within subjects analysis of real time reports. *Journal of Consulting and Clinical Psychology, 64*, 366–379.

Shiffman, S., Paty, J.A., Gwaltney, C.J., & Dang, Q.Y. (2004). Immediate antecedents of cigarette smoking: An analysis of unrestricted smoking patterns. *Journal of Abnormal Psychology, 113*, 166–171.

Snyder, D., & Kelly, W.R. (1977). Conflict intensity, media sensitivity and the validity of newspaper data. *American Sociological Review, 42*, 105–123.

Stewart, J., de Wit, H., & Eikelboom, R. (1984). Role of unconditioned and conditioned drug effects in the self-administration of opiates and stimulants. *Psychological Review, 91*, 251–268.

Stone, A.A., & Shiffman, S. (1994). Ecological momentary assessment (EMA) in behavioral medicine. *Annals of Behavioral Medicine, 16*, 199–202.

U.S. Surgeon General (1988). *The Health Consequences of Smoking: Nicotine Addiction*. Washington, DC: U.S. Government Printing Office.

Vermunt, J.K. (1997). *Log-linear Event History Analysis: A General Approach with Missing Data, Latent Variables, and Unobserved Heterogeneity*. Tilburg, Netherlands: Tilburg University Press.

Wheeler, L., & Reis, H.T. (1991). Self-recording of everyday life events: Origins, types, and uses. *Journal of Personality, 59*, 339–354.

Willet, B. & Singer, J.D. (1995). It's déjà vu all over again: Using multiple-spell discrete time survival analysis. *Journal of Educational and Behavioral Statistics, 20*, 41–67.

World Health Organization (2004). Why is tobacco a public health priority? From http://www.who.int/tobacco/about/en/, retrieved March 15, 2004.

Zheng, X., & Vere-Jones, D. (1994). Further applications of the stress-release model to historical earthquake data. *Tectonophysics, 229*, 101–121.

11

Emerging Technologies and Next-Generation Intensive Longitudinal Data Collection

Sarah M. Nusser, Stephen S. Intille, and Ranjan Maitra

In this chapter, we consider newly emerging measurement technologies for intensive monitoring of individual behaviors and physiological responses in a wide range of settings. Our goal is to introduce these wearable assessment systems and the algorithms being used to extract data summaries from massive amounts of raw multidimensional sensor data. Because the social science community is largely unfamiliar with this new class of longitudinal data, we focus on the opportunities and limitations associated with the intensive longitudinal data generated by these technologies, how they impact study design and subsequent analyses, as well as statistical issues associated with processing such voluminous datasets into meaningful forms. We refer the reader to the longitudinal data modeling approaches discussed in the preceding chapters, many of which offer the potential to generate substantive knowledge from the raw data streams and data summaries (i.e., features) that are created by these mobile measurement and processing systems.

With the technology available today, a comfortable device can be created that collects continuous streams of video data describing everything the subject sees, audio data of everything the subject hears and says, accelerometer data of the subject's limb motion, data on physiological parameters such as heart rate, data on the subject's location in the community, as well as other miscellaneous data about how the subject is feeling, as reported by the subject via a mobile computing device user interface. Emerging assessment systems essentially integrate and expand prior data acquisition approaches for social science research by creating a "digital diary" (Clarkson & Pentland, 1999; Gemmell et al., 2002) that a researcher might use for studying the behavior of people in their natural settings. The value of such an experiential memory device was realized as early as 1945, but only recently have improvements in the size and cost of technology made the vision achievable (Bush, 1945; Gemmell et al., 2002; Norman, 1992).

Today's assessment tools rely on a small computer, such as a mobile phone or personal digital assistant (PDA), carried by the study participant at all times. The device gathers input via touch-screen displays and wireless sensors (e.g., position locators, voice recognizers, cameras, and barcode scanners). Alternatively, a special watch may be used to collect and display information from another body sensor, such as a chest strap heart rate monitor (FitSense, 2002). These personal appliances are essentially *wearable computers* with small and convenient ergonomic designs that permit users to comfortably and effortlessly wear or carry them at all times. The small computer can use advanced signal-processing algorithms to infer and then respond in real time to information about a subject's context (e.g., location, physiological state, and the activity he or she is engaged in), and can request self-reported data from the subject. In compressed form, a year of these data can be stored with less than 1 terabyte of disk; by 2007, 1 terabyte of diskspace is expected to retail for approximately $300 (Gemmell et al., 2002).

The use of mobile data collection and processing devices for social science research is made possible by several recent advances. Low-cost miniaturized sensors have been developed that can be worn by a subject and effortlessly capture frequent digital readings on physiological and environmental parameters. At the same time, mobile computing devices equipped with fast processors and nonvolatile memory can retrieve, process, and store volumes of data from sensors, and use these data to trigger a request for additional information from the study participant. The trigger is inferred with pattern recognition algorithms on the mobile device that detect specific events from the subject's data. Device interoperability and communication are facilitated by ubiquitous wireless networks that transfer information among sensors, computing devices, and data servers.

When all of these sensor components are combined into a single tool, a researcher in the health and social sciences has a powerful new research assessment device. Studies can be envisioned where researchers ask questions about physical activity only when subjects are "exercising" as determined by algorithms that infer activity from heart rate and accelerometer data. Similarly, researchers could run studies where questions about diet and exercise are asked only if the system determines that someone is "walking toward a food store" or has just "dined in a restaurant and taken a walk to the car." Studies would also be possible in office settings where researchers can ask colleagues to comment on the activities of others; for example, when one person is "walking" nearby, a colleague could be asked to assess the person's mood.

As tools to record digital diaries are developed, we must anticipate the types of data that will be generated and how statistical science may be called upon to support the processing and analysis of sensor data. Each data stream and the features extracted from the raw data will have specific characteristics that must be considered when combining and analyzing such datasets. In the following sections, we highlight potential issues in study design and analysis of data obtained from wearable assessment systems, with a particular focus on data captured in coded or numeric formats (i.e., audio and video data are beyond the scope of this chapter). We begin by offering a brief example of an intensive data collection system. We then outline the nature of the data these systems generate and problems that arise in the data acquisition or processing phase. Finally, we discuss some of the statistical challenges that arise in developing algorithms and analysis approaches for analyzing data from these new systems.

11.1 Intensive Data Collection Systems

11.1.1 Example: Supervised Learning and Context-Aware Data Collection in Physical Activity Studies

Researchers studying physical activity are interested in developing better tools to understand how, when, and where people get exercise of moderate or greater intensity. In this section, we outline a research tool under development that will support future physical activity assessment studies. We describe the hardware configuration, the need for algorithms and training datasets used to detect specific forms of physical activity, as well as the methods used to sample additional data from the study participant. Our goal here is to describe the layout of such a system and use it to motivate the statistical challenges that arise via systems that use such real-time sensors and data collection technologies. Each of the components of this particular tool have already been used in a laboratory setting, and components are now being integrated into a single tool in ongoing research. In a few years, the functionality we describe will be available with a typical, low-cost mobile phone.

The physical activity measurement tool relies on a connected system of sensors. The base of the system consists of a handheld computing device, such as a personal digital assistant (PDA). The PDA acts as a data logger for several devices distributed on the body and is used to prompt subjects to complete a short questionnaire when specific data patterns appear. A global positioning system (GPS) receiver that logs the location of the subject is attached to the PDA. A wireless heart rate monitor is worn as a chest strap and transmits beat-to-beat values to the PDA in real time. The PDA also acts as a receiver for a set of small wireless accelerometer sensors that measure limb motion. These sensors are integrated into elastic bands and worn by subjects on the dominant wrist, nondominant upper arm, dominant ankle, nondominant upper thigh, or attached to the subject's belt.

Each accelerometer measures motion of $\pm 10g$ with a $0.04g$ sensitivity (Analog Devices ADXL210). Forces experienced by the body during typical everyday activities are usually below $10g$ except at limb extremities in very rigorous activity (Bouten et al., 1997). Subjects use the PDA touch-screen to answer questions when an audio or tactile vibration prompt indicates that a response is needed. Even prototype versions of wireless sensors used by researchers are small and can be easily worn under clothing. Future commercial versions will be integrated directly into watches, phones, glasses, shoes, and other clothing. Some of these sensors provide continuous streams of data (e.g., motion sensors on the body). Other useful sensors for some research provide intermittent signals owing to the invasive nature of the sensing itself (e.g., blood pressure, which requires a cuff to inflate). In many situations, it is necessary to process or chunk the raw data to minimize storage requirements or create input for subsequent statistical analysis.

One of the novel aspects of such a system is that algorithms that analyze data in real time can be used to detect specific activities, such as "walking," "cycling," "scrubbing," and "vacuuming" (Bao & Intille, 2004). Upon identifying such an event, the system can prompt the study participant to respond to specific questions during or just after the subject engages in specific physical activities. An inference algorithm might combine accelerometer data and heart rate to classify motion as either "moderate or greater physical activity" or otherwise. GPS in combination with physical activity detection might also be used to trigger self-reports only at particular types of locations and when certain activities are performed. Current algorithms can detect activities in less than one second on present-day mobile device technology.

In the following subsections, we outline expected characteristics of specific measurements and data processing methods used to acquire intensive longitudinal data in a wearable assessment system.

11.1.2 Location Data

Systems that record a location track for an individual over time rely on fixed transmitters that send signals to mobile receivers, which use the signals to determine the receiver's location. The specific technologies differ for indoor and outdoor environments.

Outdoor global positioning systems (GPS) rely on four or more satellite radio transmitters to triangulate a person's position on the earth. A GPS data point will typically consist of a time stamp and longitude, latitude, and altitude coordinates. Some devices will report an error measure on the imprecision of the location, which results from timing inaccuracies due to atmospheric conditions or wireless signal multipath bouncing. These errors are often 30–100 m, a distance that is significant given that GPS is often used to determine location at the city scale (e.g., street corner, nearby storefront, etc.). Some technologies such as differential GPS use known transition locations to improve position estimates substantially (e.g. to within 10 m).

Prior to determining a position, a GPS system must identify the four satellites. This initialization procedure can require from several seconds to over a minute on many devices when a person emerges from an indoor location. Once a stable signal has been established, sampling rates on commonly available GPS receivers can be as fast as several positional readings per second. Taken as a series, GPS data points represent the path that a subject has traveled when signals have been received (figure 11.1).

Figure 11.1 Examples of GPS track data from two subjects who were instructed to drive from a starting place (star) to specific destinations (targets). Note the differences in travel behaviors that such data can provide, as well as the gaps in the path that can create problems in the analysis phase.

Indoor positioning systems emerging from research labs are not yet widespread, but similar strategies are used to determine a person's position. For example, one system uses wireless (802.11b or "Wi-Fi") network hubs in an office to determine the position of a computing device such as a PDA (Bahl & Padmanabhan, 2000). Another system uses fixed-location transmitters with special ultrasonic tags to identify the location of a tag to within several feet (Priyantha et al., 2000).

Various kinds of problems are expected to arise in logging positional data, depending on the type of sensors. Any location sensor can be expected to generate small increments of missing data at least occasionally, as seen in figure 11.1. Larger segments of missing data will regularly result from positioning systems whenever a person is not within range of the signal transmitters used to determine a location. For GPS, this situation occurs whenever someone is either indoors or within a dense urban or forested area that does not have line of sight to four satellites. Most GPS receivers will also miss data for 10–90 seconds each time a person emerges from a building.

Finally, most GPS receivers consume too much power to run continuously on a mobile device for an entire day. For these types of studies, GPS receivers will typically be programmed to sample position either randomly, or based upon another sensor reading, such as when accelerometers detect motion or when the user turns the phone on. Power management issues for many sensors will require that the activation of the sensor be driven by the activation of another, presumably less power-hungry, sensor. Depending on the strategy used to activate data acquisition, this approach may raise issues of sampling bias.

11.1.3 Cartographic Data

Positional errors in GPS readings may be compounded when GPS data are used in combination with cartographic mapping sources, which are subject to their own errors (figure 11.2). The error inherent in these electronic map databases

Figure 11.2 An example of the potential mismatch that can occur when combining photographic imagery (or map data) and GPS track data, each with its own sources of errors. With this type of registration difference, the map may indicate a person walking on a sidewalk, when in fact the subject is walking in the park.

may be difficult to characterize in a manner that can be used within a sensing study or may simply be unknown. GPS longitude and latitude data are typically used to determine position relative to certain features obtained from cartographic information sources (e.g., location of community businesses and landmarks). Misclassification errors can create problems if a researcher wishes to trigger questions whenever a subject is "in the vicinity of a food store" or "inside a park." Prior to conducting the study it may be necessary to verify that critical map features used to make these assessments are positionally accurate and labeled correctly. While this may be possible for projects of limited geographic scope, manual map verification may not be feasible for national studies deployed on mobile phones that use commercial map databases (e.g., MapQuest.com).

11.1.4 Accelerometer Data

Miniature, low-cost, and wireless accelerometers can transmit motion data from body movement in real time to the mobile computing device (Munguia Tapia et al., 2004). Accelerometers currently in use for this purpose can measure and transmit 3-axis movement at high sampling rates (e.g., 30–100 Hz) for distances indoors of up to 30 m. A typical data point consists of a time stamp, an epoch length indicating milliseconds over which movement data is aggregated, and a measurement indicating the rate of change in the velocity of the sensor. Thus, for example, when an arm is waved, there is a pronounced sinusoidal wave in one axis, where the peaks correspond to a change in arm direction (figure 11.3).

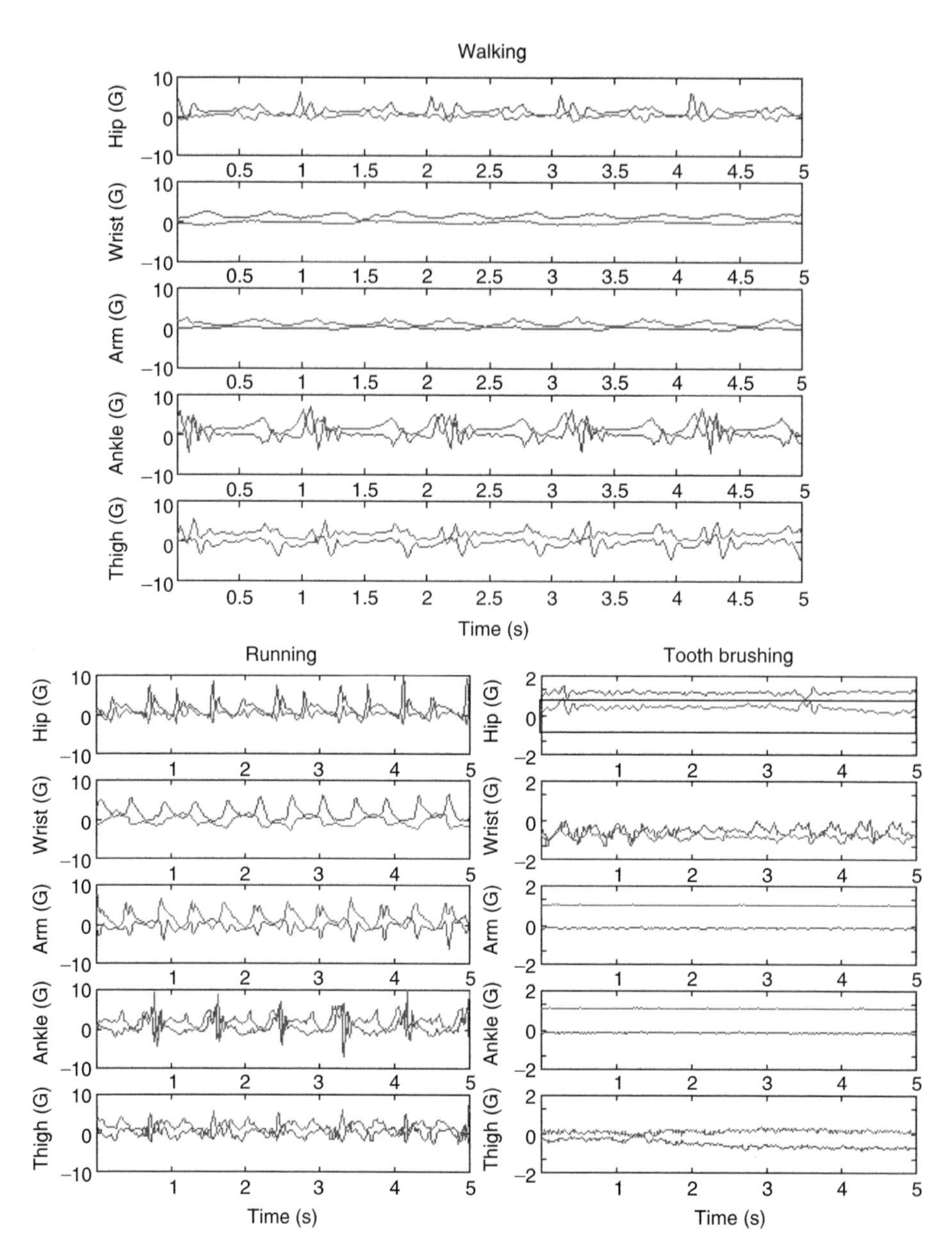

Figure 11.3 An example of data generated by placing 2-axis accelerometers on five points on the body: dominant wrist, nondominant shoulder, dominant ankle, nondominant thigh, and hip. Pattern recognition algorithms to automatically detect activities such as walking and brushing teeth can be developed that use features encoding correlations between signals from each sensor.

There is noise in the accelerometer electronics itself, but relative to the scale of human motion this noise is negligible if the accelerometers are calibrated correctly. The data are transmitted via wireless at 2.4 GHz frequency from the sensor to a receiver, a process that introduces more significant noise. These transmission errors are dependent upon the person's environment and lead to erroneous measurements and data gaps. By adjusting the number of bits used by error correction codes in the wireless transmission protocol, erroneous data points can be minimized at the cost of missing other potentially good data points. In a typical environment, we have observed that an error rate of 1 data point every 3–4 seconds is common, with missing chunks of several milliseconds of data every few minutes. Transmitter technologies with built-in error checking and retransmission capability can improve the data significantly, at a cost of a shorter battery life.

Multiple transmitters can be placed simultaneously on the body. Error rates depend upon the position of the transmitter on the body relative to the receiver on the body. For example, a receiver carried on the hip will typically generate less error from a hand sensor than a receiver carried in a backpack cluttered with objects. Environmental conditions that cause one sensor to fail (e.g., a preponderance of metal objects) will often cause all of the accelerometer transmitters to fail simultaneously.

Perhaps the most likely error when using these sensors in practice may be introduced by improper or inconsistent placement on the body by the end user. Sensors must be oriented in the same way and located on the same spot on the body in order to return consistent measurements over multi-day, week, or month periods. A sensor that is simply oriented in the wrong direction but located in the right position will provide a data stream that is correct when the axes are reoriented to the expected position, but first this error must be discovered by the researcher, potentially a daunting task for a large study where hundreds of megabytes of accelerometer data have been collected for multiple weeks or months. Even with good training procedures and materials such as video instruction and reminders, each time the subject removes and then replaces the sensor, a new opportunity for systematic error is introduced.

11.1.5 Heart Rate Data

The heart rate monitor measures beat-to-beat intervals via a chest strap worn under the clothing and transmits the data to the PDA, which can then filter the readings to obtain an instantaneous heart rate measure (figure 11.4). As with accelerometers, environmental conditions can disrupt wireless transmission.

More common, however, is that the user may inadvertently alter the sensor's position from the correct place on the body. The strap must hug the skin tightly enough to cause perspiration, which improves the electrical contact between the device and the skin. The heart rate monitor will occasionally produce no data if the strap slips or becomes loose. The chest straps are less comfortable for women to wear, which may lead to a gender-based difference in error rates. Choices such

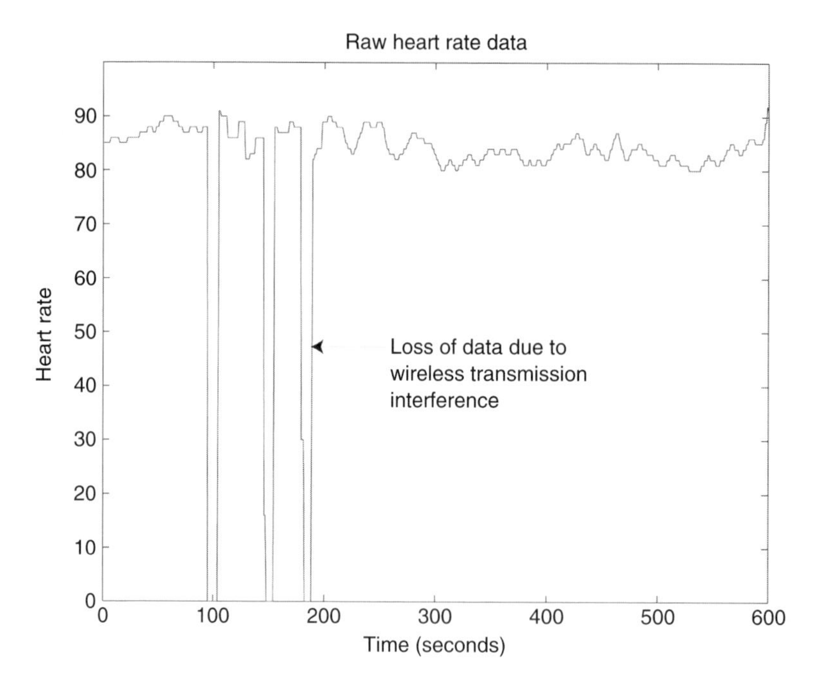

Figure 11.4 An example of data that are wirelessly transmitted from a chest strap heart rate monitor to a receiver located in a person's pocket. Short segments can be lost owing to wireless transmission dropouts or poor contact between the chest strap and the skin.

as whether sensors and the receiver are carried in a bag or pocket or worn on a belt may also lead to systematic differences in error rates between genders. Use of electrode gel would reduce but not eliminate these errors at the cost of additional burden to the user.

A different kind of problem arises with measurement errors. Some data will be erroneous, as is the case when the chest strap is slightly out of place and can not sense all of the heart beats. The algorithm that computes the instantaneous heart rate must filter these readings. Typically, data points are thrown out if they are above a hand-coded threshold or they do not appear within a series of regular readings. A simple running average is used to further smooth the data. With current technology, a kernel of several seconds is required. If a researcher requires to detect heart rates instantaneously, electrode gel and multiple on-body contacts can be used, at the cost of additional burden to the user. Filtering will still be required, however.

In some cases researchers allow the heart rate device to perform some of this filtering. However, in those cases, the algorithms the devices use (and how they deal with error) may not be available to the researcher. Direct processing of raw signals for heart rate and accelerometers ensures that the researcher is aware of various sources of error and how they are handled.

11.1.6 Data Reduction to Minimize Storage Requirements

Although eventually storage on mobile computing devices will be essentially free, today's devices sometimes require "uninteresting" data to be thrown out at the time they are received rather than using limited storage space. For example, suppose a device is recording accelerometer data and there is no movement of the device. Rather than storing megabytes of data, the device could cluster the data into chunks of "no motion." Doing so, however, requires that the researcher set thresholds for the meaning of each of the "uninteresting" states. In some cases, data may be thrown out because of readings (or lack of them) from another sensor device. For example, in some studies researchers may feel that accelerometer data without heart rate data are of little value. Therefore, they may choose to minimize disk storage requirements at the expense of possibly losing information. This has a negative impact on immediate and future analyses, for example, when a sensor's data are set to missing but could have been used to impute values for the sensor that was not functioning, or when an unanticipated analysis calls for data from the sensor that was functioning but censored, but not from the nonfunctioning sensor.

11.1.7 Inferring Events

Recent advances in pattern recognition algorithms and processor speed have led to systems that can robustly infer events from noisy sensor data. The most promising techniques typically use supervised learning algorithms, where a set of labeled examples is used to create exemplar models of each target class and then new examples are classified based on functions that assess similarity with the target classes. Using such pattern recognition, raw sensor data can be converted into more meaningful labels describing the physical activity. For example, data on limb motion collected from accelerometers worn on multiple body parts can be used to automatically determine which of a list of known activities such as "walking," "cycling," and "scrubbing" a person may be performing (Bao & Intille, 2004).

Although software applications that use pattern recognition for tasks such as speech recognition are commercially available today, at the time of this writing, software for detecting context such as physical activity from mobile sensors is still under development in laboratories. As mobile devices improve and gather data from mobile sensors such as accelerometers, however, commercial applications will become available. Commercial applications that can be exploited by researchers without a background in pattern recognition are likely to include "setup" procedures where the end users "train" the device to recognize specific contextual cues, such as specific physical activities they perform. This will not be unlike modern speech-enabled mobile phones that allow users to speak a few phrases from which the phone's software can learn in order to build good models. Exactly how these training phases will work for contextual information such as physical activities is an area of active research.

Systems that use learning algorithms to infer activities from data will introduce several types of error. Recognition algorithms will make false positive detections of specific activities, and they will make false negative errors when activities are not detected. Some algorithms will characterize their own error when making a classification, others will not. Even those algorithms that provide some measure of accuracy for each inference, however, will base that measure on a training set of data. The size and quality of that training set will have a dramatic influence on the accuracy and relevance of the misclassification rates produced.

Even in cases where the details of the inference algorithm are known to the investigator, the investigator will need to know how the algorithm was trained and how well the training examples represent the population of people to be studied. In practice, this information is unlikely to be available because detection algorithms will be "black box" systems, just as speech recognition systems are today. Black box systems may intentionally (for proprietary reasons) or unintentionally (owing to complexity) hide the internal mechanisms that make them work. Already mobile products on the market that are being used for research, such as an electronic pedometer (FitSense, 2002), rely on proprietary algorithms to infer simple events such as "walking" and "shuffling" from raw accelerometer data. Information about how inference algorithms work, let alone how they were trained and on what datasets, will not necessarily be available to the social science researcher.

Another challenge is that inference algorithms are entirely dependent upon the sensors that provide the raw data. Therefore, the performance of the inference algorithm will be strongly correlated with the availability and quality of sensor data. If the training datasets do not include examples of sensor failure similar to those experienced in a particular experiment, the inference algorithms may perform erratically (e.g., failing to recognize a pattern or indicating that a pattern occurred when it did not).

11.1.8 Eliciting Responses from Subjects

The experience sampling method (ESM) (Csikszentmihalyi, 1982) and ecological momentary assessment (EMA) (Stone & Shiffman, 1994) are techniques where a subject records quantitative or qualitative observations about his or her context, often triggered by an electronic timing device. These techniques generate longitudinal datasets that often include time series where specific measures of interest are sampled throughout a subject's everyday life experiences. A researcher who wants to avoid the use of independent observers may choose to use one of three sampling methods: (1) systematic sampling on a fixed interval schedule, such as every 30 minutes; (2) stratified sampling on a random interval schedule, such as on average once every 30 minutes or sometime randomly within every two-hour window; and (3) purposive sampling in response to user initiative, where the user is told to make a data entry whenever he or she performs a particular activity (Stone & Shiffman 1994).

Sensors that transmit data to the computing device in real time offer a new option: context-sensitive ecological momentary assessment (CS-EMA), where specific questions are asked only when a user does a particular activity or a sensor has a particular reading (Intille et al., 2003). In the simplest case, questions can be triggered using explicit, manually coded mappings between questions and sensor readings. For example, a question can be triggered each time a heart rate threshold is exceeded or each time the GPS indicates that a person is in a particular location.

The computer can also preprocess the data in real time, however, to either filter raw data or infer a user state using a pattern recognition algorithm. The data can trigger a question that is asked based on the *likelihood* that a subject has performed a particular action. Context-awareness modules permit a researcher to acquire more information about the behavior or situation of interest by sampling only during or just after the activities of interest. Contextual inputs, such as location, time, event, or biosensor data, can trigger the sampling. This approach may minimize some interruption annoyance without compromising the quality of data acquired on the target phenomena.

Sensor-triggered self-reports do not necessarily need to depend upon sensors worn by the subject. Wireless networks make it possible for mobile computing devices to share information among subjects. For example, using a Bluetooth network, two of the computing devices can detect when they are within a few meters of one another and share data from each other's sensors. Therefore, self-report readings for one subject in a study can be triggered by sensor readings from another subject in the study. A researcher interested in how interaction between people impacts activity could program the mobile computing device to ask a question just after two people "interact," where interacting is defined as any time they move within some predetermined distance from one another.

Prompting a user for information could create reactivity effects, and researchers using sensor-triggered self-report need to design sampling strategies so that the prompting minimally disrupts the naturalistic behavior that is actually being studied. Work to date suggests that reactivity to ecological momentary assessment may be small (Hufford et al., 2002).

11.2 Statistical Issues for Intensive Longitudinal Measurement

The physical activity measurement tool provides an example of a data collection tool and methodology that may be available to social scientists in the near future. Research subjects wear sensors that generate multivariate data streams indexed by time and, in many cases, space. Real-time processing of incoming data creates summaries needed to reduce storage requirements, detect activities, trigger event sampling, or conduct analyses after the study has been completed. The resulting data streams and data summaries facilitate the construction of a complete digital record of a subject's behaviors, physiological responses, and perceptions over a period of time.

While technologies have simplified the acquisition of multivariate time series of responses on individual subjects, the statistical issues introduced by the nature of these rich and voluminous data are myriad. In this section, these issues are discussed for raw data as well as techniques to reduce data or infer events from data.

11.2.1 Data Quality Problems in Raw Data Streams

Continuous data recording is an idealized concept that, in reality, is complicated by the kinds of errors that inevitably occur in any data collection process. Two major problems that arise in collecting data are the failure of a device or subject to record data (i.e. missing data), and measurement errors that arise in recorded data.

11.2.1.1 Missing Data

Many sensors will generate continuous data streams with small amounts of missing data. Often, smoothing techniques (e.g., moving average) will be sufficient to eliminate or minimize the occurrence of data gaps in the analysis dataset. Alternatively, the data may be summarized for discrete chunks of time to create a sequence of values without data gaps. Smoothing and data summarization techniques may gloss over interesting features in the data, but are simple to implement. A more complex approach is to consider methods for imputing missing data to create a complete data array (Little & Rubin, 1987). Some methods take advantage of the structure present in multivariate data streams by explicitly modeling the distribution of missing data and their relationships with other variables (Gelman et al., 2003).

Longer gaps of missing data are more troublesome. Sensors may generate potentially large data gaps when signals fall below a minimum detection level or are not received at all owing to problems with the equipment or subject compliance. For example, if a chest strap slips, no information will be recorded on heart rates. Alternatively, GPS sensors will not receive satellite signals indoors or in some outdoor environments that block signals, leading to periods of missing positional data for one sensor when others are continuing to log information; the amount of time elapsed between the last known position and the first known position after signals are regained is observable, but it may be difficult to assess where the subject has been during the interim period.

Since most sensors tend to provide somewhat erratic data streams over time (see figures 11.3 and 11.4), it may be easier to take advantage of correlations among sensor data streams to develop a model for imputing missing data. In some cases, it may be difficult to identify a sensor that generates correlated data, as is the case with large gaps in a GPS track. For this example, one option is to use a built-in camera that captures still images or audiovisual sequences of activity to infer where a subject has been. New algorithms are being developed that automatically process video streams from cameras worn on the body to determine contextual information such as type of location in areas were GPS is not available (Clarkson & Pentland, 1999).

Alternatively, study designs can be developed to minimize data gaps. For instance, sensors could be monitored for slippage and prompt subjects to refasten them, or, if a sensor fails unexpectedly, a user could be promoted to report their activity or environmental conditions. However, such efforts run the risk of increasing respondent burden and will not result in ideal data.

As noted earlier, researchers sometimes implement algorithms that cause data acquisition on multiple sensors to cease when one of the sensors in the system is not working. This approach should be avoided or at least carefully evaluated since it may induce new problems when analyzing data. Sampling bias may occur if the primary sensor failure that shuts down recording of other sensors is related to the phenomena being observed, as is the case when a GPS sensor erroneously fails to acquire data at the same location and consistently shuts down other sensors that would otherwise provide valuable information at that time. Also, data from sensors that are still functioning (but at risk of being turned off) may in fact be useful in modeling missing sensor data or in assessing the potential for systematic bias.

11.2.1.2 Measurement Errors

Error in sensor measurements is a second major data quality issue. From a statistical perspective, measurement error in the sensor data stream may take on the form of systematic bias and/or extra variation. Each sensor will have its own error characteristics, and the magnitude of the problem may or may not be worthy of attention. In many cases, the measurement error for a sensor will be correlated over time. There are also many instances where the error properties of multiple sensors will be correlated. For example, a single source of interference in wireless data transfer may simultaneously distort values generated from more than one sensor.

Measurement bias is most likely to occur because of calibration problems or sensor misplacement. It may be possible to reduce or prevent these biases with study designs that place a strong emphasis on encouraging good compliance by subjects or by paying careful attention to the calibration of sensors in relation to study conditions. In some cases, however, proprietary algorithms may be used that make it impossible to address or even detect calibration bias.

Random variation will nearly always be present. It may be reduced by good study design, but it is unlikely to be eliminated. The relative magnitude of random error and the presence of bias will vary with the sensor and with the subjects wearing the sensors.

As an example, consider a GPS receiver used to determine a subject's location. The GPS does not typically have a bias associated with it, unless parameters such as the datum or projection are incorrect. The precision of the subject's position, however, tends to be heterogeneous in space and time. For example, at any point in time and space, a GPS position is affected by the current satellite configuration (e.g., whether the four satellites are geographically dispersed or reasonably near each other) as well as the presence of physical barriers between the subject and the satellites used to determine the subject's position. If a differential GPS receiver is used, for instance, then the random error in signals received is likely to be quite small relative to other sources of variation.

Data from other sensors such as a heart rate monitor or accelerometer are more likely to incur biases from the misplacement of the sensor by the study participant. Prevention of these biases through proper training and user support is critical, and it would be wise to establish real-time data-checking algorithms for logged data. Data from accelerometers can be quite noisy, and thus may require smoothing or filtering techniques.

Data reduction steps used to create indicators for system triggers or summaries for statistical analyses transform error properties in raw sensor data streams. The impact may be positive or potentially deleterious. If smoothing algorithms are used, the impact of random errors is typically reduced. However, for sensor data that are used to trigger events, measurement errors or misclassified events can have a serious impact on data streams that depend on the sensor context to request additional data acquisition. Bias in trigger variables could also lead to context-sensitive ecological momentary assessment sampling bias. This occurs when there is a tendency to fail to detect a real event (false negative) or erroneously detect a nonevent (false positive).

Measurement errors may also be propagated when data streams are combined with other auxiliary information. For example, when using positional data from a location sensor in combination with a digital map, two sources of errors are present: positional error from the location sensor, and cartographic error. Cartographic error takes on many forms, including errors of omission or commission in map features such as buildings or streets, as well as positional errors in the location of map features. In the example discussed earlier, a map error could lead to the false trigger for a self-report or failure to gather user data when the trigger condition is in fact present.

The relative magnitude of measurement error bias or variance relative to other sources of error may be insignificant. Variance in particular may be small relative to potentially large levels of person-to-person variation that would be common for these types of studies. However, if the magnitude of the measurement error variance or bias is sufficiently large, it is useful to characterize the properties of the measurement error so that it can be adjusted for in subsequent analyses.

If information on measurement error properties is available, it is possible to use statistical methods that take error properties into account when generating data summaries or making inferences from the raw data streams and auxiliary inputs from a given study. For example, in linear regression analyses, measurement error variance is known to lead to bias in regression coefficients describing the relationship between the response variable and the covariates, which can be corrected for if the measurement error variance is known or can be estimated. Fuller (1987) and Carroll et al. (1995) discuss approaches to measurement error modeling under various settings. Cressie and Kornak (2003) present methods that address measurement error in both the sample location and the measured variable. Skinner (1998) provides an example of measurement error modeling in longitudinal data.

However, obtaining measurement error information may be extremely difficult. Data to estimate or prior estimates of measurement error properties of a sensor may not be available to the researcher because of lack of documentation or proprietary concerns, or the assessment of measurement properties may be developed in an experimental environment that is not well matched to the study setting. This is especially true of black box systems that are more likely to be used by social science researchers, and it would be beneficial if vendors could provide detailed information on misclassification rates and measurement error properties, including how these quantities were estimated.

11.2.2 Supervised Learning and Pattern Recognition

The system described in the case study above would use supervised learning to process data from the incoming accelerometer, chest band, and GPS systems. Researchers in computer vision, speech processing, machine learning, and other computational perception domains have found supervised learning algorithms to be highly effective at complex recognition of activity by computer from noisy sensor data. These algorithms use training sets of sensor data that have been annotated with activity labels to build models that capture the variability of interest in the examples and the nuisance uncertainty in the sensor measurements.

Similarity measures are used to compare new sensor readings with the models to determine the classification. Inputs to the algorithms used to classify observations often take the form of "features" (e.g., local peaks, inflection points, principal components) that are calculated from the raw data or from other features, possibly obtained from a different time point. For example, higher-level activities such as "going to work" may be made up of a loosely ordered stream of other detectable events; classification of this activity may require the modeling of relationships between features across relatively long time intervals (in minutes or perhaps even in hours).

With good training sets and good feature selection in the development of the algorithm, supervised learning techniques can provide significantly more accurate classifications than context detectors that employ hand-constructed rules (Witten & Frank, 1999). Further, supervised learning techniques can be used to create context detection algorithms that are customized to individual users by training the algorithms on user-specific datasets. In this section, we discuss factors that affect the efficacy of feature recognition, its use in context-sensitive ecological momentary assessment sampling, and computational issues that arise in applying these methods.

11.2.2.1 Using Training Sets to Derive Classification Rules

The first step in developing a physical activity recognizer is to generate labeled data for training the supervised learning algorithm. Since most user activity is strongly influenced by the setting, training data are most useful if they are generated from settings representative of the field study. Ideally, test volunteers are recruited from groups that have characteristics similar to the study's target population, and realistic test activities and similar environments used as the basis for training examples; however, for certain types of activity these data may not be available (Intille et al., 2004). In the test to create training data for the physical activity recognizer, volunteers are asked to wear sensors on the same body locations that will be used by study subjects. For each activity of interest, a set of examples are created where volunteers are performing the activities as naturally as possible. Subjects are provided descriptions of each target activity and asked to perform them.

This training step will most likely be performed during the development of the algorithm, rather than by the social scientist who uses the pattern recognition algorithms to trigger context-based queries and data collection. Typically all that will be available are general performance characteristics. For instance, the physical activity assessment tool is being developed using a supervised learning algorithm based on decision trees (Breiman et al., 1984; Quinlan, 1993), which is a tree-structured classification tool that approximates discrete-valued target functions, with a test prescribing a decision at each node. Inputs for this classifier are mean, energy, entropy, and correlation measures computed from the accelerometer data streams in real time by the PDA. The algorithm was trained on three hours of training data collected from 20 subjects, and the algorithm results in an overall accuracy of 84.3% $\pm$ 5.2 on 20 target household activities (Bao & Intille, 2004). Using a leave-one-subject-out validation process on the training data, a misclassification matrix (Kohavi & Provost, 1998) is generated by comparing actual and predicted classifications to arrive at the percentages of false positives and negatives. Recognition results of 80% to 95% for walking, running, climbing stairs, standing still, sitting, lying down, working on a computer, bicycling, and vacuuming are obtained and are comparable with recognition results using laboratory data from previous work on other types of activities (Foerster et al., 1999; Lee & Mase, 2001; Mantyjarvi et al., 2001; Randell & Muller, 2000; Uiterwaal et al., 1998; Van Laerhoven & Cakmakci, 2000). Overall recognition rates of 80–95% are considered good for detection of physical activities from accelerometer data.

In general, activities that are highly constrained owing to the shape of the human body, such as walking, will be recognized more reliably than activities that can be performed with more variability between subjects, such as stretching. For example, recognition accuracies for stretching and riding an elevator were below 50% using the same training set and algorithm that performed well recognizing ambulation and posture. Recognition accuracies for "watching TV" and "riding escalator" were 77.29% and 70.56%, respectively. These activities do not have unambiguous motion characteristics and are easily confused with other activities. User-specific training sets of sufficient size could lead to improved recognition for these activities, although obtaining these training sets from individual subjects may not be possible in many studies. Use of more complex data features (e.g., peaks, inflection points, principal components) that can model the most salient differences between these activities (and the probabilistic encoding of temporal information capturing correlations across different data streams) would probably improve the discriminatory power of most supervised learning methods.

In practice, the social science researcher may be given this type of information on performance, but in most situations will be provided only with a black box activity recognizer. The algorithm developer has a large number of options to choose from, both for computing features from raw data and classification method. In practice, however, the social science researcher will not be aware of how or why these decisions were made and how the activity detection algorithms actually work. Encoded in the result of the event detection classifier may be several layers of inference. For example, raw data such as heart rate may have been run through a filter to fill in missing data and smooth noise. Features may then have been computed by calculating correlations between multiple filtered data streams. Some of these features will compare points in the data streams widely separated in time. The feature detectors themselves may include well-defined mathematical operations (e.g., finding peaks in the frequency domain) or hard-coded rules developed directly from domain knowledge (e.g., detection of footfalls using a hand-crafted correlation operation). The classification algorithms themselves may use the results of other classification algorithms as feature input. For instance, results from classifiers for activities such as posture or ambulation (e.g., walking) may be used as feature input into other classifiers that detect longer-term activities such as "exercising," or "cooking," or "socializing." Each step in this process may use a different type of classification algorithm. Some may use probabilistic decision trees, sets of hierarchical decision rules for classifying an item of interest given a set of computable features. Others may use artificial neural networks (ANNs), which are complex, multilayered classification structures whose layers and parameters need to be fit iteratively based on observed training data (see Maitra, 2002, and the references therein). Once fitted, the classification algorithms at each stage in the hierarchy may be trained on data collected from different subjects.

11.2.2.2 Pattern Recognition in Relation to Context-Sensitive Ecological Momentary Assessment Sampling

To enable the physical activity measurement tool to collect self-reports from the study participant, a real-time version of the activity recognition algorithm is used to detect events and trigger context-sensitive ecological momentary assessment. Decision trees are generally slow to train but fast to run and are therefore amenable to real-time implementation.

A researcher would have the option of triggering a self-report question only after the system infers that a specific physical activity has taken place. However, the inference algorithm's performance varies from activity to activity and from user to user. The performance of the algorithm on a particular subject may not be known, since testing on the subject would involve a prohibitively long calibration process.

When using the physical activity assessment tool, the researcher would also be able to trigger questions based on the GPS and/or heart rate data. It is possible that one inference algorithm might combine accelerometer data and heart rate to classify motion as either "moderate or greater physical activity" or not. GPS in combination with physical activity detection might be used to trigger self-report only at particular types of locations (e.g., parks) when certain activities are performed (e.g., running or cycling).

The experimenter has several factors to consider when determining how to trigger sampling. First, are the activity training sets sufficiently close to the activity that will be observed with the target subjects so that recognition performance will match the expected performance? Second, given the error rates of the recognition algorithms and the uncertainty in all the data collected (e.g., GPS), what information must be collected to keep statistical analysis manageable and meaningful? What threshold values should be used to trigger context-sensitive questions for each specific activity of interest and how does the selection of the cutoff value impact statistical analysis? Finally, could statistical sampling techniques be used to replace deterministic rules for ecological momentary assessment sampling with probabilistic methods that protect against false positives and negatives? Each of these factors impacts the meaning of triggered data, and should be carefully weighed when designing the study and analyzing data.

11.2.2.3 Computational Issues in Supervised Learning

Intensive longitudinal data generated by emerging assessment systems will require that challenges generated by massive numbers of multidimensional observations from multiple sensors and the nature of the multivariate distribution of the sensor data or summaries be addressed. In this section, we outline potentially useful methodologies for similar problems discussed in the statistical and artificial intelligence literature.

As noted earlier, emerging systems for intensive longitudinal data collection will result in vast arrays of data arising from the different types of sensor measurements and the number of readings that are logged by each sensor. Analysis of longitudinally generated massive datasets presents a formidable practical challenge to classical and even contemporary statistical theory and methodology. In recent years, methods to analyze massive datasets have received attention in the statistical literature. For instance, one approach to reduce dimensionality to facilitate pattern recognition and supervised learning is the use of principal components (Mardia et al., 1979). The idea behind this technique is to provide uncorrelated projections of the variables in the dataset such that most of the variation in the data is captured by the first few components, thus resulting in a considerable reduction of data. Although principal components do not always result in easily interpretable quantities, this is likely not to be an issue when the goal is to detect specific activities, rather than interpreting the principal components themselves. A more serious problem is the need to decompose the severely multidimensional variance–covariance matrix given the large numbers of observations. Dimensionality of the dispersion matrix may be reduced through careful examination of the variables with high loadings for one or more components.

Another issue relates to the multivariate distribution of the original variables. Principal components give rise to uncorrelated projections, which are considered independent if the observations are multinormally distributed. If the distributions of the original variables are not close to Gaussian, a closely related technique can be used to separate signals, referred to as independent component analysis (see Hyvärinen and Oja, 2000, for an excellent survey of the technique).

Fundamental in the preceding discussion is the question of how to effectively annotate and analyze the raw data streams. Volumes of data generated over a short period of time will require computer algorithms to preprocess the raw data into quantities (i.e., features) that describe attributes or events for statistical analysis. It may also be necessary to apply data reduction algorithms to address storage and network constraints. Captured auditory and image signals from the mobile device may help the researcher initially to hand-label or semiautomatically label training examples in the raw data streams.

The selection of what features or summaries to compute from the raw data streams is critical for creating algorithms that can successfully infer higher-level activities from the data. Although poor selection of features will render even the best inference algorithms ineffective, computing all possible features for any dataset with multiple sensor data streams becomes computationally prohibitive when real-time performance is required. Typically, the algorithm designer evaluates the tradeoff between the utility of a feature and computer time and makes a judgment call, mixing and matching known techniques. Other than principal components, commonly used features in pattern recognition for data reduction are peaks and inflection points when data streams are converted to a "meaningful" space (e.g., the frequency domain), and correlations between raw signals or other features between different data streams. Correlations that are particularly important when attempting to infer activity are those computed between data streams at different periods in time. For example, to recognize an activity such as "cooking" may require comparison of data streams at intervals of time, some separated by only milliseconds but also others at intervals of time separated by minutes and possibly hours. Therefore, missing or erroneous data in one stream may have consequences for an inference algorithm at some other point in time because many key features will encode temporal information.

Supervised learning in the presence of missing or erroneous longitudinal data streams may possibly be performed using the methodology on advanced smoothing and flexible discriminant analysis of Li et al., discussed in chapter 2 of this book. Another possibility is to smooth the data with B-splines and use the basis coefficients used for simultaneous modeling of long- and short-term trends by Fok and Ramsay in chapter 5.

Finally, another issue that has connections with real-time context-sensitive ecological momentary assessment is that there may be patterns which have not been recorded in the training stage. For instance, a person may actually have been performing an activity that did not arise while training and that is important to record in the database. As an illustration, suppose that a subject in a study on physical activity performs intense yard work but that the system was never provided with an example set of this type of activity and examples that are customized to the subject being observed. The issue then becomes to test whether the observed pattern is from an activity that has already been identified or is from some task that has not been encountered yet. Likelihood-ratio-type tests can be constructed in order to first test the above hypothesis and classification or clustering done accordingly. We refer to the steps in the multistage clustering algorithm for massive datasets of Maitra (2001); after clustering a small subsample of observations, the remaining observations are tested to assess whether all groups in the dataset have been reasonably identified. Although the context is slightly different, we believe this approach can be readily extended to the case above. Once testing is done, observations that can be reasonably grouped are classified accordingly, while the rest are submitted to further pattern recognition algorithms for further processing.

11.3 Summary

In this chapter, we have described the types of measurement tools that are emerging for intensive measurement of human responses in social settings. These wearable systems incorporate new information technologies, as well as sophisticated statistical and computational tools for summarizing and extracting inferences from multivariate data streams. It will be critical for social scientists to understand the nature of the data generated by these technologies in developing approaches to study design and analysis.

Although black box systems are likely to be the main mode by which these tools are introduced, the data manipulations that occur within these systems and their impact on the summaries obtained from them should be considered prior to identifying an approach for analysis. Once the characteristics of these data are well understood, it will be possible to choose from existing and newly developed statistical methodologies to draw inference from these multidimensional data streams, many of which are discussed in this book. In addition, as further innovations in wearable assessment tools motivate new research threads in the social sciences, we anticipate the need for new statistical approaches to support these inquiries.

ACKNOWLEDGMENTS

The authors wish to thank reviewers of an earlier version of this manuscript for their insightful comments. S. M. Nusser is supported in part by National Science Foundation Digital Government Grant 9983289 and 0306855. S. S. Intille is supported in part by National Science Foundation ITR Grant 0112900 and the Changing Places/House_n Consortium. R. Maitra is supported in part by National Science Foundation Career Award DMS-0437555.

References

Bahl, P., & Padmanabhan, V.N. (2000). RADAR: An in-building RF-based user location and tracing system. In *Proceedings of IEEE Infocom 2000* (Vol. 2, pp. 775–784). Piscataway, NJ: IEEE Press.

Bao, L., & Intille, S.S. (2004). Activity recognition from user-annotated acceleration data. In A. Ferscha & F. Mattern (Eds.), *Proceedings of PERVASIVE 2004* (pp. 1–17). Berlin: Springer.

Bouten, C.V., Koekkoek, K.T., Verduin, M., Kodde, R., & Janssen, J.D. (1997). A triaxial accelerometer and portable data processing unit for the assessment of daily physical activity. *IEEE Transactions on Bio-Medical Engineering, 44*, 136–147.

Breiman, L., Friedman, J.H., Olshen, R.A., & Stone, C.J. (1984). *Classification and Regression Trees*. Belmont, CA: Wadsworth.

Bush, V. (1945). As we may think. *Atlantic Monthly, 176*, 101–108.

Carroll, R.J., Ruppert, D., & Stefanski, L.A. (1995). *Measurement Error in Nonlinear Models*. London: Chapman & Hall.

Clarkson, B.P., & Pentland, A. (1999). Unsupervised clustering of ambulatory audio and video. In *Proceedings of the 1999 IEEE International Conference on Acoustics, Speech and Signal Processing* (Vol. 6, pp. 15–19). Piscataway, NJ: IEEE Press.

Cressie, N., & Kornak, J. (2003). Spatial statistics in the presence of location error with an application to remote sensing of the environment. *Statistical Science, 18*, 436–456.

Csikszentmihalyi, M. (1982). Toward a psychology of optimal experience. *Review of Personality and Social Psychology, 2*, 13–36.

FitSense Technology (2002). FitSense FS-1 Speedometer. Southborough, MA: FitSense Technology.

Foerster, F., Smeja, M., & Fahrenberg, J. (1999). Detection of posture and motion by accelerometry: a validation in ambulatory monitoring. *Computers in Human Behavior, 15*, 571–583.

Fuller, W.A. (1987). *Measurement Error Models*. New York: John Wiley.

Gelman, A., Carlin, J.B., Stern, H.S., & Rubin, D.B. (2003). *Bayesian Data Analysis*, (2nd ed.). New York: Chapman & Hall.

Gemmell, J., Bell, G., Lueder, R., Drucker, S., & Wong, C. (2002). MyLifeBits: Fulfilling the Memex vision. *Proceedings of ACM Multimedia '02* (pp. 235–238). New York: ACM Press.

Hufford, M.R., Shields, A.L., Shiffman, S., Paty, J., & Balabanis, M. (2002). Reactivity to ecological momentary assessment: An example using undergraduate problem drinkers. *Psychology of Addictive Behaviors: Journal of the Society of Psychologists in Addictive Behaviors, 16*, 205–211.

Hyvärinen A., & Oja, E. (2000). Independent component analysis: Algorithms and applications. *Neural Networks, 13*, 411–430.

Intille, S.S., Bao, L., Munguia Tapia, E., & Rondoni, J. (2004). Acquiring in situ training data for context-aware ubiquitous computing applications. In *Proceedings of CHI 2004 Connect: Conference on Human Factors in Computing Systems* (pp. 1–9). New York: ACM Press.

Intille, S.S., Rondoni, J., Kukla, C., Anacona, I., & Bao, L. (2003). A context-aware experience sampling tool. In *Proceedings of CHI '03 Extended Abstracts on Human Factors in Computing Systems* (pp. 972–973). New York: ACM Press.

Kohavi, R., & Provost, F. (1998). Glossary of terms: Special issue on applications of machine learning and the knowledge discovery process. *Machine Learning 30*, 271–274.

Lee, S.W., & Mase, K. (2001). Recognition of walking behaviors for pedestrian navigation. In *Proceedings of 2001 IEEE Conference on Control Applications* (*CCA01*) (pp. 1152–1155). Piscataway, NJ: IEEE Press.

Little, R.J.A., & Rubin, D.B. (1987). *Statistical Analysis with Missing Data*. New York: John Wiley.

Maitra, R. (2001). Clustering massive datasets with applications to software metrics and tomography. *Technometrics, 43*, 336–346.

Maitra, R. (2002). A statistical perspective on data mining. *Journal of the Indian Society for Probability and Statistics, 6*, 28–77.

Mantyjarvi, J., Himberg, J., & Seppanen, T. (2001). Recognizing human motion with multiple acceleration sensors. In *Proceedings of the IEEE International Conference on Systems, Man, and Cybernetics* (pp. 747–752). Piscataway, NJ: IEEE Press.

Mardia, K.V., Kent, J.T., & Bibby, J.M. (1979). *Multivariate Analysis*. New York: Academic Press.

Munguia Tapia, E., Marmasse, N., Intille, S.S., & Larson, K. (2004). MITes: Wireless portable sensors for studying behavior. *Proceedings of Extended Abstracts Ubicomp 2004: Ubiquitous Computing*. Available online: http://ubicomp.org/ubicomp2004/adjunct/demos/tapia.pdf

Norman, D. (1992). *Turn Signals Are the Facial Expressions of Automobiles*. Reading, MA: Addison Wesley.

Priyantha, N.B., Chakraborty, A., & Balakrishnan, H. (2000). The Cricket location-support system. In *Proceedings of the Sixth Annual ACM International Conference on Mobile Computing and Networking* (*MOBICOM*) (pp. 32–43). New York: ACM Press.

Quinlan, J.R. (1993). *C4.5: Programs for Machine Learning*. San Mateo, CA: Morgan Kaufmann.

Randell, C., & Muller, H. (2000). Context awareness by analyzing accelerometer data. In B. MacIntyre & B. Iannucci (Eds.), *The Fourth International Symposium on Wearable Computers* (pp. 175–176). Piscataway, NJ: IEEE Press.

Skinner, C.J. (1998). Logistic modeling of longitudinal survey data with measurement error. *Statistica Sinica, 8*, 1045–1058.

Stone, A.A., & Shiffman, S. (1994). Ecological momentary assessment (EMA) in behavioral medicine. *Annals of Behavioral Medicine, 16*, 199–202.

Uiterwaal, M., Glerum, E.B., Busser, H.J., & van Lummel, R.C. (1998). Ambulatory monitoring of physical activity in working situations, a validation study. *Journal of Medical Engineering and Technology, 22*, 168–172.

Van Laerhoven, K., & Cakmakci, O. (2000). What shall we teach our pants? In B. MacIntyre & B. Iannucci (Eds.), *The Fourth International Symposium on Wearable Computers* (pp. 77–83). Piscataway, NJ: IEEE Press.

Witten, I.H., & Frank, E. (1999). *Data Mining: Practical Machine Learning Tools and Techniques with Java Implementations*. San Francisco: Morgan Kaufmann.

Index